POWERLESS SCIENCE?

The Environment in History: International Perspectives

Series Editors: Dolly Jørgensen, *Umeå University*; David Moon, *University of York*; Christof Mauch, *LMU Munich*; Helmuth Trischler, *Deutsches Museum, Munich*

ENVIRONMENT AND SOCIETY

Volume 1
Civilizing Nature
National Parks in Global Historical Perspective
Edited by Bernhard Gissibl, Sabine Höhler, and Patrick Kupper

Volume 2
Powerless Science?
Science and Politics in a Toxic World
Edited by Soraya Boudia and Nathalie Jas

Forthcoming Titles:

Volume 3
Managing the Unknown
Essays on Environmental Ignorance
Edited by Frank Uekötter and Uwe Lübken

Volume 4
Creating Wildnerness
A Transnational History of the Swiss National Park
Patrick Kupper
Translated by Giselle Weiss

Powerless Science?
Science and Politics in a Toxic World

Edited by
Soraya Boudia and Nathalie Jas

Published in 2014 by

Berghahn Books

www.berghahnbooks.com

© 2014 Soraya Boudia and Nathalie Jas

All rights reserved. Except for the quotation of short passages for the purposes of criticism and review, no part of this book may be reproduced in any form or by any means, electronic or mechanical, including photocopying, recording, or any information storage and retrieval system now known or to be invented, without written permission of the publisher.

Library of Congress Cataloging-in-Publication Data

Powerless science? : science and politics in a toxic world / edited by Soraya Boudia and Nathalie Jas.
    pages cm. — (The environment in history ; volume 2)
    Includes bibliographical references and index.
    ISBN 978-1-78238-236-2 (hardback : alk. paper) —
    ISBN 978-1-78238-237-9 (institutional ebook)
  1. Science and state. 2. Science—Political aspects. 3. Science—Moral and ethical aspects. I. Boudia, Soraya, editor of compilation. II. Jas, Nathalie, editor of compilation.
    Q125.P923 2014
    338.9'26—dc23

2013017868

British Library Cataloguing in Publication Data

A catalogue record for this book is available from the British Library

Printed in the United States on acid-free paper

ISBN 978–1-78238-236-2 hardback
ISBN 978-1-78238-237-9 institutional ebook

Contents

List of Figures and Tables	vii
Acknowledgments	ix
Introduction. The Greatness and Misery of Science in a Toxic World *Soraya Boudia and Nathalie Jas*	1

PART I. KNOWLEDGE, EXPERTISE, AND THE TRANSFORMATIONS IN REGULATORY SYSTEMS

1. Precaution and the History of Endocrine Disruptors *Nancy Langston*	29
2. The Political Life of Mutagens: A History of the Ames Test *Angela N.H. Creager*	46
3. DES, Cancer, and Endocrine Disruptors: Ways of Regulating, Chemical Risks, and Public Expertise in the United States *Jean-Paul Gaudillière*	65
4. Managing Scientific and Political Uncertainty: Environmental Risk Assessment in a Historical Perspective *Soraya Boudia*	95

PART II. ACTIVISM AND NONACTIVISM: ALTERNATIVE USES OF KNOWLEDGE

5. Work, Bodies, Militancy: The "Class Ecology" Debate in 1970s Italy *Stefania Barca*	115
6. What Kind of Knowledge is Needed about Toxicant-Related Health Issues? Some Lessons Drawn from the Seveso Dioxin Case *Laura Centemeri*	134

7. **From Suspicious Illness to Policy Change in Petrochemical Regions:** Popular Epidemiology, Science, and the Law in the United States and Italy 152
    Barbara L. Allen

8. **Guinea Pigs Go to Court:** Epidemiology and Class Actions in Taiwan 170
    Paul Jobin and Yu-Hwei Tseng

PART III. PUTTING KNOWLEDGE, IGNORANCE, AND REGULATION INTO PERSPECTIVE

9. **Reckless Laws, Contaminated People:** Science Reveals Legal Shortcomings in Public Health Protections 195
    Carl F. Cranor

10. **Untangling Ignorance in Environmental Risk Assessment** 215
    Scott Frickel and Michelle Edwards

11. **Low-Dose Toxicology:** Narratives from the Science-Transcience Interface 234
    Sheldon Krimsky

12. **Unruly Technologies and Fractured Oversight:** Toward a Model for Chemical Control for the Twenty-First Century 254
    Jody A. Roberts

Notes on Contributors 269

Index 274

Figures and Tables

FIGURES

2.1. Pictures of Ames Tests for (A) Spontaneous Revertants and exposure to (B) Furylfuramide, (C) Aflatoxin B1, and (D) 2-Aminofluorene. The mutagenic compounds in B, C, and D were applied to the 6 mm filter disk in the center of each plate. Each petri plate contains cells of the tester strain in a thin overlay of top agar. (The strain used here is TA98, derived by adding a resistance transfer factor to a *Salmonella* tester strain, mutant *hisD3052*, that scores frameshift mutations.) Plates C and D contain, additionally, a liver microsomal activation system isolated from rats. The spontaneous or compound-induced revertants, each of which reflects a mutational event, appear in a ring as spots around the paper disk. Ames, McCann, and Yamasaki 1975. © Elsevier 51

3.1. Co-occurrence of keywords, 1992–1999. 84

3.2. Co-occurrence of keywords, 2000–2005. 85

3.3. Co-occurrence of keywords, 2006–2009. 86

3.4. Association between keywords and journals, 1992–1999. 87

8.1. TAVOI's General Secretary Hwang Hsiao-ling addressing the media in front of the Taipei District Court on the day of the first court hearing, 11 November 2009. On her right is Lin Yong-song, lead counsel for the plaintiffs. © Paul Jobin 176

8.2. Placard forbidding the use of the fish farms, Anshun, August 2008. © Paul Jobin 183

8.3. Containers of soil contaminated by the dioxin, stored in a former factory of Taiwan Alkali Industry. © Paul Jobin 184

8.4. Pilgrims at Mazu Temple, Anshun, October 2012. The health checks take place in the temple. © Paul Jobin 186

10.1. Post-Katrina Flooding in Greater New Orleans. 218

10.2.	"Epistemic Efficiency" in Risk Assessment.	223
10.3.	"Epistemic Reach" in Risk Standards. The EPA Hierarchy of Human Toxicity Values.	227
11.1.	Case Control Study. Hidden Genetic Effects.	240
11.2.	Causal Chain of Endocrine Receptor Mediated Effects.	244
11.3.	A Two-Range Dose Response Curve Reflecting Two Mechanisms of Action.	245

TABLES

3.1.	Four Ways of Regulating Health-Threatening Food and Drugs	91
8.1.	Carcinogens Implicated in the RCA Case	179
10.1.	Information Gaps in the EPA's Hierarchy of Human Toxicity Values	227

Acknowledgments

As editors of the book, we are indebted to many people without whom this project would not have been possible. We are thankful for the many stimulating and friendly discussions we had with Francis Chateauraynaud, Josquin Debaz, Antoine Blanchard, Axel Meunier, and Elifsu Sabuncu of the Research Project "FADO"—Low Doses and Expertise: Historical and Sociological Approaches. We also would like to thank Amy Dahan, Jean-Paul Gaudillière, Dominique Pestre, and Sezin Topçu for the sometimes lively, but always heavily inspiring, discussions we had during the three years that the seminar on the Global Government of Technosciences (GATSEG) lasted. The debates we had within the FADO and GATSEG projects greatly improved the conception of the book and helped us strengthen our main thesis.

We would like to thank the commentators on earlier versions of the first chapter of the book. In this respect, Beate Bächi, Christophe Bonneuil, Olivier Borraz, Yves Cohen, David Demortain, Jean-Baptiste Fressoz, Claude Gilbert, Emmanuel Henry, Jean-Noël Jouzel, Carsten Reinhardt, Alexander von Schwerin, and Didier Torny have been especially helpful and stimulating. We are also grateful to the anonymous referees for their insightful comments.

The generous support of the Agence Nationale de Recherche (ANR) for the FADO and GATSEG projects facilitated our own research for this book.

Introduction
The Greatness and Misery of Science in a Toxic World

Soraya Boudia and Nathalie Jas

> Most of the necessary knowledge is now available but we do not use it.
> —Rachel Carson, *Silent Spring*

Twenty-five years after the Chernobyl disaster, the Fukushima catastrophe once again brings into sharp focus the risks imposed on all of humanity by certain technologies. An earthquake, followed by a tsunami, triggered a major international crisis, arousing fears of an unprecedented technological disaster. The nuclear explosion ultimately did not take place, and the worst seems to have been avoided. But significant quantities of radioactive material, iodine 131 and caesium 137 in particular, were released into the atmosphere by three of the six reactors that partially melted. Moreover, large quantities of seawater that had served to cool down the reactors were released into the environment. This event highlights a number of problems linked to the dangers of technoscience. It shows that even in one of the richest and safest countries in the world—and one of the most economically and technologically developed ones—in a high-tech sector that mobilizes a large community of experts and is subject to a whole range of very strict international regulations, and in spite of decades of experience, the management of technoscientific risks—particularly environmental contamination by dangerous chemical substances—is still a major scientific, technological, social, and political problem.

Fukushima is a perfect illustration of the observation that underpins this book and that presents itself as a paradox. Throughout the twentieth century, scientific knowledge and expertise were constantly mobilized to develop public policies designed to prevent or manage the effects of toxic substances on health and the environment. Science has thus served as the guarantor of the effectiveness of systems regulating dangerous chemical substances and physical agents. Yet today, in spite of decades of development in research on toxicants, along with the growing role of scientific expertise in public policy making and the unprecedented rise in the number of national and international institutions dealing with environmental health issues, problems surrounding contaminants and their effects on health are far from being resolved. Indeed, they are often at the heart of new public crises and advocacy movements denouncing the shortcomings or even failure of policies implemented. These problems therefore remain a major issue for Western societies and international institutions. However, while scientific knowledge has not made it possible to truly protect populations, it has retained a key position within all public debates—particularly because it is still essential in the identification and characterization of toxicants as well as in public legitimization of different policies related to toxicant-related issues. Scientific knowledge and techniques thus have played and continue to play a determining role in rendering the toxic world visible and in making the resulting issues public.

This statement calls for a reconsideration of the roles of scientific knowledge and expertise in the definition and management of toxicant-related health issues. That is the aim of this book, which seeks to shed light on the way environmental health problems posed by toxicants have been conceived and governed since the 1940s. The different chapters analyze the historical, social, and political trajectories that have structured and continue to structure the statuses and functions of scientific knowledge in toxicant-related issues, whether in toxicant regulation regimes or in the different advocacy movements surrounding them.

The analysis in this book is founded upon three methodological choices. First, it encompasses various approaches, both in its questions and methods of investigation, stemming from environmental history, science and technology studies, political science, sociology, and the philosophy of law. By drawing on very different yet complementary perspectives, we can highlight a much broader range of mechanisms, which have governed and organized the production and use of scientific knowledge, expertise, and counter-expertise for the management of problems posed by toxicants. Second, together, the contributions in this book cover a sufficiently long period of time to account for the important transformations of the role of knowledge in the regulation of toxicants, as well as for the diversity of ways in which knowledge has been produced and mobilized in toxicant policies since 1945. Third, the proposed analysis considers several spatial scales, namely, local, national, and transnational, with a diversity of case studies covering different geographic areas.

As a result, this book analyzes the official and alternative statuses and uses of scientific knowledge in the social and political handling of the issue of toxicants at different times from the late 1930s until today, at different levels, from the most local level to international institutions. A significant part of the chapters are focused on the United States, as that is where the design, experimentation, and transformations surrounding the ways toxicants have been governed historically took place, to then spread to the rest of the world. However, that is not to say that we neglected other parts of the world; we selected case studies through which a much broader host of configurations could be addressed. Thus the Italian case, that of a country that industrialized rapidly in the 1960s and 1970s and witnessed a substantial number of major industrial incidents, the best-known one being Seveso, offers a national configuration very different from that of the U.S. case. The presence at the time of a powerful left wing and trade unions that had found original ways of integrating health and environmental concerns also produced forms of mobilization and counter-expertise worth discussing. Finally, we selected Taiwan in Asia, as it offers yet another configuration, insofar as the contaminated sites result from a long history, related to both colonialism and Western industrial relocations, that further complicates both the production of knowledge on contaminations and advocacy. Through these choices, this book thus offers original perspectives and renewed insights into the issues and processes involved in the management of toxicants.

This book is organized into three parts. Each of them explores a particular aspect of the roles of science in the definition and management of toxicant-related health issues. In this Introduction we discuss each of these main themes. First, we present the various changes in the scientific conceptualization of toxicants since the 1940s, and the ways in which these changes have shaped expertise on and the regulation of toxicants and the problems they pose. We thus show how the production of scientific knowledge and expertise on toxicants and their effects evolved alongside the modes of toxicant regulation. In the second part, we examine the production and uses of scientific knowledge in advocacy movements and in the gradual construction of counter-expertise. We analyze the appearance of counter-expertise in the 1970s and describe the different forms it took on, whether stemming from scientific academia, from the work of scientists working for regulatory agencies, or from lay persons involved in advocacy movements. We identify the diverse roles that the different forms of production of scientific knowledge have played and continue to play in social and political movements surrounding toxicant issues. We emphasize the complex, nonmechanistic relations that subsist between advocacy, nonadvocacy, and knowledge—whether extensive or poor—or ignorance about toxicants. In so doing we highlight that while advocacy movements may involve dynamics of production of knowledge, the existence of significant knowledge on contamination does not necessarily ensure the success of movements, nor even the strengthening of movements.

Finally, in the third part, we consider the role of the social sciences and humanities in the production of knowledge about the ways toxicants have been regulated and as resources for action, whether for regulatory systems or as part of advocacy movements. We first turn back to the main frameworks of analysis that have been developed, such as the propositions formulated by the social sciences and humanities since the end of the 1960s—when they began to consider ways in which toxicant regulatory systems could be improved. We then present a series of current approaches emanating from the social sciences and humanities after decades of toxicant policies and at a time when regulatory systems in Europe, the United States, and international organizations are being reconfigured. The propositions made seek to define the conditions of production and mobilization of knowledge in regulation, so as to develop systems that can deal more effectively with the public health and environmental problems generated by toxicants.

Knowledge, Expertise, and the Transformations in Regulatory Systems

The issues underlying the problems posed by environmental health risks have a long history that has significantly shaped their role in current expert and decision-making communities, as well as in the public sphere. The current ways of managing the environmental health problems posed by toxicants and the roles that scientific and technical knowledge have played in these are the result of an historical accumulation of actions, responses, and institutional configurations and reconfigurations that are rooted in long-term processes about which more needs to be said (Boudia and Jas 2007; Boudia and Jas 2013).

The scientific understanding and study of environmental health problems and the regulatory and public policy systems dealing with them are the product of changes that began back in the nineteenth century and that are closely intertwined with the history of capitalism. Already in the nineteenth century, galloping industrial change profoundly altered the environment, at the cost of chemical pollution, technical accidents, and the poisoning of the bodies of workers, residents, and consumers alike. These multiple effects were not overlooked. They triggered numerous debates and controversies as well as the implementation of a wide range of management mechanisms: expert commissions, especially within academia, court cases, insurance policies, compensation, improvements to technical systems to limit emissions or their effects, the development of sets of regulations to frame the use of toxicants, and new administrations dedicated to the management of potentially dangerous substances (Young 1986; Bernhardt and Massard-Guilbaud 2002; Dessaux 2007; Fressoz 2012; Massard-Guilbaud 2010). Regulation of the activities generating pollution found itself caught up between contradictory logics with, on the one hand, the struggle against visible environmental damage and long-term concerns regarding such damage, and, on the other, the desire to legitimate sustained industrial growth by states concerned first and foremost with ensuring economic development.

Holding these contradictory logics together has constituted a major issue for the administrations in charge of managing pollutants and the dangers caused by industrial activity. These administrations primarily resorted to science and technology as solutions to hold often contradictory imperatives together: to simultaneously ensure industrial and economic development and manage the concerns and protests that could arise—and to provide forms of health and environmental protection. A doctrine of management of industrial excesses developed in the nineteenth century. It elaborated a logic and rhetoric of intervention that gave scientific knowledge and expertise a central role. Thanks to these, it was possible to regulate the dangers posed by industrial pollutants, by precisely determining danger thresholds and elaborating tools of effective control, management, prevention, remediation, and reparation. As a result, the constant progress of science and technology also allowed for regular improvement of the systems regulating the deleterious effects of industrial activities.

Although laws in this respect were inherited from the early nineteenth century, from 1870 on the implementation of regulatory systems accelerated. This corresponded to a period during which, in general, the state was expanding its ambit and simultaneously changing its methods, notably by developing new administrations in which scientific expertise played an essential part. The last third of the nineteenth century and the early twentieth century was thus a period in which the foundations were laid for many national regulatory systems, namely, with regard to foodstuffs, medicines, professional medicine, toxic substances, and industrial pollution. Science played a crucial role in these changes, in several respects. From the growth of chemical analysis to the rise of the hygienist paradigm, and from the development of toxicology to the increasing normalization and security standards on technological facilities, science and technology, through the knowledge and instruments they produce, contributed to building and ensuring the functioning of systems regulating dangerous activity. But although these regulatory systems became stronger during the interwar period, they failed to prevent sanitary scandals resulting from the development of certain sectors of activity: pollution through industrial accidents and collective poisoning through pesticides, medicines, cosmetics, paintings, foodstuffs, etc. (Kallet and Schlink 1933; Whorton 1974; Sellers 1997). These numerous scandals pointed to regulatory systems' incapacity to prevent the dangers posed by the unfolding Chemical Age. They sometimes brought to light regulatory systems' functioning mechanisms and showed their limits. In most cases the regulatory policies implemented seemed to result from negotiated compromises that were acceptable for industrial actors, not from a desire to encourage the production of scientific expertise on the health and environmental effects of toxicants with the goal of elaborating regulatory measures centered on the protection of public health.

Right at the end of the 1930s, these repeated scandals led to the creation of a movement to amend legislation on toxic substances, which remained active throughout World War II and after it ended. The transformations of regulatory systems that took place between the late 1930s and early 1950s gave an even more explicit role to scientific knowledge and expertise. During this period, the principle of toxicity evaluation prior to issuing a product license, namely, was imposed in a number of countries and for a number of substances (medicine, pesticides, food additives). The aim of these evaluations was to decide whether the substances could be authorized or not, and to set the conditions for their use so that they did not present a danger for public health. The designers and promoters of these new regulations argued that the objective was the complete elimination of "hazards." The Food, Drug, and Cosmetic Act passed by the U.S. Congress in 1938, discussed in the first chapter of this book by Nancy Langston, offers a paradigmatic example of this new approach. Langston shows that this law was based on precaution, but that that was not enough to prevent the dissemination of a substance that is as dangerous as diethylstilbestrol (DES). She analyzes how during the 1940s, three instances of industrial lobbyists' political work achieved the reversal of a decision by the Food and Drug Administration (FDA) that, for precautionary reasons and within the framework of the 1938 Food, Drug, and Cosmetic Act, had demanded that DES be banned.

Among other things, the emblematic case of DES, discussed by Langston, shows that laws on toxicants in the late 1930s, the 1940s, and the 1950s, while theoretically very protective, were not able to deal with the radical change of scale in the problems posed by toxicants from the end of World War II on. First, the numerous biases toward industry did not disappear with these new regulatory systems, and the development of economic activity remained a major concern that justified public health protection systems being virtually systematically bypassed. This was made all the more easy by the rise of potentially dangerous industries like the petrochemistry, synthetic chemistry, and nuclear industries, which stood as emblems of a modernity that promised wealth and a new well-being. These industries developed at such speed that regulatory systems, with far more limited means, could hardly be effective. These industries were socially, economically, and politically far too powerful for public health or environmental protection to have been considered by political authorities as a sufficient reason to restrict their expansion. As a result of the development of these industries, the world witnessed an unprecedented increase in the quantities of chemical substances put into circulation and onto the market, and some of those substances started to be found in the atmosphere, the soil, and water. Although the regulatory systems did rely on scientific expertise, they did not have the means to carry out in-depth examinations of the numerous new substances brought onto the market (Davis 2001; Ross and Amter 2010; Vogel 2012). In fact, most of them were not evaluated or regulated in any way whatsoever.

This book shows that it is crucial to understand and analyze the changes that took place between the late 1960s and the early 1980s if we are to make sense of the way the regulation of toxicants is structured and functions at present. The most significant change during this period was the unprecedented growth of environmental issues and the long-term inscription of environmental health issues within the different public and professional arenas (Hays 1989; Brooks 2009). At the end of the 1960s, in the wake of the social and political movements of the time, the environment gradually became a major theme of radical criticism. There was a proliferation of environmental health issues making their way onto the political agenda: various types of chemical pollution, air pollution, water contamination, and food contamination were denounced and associated with an unrestrained capitalist economic development.

There was a shift in the way the nature of the issues raised was represented, as evidenced in several chapters of this book. The crisis of the 1970s brought to light the rise of problems whose scale and potential consequences were unprecedented. These new problems were partly defined by the greater scales of space and time within which they existed. Pollution was no longer local but could affect the entire planet. It affected not only health but the entire ecosystem. The consequences were not only immediate; they could be felt decades after exposure or contamination, and over several generations. Due to their unprecedented scale, from the infinitely small to the infinitely big, health and environmental issues raised a host of new questions that experts and institutions had to deal with. Various types of answers were provided. They were both political and administrative, involving regulatory and institutional reconfigurations. At national level, in the United States and certain European countries, this translated into the creation of agencies to manage environmental problems, and/or the reconfiguration of systems regulating toxicants, as symbolized by the creation of the Environmental Protection Agency (EPA) in the United States in 1970 or the development of environmental regulations by the European Economic Community and in European countries from the late 1960s on. At transnational level, new initiatives proliferated. The United Nations Conference on the Human Environment held in Stockholm in 1972, for instance, was organized to discuss the general state of the environment and to identify problems requiring international collaboration. One of the memorable initiatives to come out of this conference was the creation that same year of the United Nations Environment Programme (UNEP).

These different transformations that took place in the late 1960s and early 1970s reflect, and themselves induced, important changes in the role and place of scientific knowledge in dealing with toxicant issues. In the context of questioning, criticism, and activism, science, along with its actors, products, and methods, came to occupy a central position. The keener attention paid to environmental issues gave a whole new standing to researchers working in the field of environmental health. In the alarms that they sounded these researchers identified the extensive presence of chemical contaminants in the environment as being responsible for the development of new health problems, such as genetic mutations and effects on reproductive problems, which thereby acquired unprecedented public visibility. A large volume of scientific work was produced. After studies on carcinogenesis came those on ecotoxicology and environmental mutagenesis (Frickel 2004). Hence, for a whole host of substances, the lack of greater precautions surrounding their use and regulation in the 1970s could not be explained by uncertainty or a lack of knowledge regarding their pathogenic effects. The absence of significant mitigation of the problems caused by toxicants, following the explosion of knowledge production in the 1970s, began to highlight the fact that, contrary to the public discourse developed for decades, "science alone cannot solve the problems posed by contaminants"—to take Langston's words.

With the proliferation of substances in circulation and the multiplication of denunciations of their effects by activist movements, the screening of dangerous substances and the precise definition of their effects became a core part of the work of researchers, experts, and new institutions in charge of managing contaminants. The U.S. agencies, such as the Environmental Protection Agency and the Food and Drug Administration (FDA), and international organizations like the International Agency for Research on Cancer, created in 1968 under the World Health Organization (WHO), all took on the role of leaders in the field. The multiplication of regulatory and expertise agencies allowed for the growth of research on testing and screening methods. Another feature characterizing the work that developed in the 1970s was the classification of chemical substances' effects. As shown in Angela N.H. Creager's and Jean-Paul Gaudillière's chapters, several research projects and institutional initiatives were dedicated to identifying a relationship between carcinogenicity and mutagenicity or reproductive effects.

Creager's chapter evidences the rise of research focusing on the screening and characterization of chemical substances' toxicity during the 1970s, an explosion that has so far been studied very little. Creager studies the evolution of the work of biochemist Bruce Ames to show the importance given to the development of dangerousness tests, both by industrial actors and by regulatory agencies and environmentalists. In 1973, Ames devised a test to determine the carcinogenicity of chemical substances, which generated strong interest given the possibility of applying it to a host of chemical substances on the market. The test stirred real enthusiasm among environmentalist groups and was rapidly adopted by industrial actors due to its simplicity and the lower costs involved compared to animal testing. It was based on the assumption that any carcinogen was a mutagen, and that a microorganism was an adequate model for testing mutagenicity as it can develop in human cells. Since the 1970s, the nature and results of this type of test—those by Ames and many others that have been put forward over the years—have played and still do play a crucial role in the definition of regulatory systems. They generate stormy controversies among scientific experts, which are visible to varying degrees in the public sphere. The movement that developed in the 1970s around the Ames test is currently at the heart of proposals to overhaul and elaborate a "new toxicology," formalized in a 2007 report by the U.S. National Research Council (NRC), and seeks to ensure that regulatory toxicology no longer relies essentially on animal testing, but on in vitro tests and computer modeling.

Research on the relationship between carcinogenic effects and toxic effects on reproduction is addressed in Gaudillière's chapter. Since both look at the DES case, comparing Langston's and Gaudillière's contributions sheds light on the nature of the transformations that took place between the 1950s and the 1970s. Gaudillière analyzes the multiple transformations, both legal and scientific, that took place throughout the American court cases on DES in the 1970s. He shows how the confrontation of experts over the course of the court cases led to the production of new knowledge on toxicants. Although this chapter contributes to highlighting an important phenomenon of the transformations that took place starting in the 1970s and that is analyzed in detail in the second part of this book, that of the diversification of the sources and places of production of knowledge on toxicants with the rise of counter-expertise, it also contributes to another very important aspect. It allows us to grasp the crucial issue of the categorization of dangerous substances in regulatory systems. While in the 1950s carcinogenic substances motivated continued investigation and classification work, in the 1970s two other categories of particularly hazardous substances were formalized: mutagens and reproductive toxicants. Later on the CMR category (Carcinogens, Mutagens, Reproductive Toxicants) was developed with a view to adopting a more holistic approach to effects, to establishing links between them, and to classifying chemical substances according to their effects. This classification comprised the substances considered to be the most dangerous, in terms of both their effects and their capacity to have a delayed effect in low doses. It has formed the basis for the development of systems of regulation of toxicants since the 1970s and, in modified versions, is still highly influential in current regulatory systems. Gaudillière's account shows how during a court case, through the confrontation of experts, some of the characteristics of DES which did not fit in with the then prevailing conceptions of toxicants' effects were highlighted. The deleterious effects of DES could be more significant in low doses than in higher doses, and the timing of exposure could play a crucial role in the type of effects obtained. Gaudillière ultimately shows how instrumental the DES case was in the early 1990s, as during the Wingspread Conference (1991) scientists linked to U.S. health and environmental activism formulated the endocrine disruptors (EDs) hypothesis, and with it a new category of highly hazardous chemicals. Activists currently use EDs characteristics to call for the overhaul of the CMR classification system and for regulatory systems implemented in the 1970s to be scrapped. They consider these both out of date and incapable of protecting populations from the deleterious effects of what they see as the "new toxic substances" (Krimsky 2000; Vogel 2012).

As well as the transformations in the scale of the problems and in the way toxicants were conceptualized and categorized, this book highlights another type of change in the 1970s. It pertains to the ways in which public policies on contaminants are managed and legitimated, as analyzed by Soraya Boudia in this book. Her chapter shows that the growth of work and the accumulation of data on contaminants and their effects led to the challenging of the threshold paradigm that had structured the perception as well as the regulation of toxicants since the end of the nineteenth century. To fully grasp these changes, it is useful to remember that environmental health problems were approached essentially through the dogma of toxicology, which holds that "the dose makes the poison," in other words, that for each toxicant it is possible to determine a threshold below which no deleterious effect is observed, or below which risks are perfectly negligible. Until the 1970s, all regulations on toxicants were based, officially at least, on this dogma. This meant that from the 1940s on, threshold values were increasingly used, with denominations specific to each domain and the creation of a host of labels, such as tolerable dosage, permissible dosage, Maximum Allowed Concentration (MAC), or Acceptable Daily Intake (ADI). These threshold values made it possible to use substances without their having—at least in theory—too significant or irreversible an effect on health. Nevertheless, from the early 1970s on, suspicion began to grow regarding this approach, through discussions on the effects of low doses of radioactivity and many carcinogens. The accumulation of results concerning the effects of exposure to carcinogens in the workplace or in the environment, along with a number of experimental studies, tended to show that, for numerous substances, nothing permitted the definition of a threshold below which no deleterious effects could be observed.

The question of low doses was a major political issue. It cast doubt on a host of activities that until then had been considered safe or seen as presenting negligible risks. Raising this issue amounted to claiming that innovations could have negative sanitary and environmental effects not only in exceptional situations like accidents, but also in "ordinary" situations, in their normal use. This was inherently a critique of various scientific and industrial domains: without generating major threats, they contributed to spreading in the air, water, and ground proportions of toxicants considered negligible. The issue of exposure to low doses undermined regulatory systems, for which defining thresholds and threshold values was a major activity. The recognition of the potential problem of exposure to low doses of pollutants de facto generated a contradiction in the practices of regulatory systems. On the one hand, this meant admitting that there is no threshold below which one can assert the innocuousness of a substance; on the other, setting threshold values remained central to regulatory systems (Bächi 2010).

As a result, starting in the 1960s the discourses legitimating regulatory policies began to change noticeably (Jasanoff 1990). To overcome the contradictions generated by the issue of low doses, the procedures used to determine these threshold values were increasingly presented as seeking not to guarantee the absolute innocuousness of the use of certain substances under certain conditions, but to establish "socially acceptable" levels of risk. It was thereby recognized that exposure norms did not result from a scientific decision only, but incorporated economic and political considerations as well. The institutional changes in the 1980s and 1990s fully took into account this new dimension, which was expressed in the desire to separate the "assessment" of substances from their "management." This was formalized in the NRC's Red Book on risk management published in 1983, as Boudia points out in her chapter of this book. The separation between "assessment" and "management" subsequently became widespread; it was adopted in both national and transnational regulatory institutions. A paradoxical situation was thereby officialized in the second half of the twentieth century, in which the way toxicants are governed is still rooted. Regulatory systems recognize that standards of exposure, and more generally, the regulation of toxicants, result from scientific as well as economic and political processes. Yet at the same time, expertise and scientific knowledge are still publicly referred to in order to legitimate decisions on toxicants and their effects.

Activism and Nonactivism: Alternative Uses of Knowledge

The rise of environmental concerns, the unprecedented accumulation of scientific work on the effects of toxicants, and the multiplication of regulatory systems as sophisticated as the ones implemented in the 1970s have not led to the disappearance or significant decline of contaminants' impact on health and the environment. On the contrary, the number and quantities of toxic or potentially toxic chemical substances disseminated since the 1950s has continued to increase, resulting in a proliferation of contaminated sites and the growth of a broad range of deleterious effects on an unprecedented scale.

The lived experience of this materiality, be it in terms of environmental degradation or damage to human health, has played a large part in the transformation of social movements surrounding the issue of toxicants and their effects since World War II. Like environmental health problems, these movements are the outcome of a long history. Industrial pollution and its effects on human health, forests, agriculture, and animal husbandry generated multiple forms of protest throughout the nineteenth century and between the two world wars, ranging from trade union movements to court cases initiated by locals, or press campaigns. In the United States in the 1930s, in the middle of an economic crisis and following numerous scandals triggered by collective toxic contamination, the chemical industry was even accused by the first consumer movements, using a highly successful book, *100,000,000 Guinea Pigs* (Kallet and Schlink 1933). From the mid 1950s on, the idea that human beings had contributed to making their environment toxic consistently gained currency. Following the wave of controversies on the effects of radioactivity, chemical pollution—particularly that linked to pesticides—became a widely debated issue. These concerns originated from certain professional circles, particularly those of cancer specialists, but also from the everyday experiences of the middle classes settling in rapidly expanding suburbs, close to fields where pesticides were used on a large scale. During the 1960s, scientific and civil society actors in the large movements of the time fully embraced the issues underpinning environmental health. The publication in 1962 of *Silent Spring*, which soon became a best seller worldwide, by a marine biologist, Rachel Carson, effectively marked the beginning of a movement that gained importance in the second half of the 1960s (Carson 1962).

The environmentalism that developed from the late 1970s highlighted a number of new questions being raised regarding the place of human beings in the biosphere, the depletion of natural resources, and environmental pollution and its immediate and long-term effect on humankind. These themes were recurrent in a number of actions and movements, led by figures such as Ralph Nader. Health was a pivotal and even structuring dimension of their interventions and a recurrent feature of activism at the time. This movement was supported by activist organizations that later became important, such as the American Environmental Defense Fund, created in 1967 to support anti-DDT movements (Dunlap 1983). These activist organizations did not spring up only in the United States. The numerous preparatory conferences between 1969 and 1972 leading up to the United Nations Conference on the Human Environment held in Stockholm in June 1972 also show the existence of this type of activism in countries of northern Europe. During the 1970s and 1980s, local and national organizations expanded their activities outside their territories of origin, as in the case of Friends of the Earth created by David Brower in the United States, which spread to 76 countries, or Greenpeace, founded in Vancouver, Canada, in 1971 by a small group of anti-nuclear activists.

The growth of these large activist organizations in the 1970s and 1980s went hand in hand with the rise of other types of organizations for which issues of environmental contamination were a major concern. Older organizations that previously focused on nature conservation reoriented their activities. In North America at least, movements for women's health engaged with the issue of the effects of toxic substances on health, initially with the question of synthetic hormones. Local victims' associations were created in long-term struggles against industrial actors responsible for the contamination of certain sites (Brown and Mikkelsen 1990; Kroll-Smith et al. 2000; Allen 2003; Brown 2007). Certain scientists involved in the production of official expertise, outraged by certain practices, founded independent research and expertise institutions, as in the case of the toxicologists and epidemiologists who founded the Italian Foundation, the Instituto Ramazzini. With a view to forming alliances, pooling their resources and increasing their capacity for action, some national organizations also federated and developed large transnational networks. Thus, over the last four decades, extremely complex webs of activist organizations have formed, including small and large organizations wielding varying degrees of power, with varied and sometimes contradictory objectives. All agree, however, on the existence of unacceptable threats to health and the environment caused by the uncontrolled excesses of the chemical era (Pellow 2007).

Scientific knowledge has played a growing role in the actions of the different advocacy movements (Ottinger and Cohen 2011). With industrial actors and political and administrative authorities denying the existence of problems related to toxicants, it became necessary to provide scientific proof of the existence of dangerous effects and to assess the extent of environmental pollution. Alternative production of scientific knowledge and counter-expertise therefore began to grow in the second half of the 1960s. The aim of such production was and still is not only to prove the existence of contaminations and deleterious effects, but also to reveal them and make them visible. It was expected that this would trigger or strengthen mobilization, thus prompting industrial actors and government authorities to deal with the problems at hand. This alternative production of scientific knowledge and counter-expertise unfolded in three interdependent processes.

The first was the involvement of established scientists—some of whom were renowned—in environmental causes in the name of science. Based on the results of research that they or others had carried out, several scientists became whistle-blowers. They decided to make facts and concerns public and to call for the implementation of prevention and remediation policies. During the 1960s and especially the 1970s, the number of renowned and less-known scientists adopting this kind of attitude multiplied. Apart from emblematic figures such as Rachel Carson (Lear 1997) or Barry Commoner (Egan 2007), many scientists, presented in a number of chapters in this book, embraced the issue of the effects of toxicants. The generalized contamination of the environment, the fauna, and human beings by PCBs (Polychlorinated Biphenyls) was revealed for instance through the relentless work of a Swedish researcher, Soren Jensen, between 1966 and 1968. His work was rapidly circulated within international arenas and contributed to launching an important movement, particularly in the United States, to reveal numerous contaminations from these substances. Despite massive lobbying by the company producing them, Monsanto, and those that used them, such as General Electric (McGurty 2009), this movement achieved a total ban on PCBs in 1979 in the United States, and in the mid 1980s in most European countries—but the ban did not resolve the problems caused by these very persistent substances.

This unprecedented involvement of scientists, whether they were well-known or not, was accompanied by a move toward the redefinition or even the creation of new disciplines to address the wide range of questions raised by the breadth and complexity of contaminations. From the 1970s, the rise of new fields such as "chemical mutagenesis," "environmental hormones," and ecotoxicology reflected the desire to articulate the promotion of new research subjects and approaches not yet recognized in the academic world, with the need to bring to light and study the problems generated by the massive circulation of potentially toxic chemical substances. This involvement motivated by professional concerns may have been complemented by a more political type of involvement. Laura Conti's scientific work in Italy in the 1970s, discussed in Stefania Barca's chapter in this book, is a particularly interesting illustration of the different types of scientific and political activism. A doctor by training and a communist, Conti developed a form of environmentalism that placed toxicants and human beings at its center. This environmentalism insisted on the multiple and complex relationships between the living and the nonliving, and showed the irreversible effects of the constant release of petrochemical waste into nature, which could not be controlled by simply resorting to thresholds on toxic concentration. Conti's scientific work was therefore nurtured by her political commitment, just as her political involvement was deeply influenced by her scientific work.

Other forms of knowledge production emerged in addition to the production of new knowledge on toxicants by academic researchers or researchers working for activist movements. Local action surrounding contaminated sites, studied in this book by Paul Jobin and Yu-Hwei Tseng as well as by Barbara L. Allen, increased exponentially starting in the second half of the 1960s, first in the United States and then in other parts of the world. The administrative and legal proceedings that took place as part of these mobilizations, providing scientific evidence of contaminations and their deleterious effects, proved to be a significant factor of success. Calling on academic researchers—even specialists able to demonstrate the existence of deleterious effects—has not always proved easy or effective. Some scientists, such as the Taiwanese epidemiologist Lee

Ching-Chang, described by Jobin and Tseung, refused to reveal their results beyond narrow academic circles. Others, such as the epidemiologist Patricia Williams, a chemical contamination specialist discussed by Allen, were first and foremost concerned with conforming to the scientism criteria of their professional community. Yet the time frame of academic research that eliminates any possible bias and the time frame of protest mobilization do not always coincide, and results can be made available too late to support the cause of activists. Moreover, the expectations inherent in academic research do not always correspond to activists' expectations, as each world has its own motivations.

Due to the inappropriateness or shortcomings of academic research in producing sufficiently conclusive scientific evidence, activist or victims' groups began developing other types of knowledge production, sometimes turning to actors other than established academic researchers. The victims themselves, relatives, and doctors or scientists who were not too concerned about their careers were able to organize themselves, identify patients, and gather data on exposure to finally show correlations between local exposure and the abnormal increase of certain serious pathologies. Patients' organizations and the cartographic work carried out at many contaminated sites gave rise in the 1980s to what Phil Brown calls popular epidemiology (Brown 2007). This is based on the elaboration and implementation of techniques that differed from those used by government authorities and regulatory bodies. It has allowed scientists allegedly less specialized in a certain subject, doctors without a research activity, retired engineers, laborers, office workers, mothers, etc., not only to produce data, but also to become experts on certain health and environmental problems. In this perspective, Allen discusses the case of Gabriele Bortollozzo, a worker from 1956 to 1990 at the highly contaminated site of Montedison in Italy, while Jobin and Tseng consider that of former workers from the Taiwanese factories of Radio Corporation of America. Both cases are highly representative of this bottom-up knowledge production by the victims themselves or their relatives—with the support, over time and depending on the locations, of activist organizations and committed scientists.

From the early 1990s and with varying time frames in different countries, a third form of change occurred through which the development of counter-expertise within local movements and national and transnational organizations, by scientists and nonscientists, took on a new dimension. From the early 1980s, the supposedly profound transformation of systems regulating toxicants that took place at national and international level during the 1970s following environmentalist activism proved to have been a failure. During the 1990s, the multiplication of highly visible issues and scandals surrounding the deleterious effects of technoscience stressed the fact that science was not in a position to provide clear answers and precise information on the dangers incurred. Yet in situations of uncertainty, decisions concerning the regulation of technoscientific practices had been taken behind closed doors by small groups of experts. Strong mobilization, defiance of certain innovations, and the discrediting of certain administrative and statal systems led policy makers to develop new modes of government, underpinned by new systems under the banners of "participation" and "transparency" (Pestre 2008). In this new context, activist organizations and committed scientists were encouraged to participate as "stakeholders," or even as experts on certain committees in order to represent "citizens'" point of view. While the shortcomings of participatory systems had become fully visible by the late 2000s (Irwin 2006; Pestre 2008), the presence of civil society representatives and alternative scientists as "experts" or "stakeholders" in current official expertise processes seems to be a given in many national and international contexts. This is closely monitored and activist organizations' representatives have a say in decision making, or have means similar to those of other interest groups—particularly industrial lobbies. But, apart from the context of the 1990s, which opened a window of opportunity for counter-expertise to get closer to official expertise processes, what made activist organizations legitimate experts within these committees was their grasp of the cases and scientific competences that they had developed in various ways, over the previous two or three decades.

While the resulting production of alternative knowledge in various contexts played a significant part in shaping the development of movements around toxicants over the last four decades, many difficulties were encountered. Providing evidence that meets scientism criteria of damage or potential damage, even serious damage, has often not been enough to obtain the compensation, remediation, or prevention demanded by activist movements or victims' organizations. The chapters in this book offer more nuanced positions regarding the role of scientific knowledge and counter-expertise in mobilizations surrounding problems related to toxicants. Numerous cases show that balancing health-related and environmental risks with the disappearance of economic activities that are essential to certain regions presents an important dilemma that even the production of irrefutable scientific knowledge cannot resolve (Auyero and Swistun 2009). In other situations, legal and administrative systems function in such a way that the production of knowledge on contaminations is far from sufficient to produce a decision in the victims' favor, or the decision provides far less than the victims had expected. Laura Centemeri's analysis of the inhabitants of the Seveso site that was contaminated by dioxins following a major industrial accident in 1976 highlights how knowledge is not sufficient motivation for taking a stand. Even though this site attracted much attention in the study of the effects of dioxins on human health, and the research results tended to show the extent of the damage caused, these data did not spur the inhabitants of this Seveso region into action. Centemeri identifies many factors to explain this paradox, including the inhabitants' attachment to the territory in which they live and their refusal to see it stigmatized by activists and scientists highlighting major pollution. The overall context made it impossible for the necessary alliances to form in Seveso to mobilize the most affected people. Thus, certain movements have failed in spite of undeniable proof of the contaminations and their effects, while others that rested on far more tenuous and debatable causality links have succeeded. This points to the fact that the success of activist movements is contingent upon their capacity to build effective alliances and apply political pressure. From this perspective, alternative scientific knowledge and counter-expertise are indeed essential but certainly not sufficient; sometimes they are not even indispensable to the success of a social movement against toxicants.

Ultimately, the important movements that have developed since the 1960s have certainly not managed to reverse the trend that began in the late eighteenth and early nineteenth centuries, which saw Western societies choosing a capitalist model of development relying on ever-increasing industrialization at the expense of the environment and human health. At the local level, however, they have managed to win trials, to prevent the creation of a rubbish dump or a waste management center, to close a factory, to clean up contaminated sites, or to compensate victims. At national and international level, they have obtained lower standards of exposure, bans on polluting substances and technologies, amendments of laws, and overhauls of regulatory systems. They have even managed to highlight unanticipated toxic effects and to introduce new issues within scientific and public arenas. The alternative production of scientific knowledge and expertise may have been essential to these achievements, but it has never been the only determining factor. The effective use of this production was possible only because it was embedded within political strategies that, for various reasons, have allowed "subrogate interests" to, at least temporarily, override "dominant interests" (Bosso 1990).

Putting Knowledge, Ignorance, and Regulation into Perspective

The multiple health and environmental problems posed by toxicants are not behind us—far from it. The number of chemical substances in circulation continues to grow. To the toxic legacy of banned or regulated substances like DTT and PCBs as well as the many unregulated ones, new substances whose effects are still relatively unknown, such as nanocarbons, are being added. Faced with this situation, many actors are currently calling for a profound reform of expertise systems and modes of regulation surrounding toxicants. Many social scientists, without all sharing the same point of view, are directing severe criticism toward existing expertise and regulatory systems, some adding their voices to different nongovernmental organizations to demand an overhaul of these systems. This attitude is not new. Since the 1970s, when many social movements highlighted the significance and the extent of contaminations that existed since the end of World War II, certain fields within the humanities and social sciences, namely, law, sociology, political science, history, anthropology, and psychology, have taken an interest in the functioning of scientific expertise and systems regulating toxicants and the technosciences and, since then, have been offering different types of analyses that have sometimes led to normative positions proposing given types of change.

Certain cross-country comparative studies have sought to bring to light the social, institutional, and cultural factors explaining the nature of the expertise produced and the way regulatory systems are organized (Brickman et al. 1985; Vogel 1986). In doing so their aim has been to define norms and strategies to improve the functioning of these systems. Extensive work in the humanities and social sciences has called for greater transparency in the procedures underpinning scientific expertise and decision making. One of the concepts that has stemmed from this work and has been taken up in the different public policies is "sound science." Such analyses, produced mainly in the 1980s, were based on the more or less explicit assumption that science is able to effectively inform public decision making, provided that systems of expertise offer experts the means to draw on "state-of-the-art science" and to make the different points of view public.

Starting in the 1980s this approach was heavily criticized by other researchers whose work insisted on two interdependent issues. First, there are many moments of significant scientific uncertainty in processes of expertise, for which no "sound science" is available. Second, drawing on several case studies, these researchers stressed that in these situations of uncertainty, experts tend to make decisions that are rather in favor of industrial actors, at the expense of consumers, citizens, or patients (Hood and Rothstein 2001; Abraham and Reed 2002). In other words, a bias in favor of industrial actors and economic imperatives exists in expertise and regulatory systems. These authors argued that reforms of systems of expertise were needed, not to guarantee the use of a "sound science," which did not necessarily exist, but to reduce the bias in favor of industrial actors and to ensure that the interests of consumers, citizens, and patients are taken into account.

Work stemming from a different perspective has also sought to promote lay or alternative knowledge as opposed to expert knowledge. It emphasizes that "lay people" have knowledge, interests, and concerns other than those of "scientific experts" regarding important issues about technoscience and its sanitary, environmental, and social impacts (Irwin and Wynne 1996; Wynne 1996; Pestre 2008). Their knowledge, interests, and concerns are no less valid; they stem from different perspectives that deserve to be taken into account in the production of expertise and in public policy making. If science, especially in situations of uncertainty, is not able to provide sure answers to the problems raised by technoscience, then it is important for public decision making to rely not only on expert claims, but to fully integrate the knowledge, concerns, and interests of "lay people." To promote a more democratic management of technoscience and the problems it poses, these researchers have often been involved in the development of participatory procedures encouraging the growth of counter-expertise and its integration into regulatory systems.

These various sets of works have gradually shown the limits of scientific knowledge in resolving the issues raised by toxicants and the often political nature of decisions regarding these substances. These two features have been emphasized in four types of work. First, certain studies, namely, in environmental history or the history of environmental health, have emphasized the materiality of the problems of environmental degradation and pathologies (Markowitz and Rosner 2002; Blum 2008). In doing so, they have highlighted the numerous instances of reductionism and downplaying in official expertise. Indeed, a deep rift exists between the materiality of damage and the existence of exposure norms, between the reality of chemical cocktails to which certain populations are exposed and substance-based approaches, between the years of illness, the individual, family or collective tragedies, and the slow pace of court cases and regulatory processes, and between situations of potential or immediate danger and the time needed to validate scientific knowledge. These studies have also shown how different social movements—economic and/or political interest groups—have sought to mend or maintain this rift, triggering numerous confrontations. A second type of work in the fields of law and political science has paid attention the construction of systems to regulate toxicants as a whole (Bosso 1990; Cranor 1997). By showing both the complexity of these systems and the extent to which they are shaped by political choices, this type of work has helped bring to light how little weight science and expertise may have in decision making—even though more often than not these systems claim to be "science-based." Such work, which often has normative objectives, has contributed to many analyses since the 1970s, analyses that have a view to inventing other, more effective, regulatory systems and that have also involved rethinking the place and role of science and expertise in systems of expertise. A third type of work, stemming from sociology and political science, stresses the impossibility of building consensus and public policies based on scientific knowledge alone. These works consider that in most risk situations, technical uncertainty is too great for robust social consensus to be built. They therefore call for new modes of discussion, decision making, and policy making to be imagined and implemented, based on the aim of building consensus between the different actors concerned. These works, looking at consensus conferences, participatory democracy, or hybrid forums, have been very successful with policy makers, namely, in Europe (Callon, Lascoumes, and Barthe 2009).

Many sociologists and political scientists thus play an important role in advising and defining regulatory policies. Developing compromise among different actors is central to this literature, which praises the many benefits of participatory systems, including overcoming profound social asymmetries through debate. A fourth and last type of work, "environmental justice studies," is particularly developed in the United States. Openly contributing to research for action, it seeks to highlight that the burden of toxic contamination is primarily borne by certain social groups that are particularly poor and discriminated against: black minorities, Mexican migrant workers, "native" populations. In so doing, this type of work associates social inequality—based on race, class, gender—with greater toxic contaminations, and the struggle against these contaminations is presented as a source of empowerment and as attempting to implement a failing social justice. To do so, it seeks to identify the most effective advocacy strategies and ensure the success of movements. In this context, particular attention is paid to the production of knowledge, whether that production is academic, stems from regulatory systems, or comes from grassroots movements ("street science"). One of the important objectives is to counter efforts that official systems may pursue to make contaminations and their effects invisible and to identify ways of transforming these systems so that they may contribute to making toxicants and their consequences more visible.

Thus for several decades now, the humanities and social sciences have not been working from an exclusively analytical perspective, but one that is also normative and aiming at transforming regulatory and expertise systems surrounding toxicants. Following several reconfigurations and attempts at transforming these systems, certain analysts are currently shifting their positions, sometimes significantly, from what their colleagues or they themselves may have proposed in the past. Carl F. Cranor's chapter offers a perfect illustration of this shift. From a philosophy of law perspective, Cranor has contributed to a lot of reflection on the use of scientific evidence in legal decisions and how society might approach the regulation of toxicants. Through his chapter in this book, Cranor's approach clearly seeks to influence public decision making in the context of the current U.S. reform of the Toxic Substances Control Act (TSCA). He tries to explore not what science is unable to know or do, but what law and regulatory systems have been or are unable to achieve. More importantly, Cranor looks at science and what it is able to show, to emphasize the ineffectiveness of law and to shed light on how inhabitants of the United States are "legally poisoned." Cranor's work shows the shift of position that some of its representatives have made. While the objective of these studies is always to think about and propose a legal framework and regulatory system with the aim of protecting human health and the environment, an explicitly activist dimension is emphasized.

The idea that the strengthening of expertise and regulatory systems does lead to greater protection can be questioned from several perspectives. One of these could point, as researchers studying the tobacco industry have done, to the importance of the economic interests at stake, and to the significant political and public work that contributes to invisibilizing or minimizing the ensuing problems. Several strategies have been studied from this perspective, from lobbying to instilling public doubt. The weight of economic interests is of course a crucial parameter in issues of expertise and regulation. And this weight is what leads certain actors to call for greater regulation. However, more regulation does not necessarily mean that toxicant problems will be resolved. The major problem is a systemic one, which lies in the very functioning of these systems. Through the long-term analysis offered in this book on the role of science in expertise and regulation, one aspect stands out: despite the immensity of the activity they have generated, these systems have not allowed for the production and accumulation of real knowledge on toxic substances, as, on the contrary, through their very functioning they have contributed to producing and disseminating ignorance. Producing ignorance does not just involve hiding certain knowledge, ignoring certain questions, minimizing certain effects, or deliberately producing public uncertainty (Proctor 1995; Oreskes and Conway 2010), even when knowledge is available to form a verdict. It is another type of production of ignorance that some of the authors of this book are concerned with. In their chapter, Scott Frickel and Michelle Edwards, through a detailed analysis of the risk assessment process for soil contamination in New Orleans after Hurricane Katrina, reflect on expertise in terms of its ability to produce not knowledge but ignorance. They also show that this ignorance then circulates and not only forms the basis of certain political decisions but is also integrated into other types of expertise. The two authors thus offer a new perspective on expertise and regulatory systems that invalidates the idea of an optimization of knowledge production in current settings. The significance of this perspective reversal is twofold. First, it is embedded in and contributes to an important theoretical shift in science studies, known as the New Political Sociology of Science, to which Frickel is an active contributor and which seeks to reposition the political at the heart of the analyses produced by science studies (Frickel and Moore 2005). Second, this reversal allows for new perspectives to shed light on processes that have not been noticed or studied much until now, and through which science-based regulatory systems are not able to protect public health and the environment.

This book therefore points toward a conclusion with important consequences: not so much a call to strengthen expertise and regulation but a call to profoundly overhaul the world of knowledge production in these systems. For such an overhaul to take place, particular attention should be placed on a careful and multidisciplinary examination of the instruments and modes of production of knowledge and rules. Yet this does not mean that scientific knowledge should form the core basis of decision making. This raises the question of knowing what should be at the heart of these systems. This book explores several possibilities that seek to subvert the very logic of these systems. Thus Sheldon Krimsky's chapter suggests the importance of working in a precautionary framework. The chapter's starting point is the study of scientific production through an analysis of the way the effects of low doses of endocrine disruptors are scientifically studied. He identifies many factors, ranging from the complexity of the issue to the actions of industrial actors, which cause a number of questions to remain without a stable answer. While the argument that science's incapacity to produce the expected knowledge has already been widely discussed, Krimsky's analysis makes two different contributions. First, as other works have done, this study shows the value of delving into the production of scientific knowledge and analyzing both the potential and the limits of such production. Second, this analysis leads to a valuable consideration, both in heuristic and political terms: if science is not able to provide the expected answers, how can we make sense of its role and of the constant rise of "science-based" regulatory systems? Krimsky's answer is unequivocal: if science cannot provide all the answers expected from it, then it should no longer be the only central frame of reference of regulatory systems; these systems must also rely on other approaches. The shift he calls for is one that grants less importance to scientific knowledge and expertise and more to other approaches, such as precaution. It is central to a current broader movement involving both scholars and activists.

What thus becomes apparent is that reinforcing expertise and regulation, without calling for a profound overhaul of all the foundations of expertise, is necessarily bound to fail. However, it is no easy task to simply enumerate what should be done. This is the difficult exercise Jody A. Roberts tackles in his chapter. His contribution is an analysis of what could be an effective regulation of the chemicals that he qualifies as "unruly technologies." Roberts first looks back on half a century of chemical regulation and reviews the reasons why these regulations never really worked. From the materiality of chemicals that never behave as anticipated, to the practices of industrial actors, through the limits of science and technology: a host of combined factors has ultimately led to the recurrent failure of regulatory systems. Roberts then discusses what could be an effective regulation of chemicals: for him, the answer lies in the diversity and multiplication of approaches. He thus explores solutions such as encouraging economical consumption, substituting, and developing green chemistry, while also recognizing their limits. Like all the other authors in this section, Roberts insists on the need to shift the center of gravity of regulatory systems. He suggests placing justice, not science, at the heart of regulatory systems as a means of guaranteeing their effectiveness in terms of health and environmental protection. In doing so he draws on and points to the value of work studying environmental justice movements. As well as opening this new perspective, Roberts's contribution reminds us just how important it is to integrate a historical dimension into any reflection on the future of the regulation of toxicants.

By showing that expertise within current regulatory frameworks rests more on ignorance than on knowledge, by offering to place precaution and social and environmental justice at the heart of policies on the management of toxicants, these chapters both reject the centrality publicly granted to science in regulatory systems and call for a reconsideration of the past and current implications of upholding this centrality. This type of approach does not discredit science in any way. On the contrary, it seeks to give it its rightful place in our societies. Above all, it seeks to remind us that while the toxicants and environmental contaminations that a society produces do constitute scientific and environmental issues, they are first and foremost political issues, involving economic and societal choices.

Conclusion

The problems caused by environmental contaminations and their effects on health are currently a major concern for many actors: scientists, activist organizations, policy makers, regulatory agencies, and industrial actors. They all stress how important these questions have become for research as much as for public policy and for the way industrial activity is performed. Reforms and new public policies like Registration, Evaluation, Authorisation and Restriction of Chemical substance (REACH) in Europe, the TSCA in the United States, or the creation of a Global Chemicals Regime, as well as industrial actors' growing references to sustainable and responsible development and to ethics, all provide an indication of unprecedented awareness and a collective desire to finally break away from past practices (Sachs 2009; Selin 2010). However, analysis of the production and use of scientific knowledge in the regulation of toxic issues as well as in advocacy movements paint a much more contrasted picture, which departs from the sometimes naive optimism demonstrated by certain social scientists. On the contrary, they call for a review and in-depth examination of past and current policies and movements and of their contributions and impasses.

The conclusion reached in this book is very dire: while science plays a determining role in defining dangerous health and environmental effects and making them visible, and while it has sometimes provided resources for advocacy movements and contributed to the adoption of new regulatory systems offering greater protection, it has also largely contributed to developing situations of invisibilization and accommodation. It has done so by conferring upon these the seal of objectivity, by producing and putting forward certain results at the expense of others and by giving the policies adopted the air of choice when in fact renouncement was primarily at stake. As result, science contributes to the development of regulatory systems producing and spreading ignorance and scientizing and legitimizing public policies that naturalized the asymmetries between those affected by the contaminations and those benefiting from them—whether financially or in terms of comfort of living.

This conclusion does not discredit science in any way. On the contrary, it seeks to give it its rightful place in our societies. Above all, it seeks to remind us that while the toxicants and environmental contaminations that a society produces do constitute scientific issues, they are first and foremost political issues, involving economic and societal choices. The new wave of regulatory reforms currently taking place makes this observation all the more important. These reforms—from REACH in Europe to the reform of TSCA in the United States—are taking place during a period of intensification of a global economic crisis, which can only make the economic dimension of the governance regarding toxic issues more significant—a dimension that played a structuring role throughout the twentieth century. Just like the climate change policies that led the way, health-environmental policies must also deal with dilemmas that are difficult to resolve. In a society where asymmetries of power and of situations are strong and play a structuring role, science is also caught up in these asymmetries it is not able to overcome—and which in many cases render it powerless. However, recognizing these difficulties, attempting to identify and enumerate them, does not mean refraining from criticizing the choices made, and certainly not giving up on the long-term transformation of a society slowly poisoning itself.

Bibliography

Abraham, John, and Tim Reed. 2002. "Progress, Innovation and Regulatory Science in Drug Development: The Politics of International Standard-Setting." *Social Studies of Science* 32, 3: 337–69.

Allen, Barbara L. 2003. *Uneasy Alchemy: Citizens and Experts in Louisiana's Chemical Corridor Disputes*. Cambridge, M.A., and London: The MIT Press.

Auyero, Javier, and Debora A. Swistun. 2009. *Flammable: Environmental Suffering in an Argentine Shantytown*. Oxford: Oxford University Press.

Bächi, Beate. 2010. "Zur Krise der westdeutschen Grenzwertpolitik in den 1970er Jahren. Die Verwandlung des Berufskrebses von einem toxikologischen in ein sozioökonomisches Problem." *Berichte zur Wissenschaftsgeschichte* 33: 419–35.

Bernhardt, Christoph, and Geneviève Massard-Guilbaud, eds. 2002. *The Modern Demon: Pollution in Urban and Industrial European Societies*. Clermont-Ferrand: Presses Universitaires Blaise-Pascal.

Blum, Elisabeth. 2008. *Love Canal Revisited: Race, Class, and Gender in Environmental Activism*. Lawrence: University of Kansas Press.

Bosso, Christopher. 1990. *Pesticides and Politics: The Life Cycle of a Public Issue*. Pittsburgh, P.A.: University of Pittsburgh Press.

Boudia, Soraya, and Nathalie Jas. 2007. "Risk and Risk Society in Historical Perspective." *History and Technology* 23, 4: 317–33.

Boudia, Soraya, and Nathalie Jas, eds. 2013. *Toxicants, Health and Regulation since 1945.* London: Pickering and Chatto.

Brickman, Ronald, Sheila Jasanoff, and Thomas Ilgen. 1985. *Controlling Chemicals: The Politics of Regulation in Europe and the United States.* Ithaca: Cornell University Press.

Brooks, Karl. 2009. *Before Earth Day: The Origins of American Environmental Law, 1945–1970.* Lawrence: University of Kansas Press.

Brown, Phil. 2007. *Toxic Exposures: Contested Illnesses and the Environmental Health Movement.* New York: Columbia University Press.

Brown, Phil, and Edwin Mikkelsen. 1990. *No Safe Place: Toxic Waste, Leukemia, and Community Action.* Berkeley: University of California Press.

Callon, Michel, Pierre Lascoumes, and Yannick Barthe. 2009. *Acting in an Uncertain World: An Essay on Technical Democracy.* Cambridge, M.A.: The MIT Press.

Carson, Rachel. 1962. *Silent Spring.* Boston: Houghton Mifflin.

Cranor, Carl F. 1997. *Regulating Toxic Substances: A Philosophy of Science and the Law.* Oxford: Oxford University Press.

Davis, Frederick. 2001. "Pesticides and Toxicology: Episodes in the Evolution of Environmental Risk Assessment (1937-1997)." Ph.D. diss., Yale University.

Dessaux, Pierre-Antoine. 2007. "Chemical Expertise and Food Market Regulation in Belle-Epoque France." *History and Technology* 23: 351–68.

Dunlap, Thomas R. 1983. *DDT: Scientists, Citizens, and Public Policy.* Princeton: Princeton University Press.

Egan, Michael. 2007. *Barry Commoner and the Science of Survival: The Remaking of American Environmentalism.* Cambridge, M.A.: The MIT Press.

Fressoz, Jean-Baptiste. 2012. *L'Apocalypse joyeuse, une histoire du risque technologique.* Paris: Éditions du Seuil.

Frickel, Scott. 2004. *Chemical Consequences: Environmental Mutagens, Scientist Activism, and the Rise of Genetic Toxicology.* New Brunswick, N.J.: Rutgers University Press.

Frickel, Scott, and Kelly Moore, eds. 2005. *The New Political Sociology of Science: Institutions, Networks, and Power.* Madison: The University of Wisconsin Press.

Hays, Samuel. 1989. *Beauty, Health, and Permanence: Environmental Politics in the United States, 1955-1985.* Cambridge: Cambridge University Press.

Hood, Christopher, and Henry Rothstein. 2001. "Risk Regulation Under Pressure: Problem Solving or Blame Shifting?" *Administration & Society* 33: 21–53.

Irwin, Alan. 2006. "The Politics of Talk: Coming to Terms with the 'New' Scientific Governance." *Social Studies of Science,* 36, 2: 299–320.

Irwin, Alan, and Brian Wynne, eds. 1996. *Misunderstanding Science? The Public Reconstruction of Science and Technology.* Cambridge: Cambridge University Press.

Jasanoff, Sheila. 1990. *The Fifth Branch: Science Advisers as Policymakers.* Cambridge, M.A.: Harvard University Press.

Kallet, Arthur, and Frederick John Schlink. 1933. *100,000,000 Guinea Pigs: Dangers in Everyday Foods, Drugs, and Cosmetics.* New York: The Vanguard Press.

Krimsky, Sheldon. 2000. *Hormonal Chaos: The Scientific and Social Origins of the Environmental Endocrine Hypothesis.* Baltimore: The Johns Hopkins University Press.

Kroll-Smith, Steve, Phil Brown, and Valerie Gunter, eds. 2000. *Illness and the Environment: A Reader in Contested Medicine.* New York: New York University Press.

Langston, Nancy. 2009. *Toxic Bodies: Hormone Disruptors and the Legacy of DES*. New Haven: Yale University Press.
Lear, Linda. 1997. *Rachel Carson: Witness for Nature*. New York: Henry Holt and Company.
Markowitz, Gerald, and David Rosner. 2002. *Deceit and Denial: The Deadly Politics of Industrial Pollution*. Berkeley: University of California Press.
Massard-Guilbaud, Geneviève. 2010. *Histoire de la pollution industrielle en France, 1789–1914*. Paris: Éditions de l'EHESS.
McGurty, Eileen. 2009. *Transforming Environmentalism: Warren County, PCBs, and the Origins of Environmental Justice*. New Brunswick, N.J.: Rutgers University Press.
National Research Council. 1983. *Risk Assessment in the Federal Government: Managing the Process*. Washington, D.C.: National Academy Press.
Oreskes, Naomi, and Erik M. Conway. 2010. *Merchants of Doubt: How a Handful of Scientists Obscured the Truth on Issues from Tobacco Smoke to Global Warming*. New York: Bloomsbury.
Ottinger, Gwen, and Benjamin R. Cohen, eds. *Technoscience and Environmental Justice: Expert Cultures in a Grassroots Movement*. Cambridge, M.A.: The MIT Press.
Pellow, David. 2007. *Resisting Global Toxics: Transnational Movements for Environmental Justice*. Cambridge, M.A.: The MIT Press.
Pestre, Dominique. 2008. "Challenges for the Democratic Management of Technoscience: Governance, Participation and the Political Today." *Science as Culture* 17, 2, 101–19.
Proctor, Robert N. 1995. *Cancer Wars: How Politics Shapes What We Know and Don't Know about Cancer*. New York: Basic Books.
Ross, Benjamin, and Steven Amter. 2010. *The Polluters: The Making of Our Chemically Altered Environment*. Oxford: Oxford University Press.
Rothstein, Henry, Alan Irwin, Steven Yearley, and Elaine Mc Carthy. 1999. "Regulatory Science, Europeanization, and the Control of Agrochemicals." *Science, Technology and Human Values* 24, 2: 241–64.
Sachs, Noah. 2009. "Jumping the Pond: Transnational Law and the Future of Chemical Regulation." *Vanderbilt Law Review* 62: 1817–69.
Sellers, Christopher. 1997. *Hazards of the Job: From Industrial Disease to Environmental Science*. Chapel Hill: The University of North Carolina Press.
Sellers, Christopher, and Joseph Melling, eds. 2011. *Dangerous Trade Histories of Industrial Hazard across a Globalizing World*.
Selin, Henrik. 2010. *Global Governance of Hazardous Chemicals: Challenges of Multilevel Management*. Cambridge, M.A.: The MIT Press.
Vogel, David. 1986. *National Style of Regulation: Environmental Policy in Great Britain and the United States*. Ithaca: Cornell University Press.
Vogel, Sarah. 2012. *Is It Safe? BPA and the Struggle to Define the Safety of Chemicals*. Berkeley: University of California Press.
Whorton, James. 1974. *Before Silent Spring: Pesticides and Public Health in Pre-Ddt America*. Princeton: Princeton University Press.
Wynne, Brian. 1996. "May the Sheep Safely Graze?" In *Risk, Environment and Modernity: Towards a New Ecology*, ed. Scott Lash, Bronislaw Szersynski, and Brian Wynne, 44–83. London: Sage.
Young, James. 1986. *Pure Food: Securing the Federal Food and Drugs Act of 1906*. Princeton: Princeton University Press.

 PART I

Knowledge, Expertise, and the Transformations in Regulatory Systems

CHAPTER 1

Precaution and the History of Endocrine Disruptors

Nancy Langston

On 6 May 2010 the American President's Cancer Panel released a bombshell in its annual report, stating that 41 percent of Americans will get cancer in their lifetimes. While efforts to fight cancer have focused on genetics, the report noted, "the true burden of environmentally induced cancers has been grossly underestimated." Carcinogens and other toxic chemicals "needlessly increase health care costs, cripple our nation's productivity, and devastate American lives."[1] The report recommended a precautionary approach to environmental carcinogens that would shift the burden of proof to industry. Rather than requiring the government or consumer to prove harm after a chemical is on the market, industry would have to demonstrate that a chemical is safe before approval.

Within days of the report's release, environmental groups applauded what they viewed as a new approach, while industry-supported voices rose in protest, claiming that no one can prove that any given case of cancer has been caused by an environmental exposure. According to some industry advocates, acting in a precautionary fashion would violate the scientific process. Because scientists have not proven that low-level exposure is the cause of reproductive problems in humans, too much scientific uncertainty remains for regulators to act. Above all, critics claimed, precaution is a novel, even radical idea, one likely to stifle innovation and destroy profits.[2]

Do these claims have any historical validity? Is precaution new, and will it destroy scientific process, innovation, and profit? This chapter explores debates over precaution in the context of endocrine disruptors, which are synthetic chemicals that alter hormone systems.

The Case of Diethylstilbestrol

Diethylstilbestrol (DES) offers a useful historical case study for understanding conflicting claims over the role of precaution in regulating synthetic chemicals. DES, synthesized in 1938 by the English biochemist Charles Dodds, was the first synthetic estrogen to be marketed and the first chemical known to act as an endocrine disruptor. Beginning in the 1940s, millions of American women were prescribed DES by their doctors, initially to treat the symptoms of menopause. In 1947 the Food and Drug Administration (FDA) approved DES for pregnant women with diabetes, and drug companies advertised it widely, promoting the use of DES in all pregnancies as a way to reduce the risk of miscarriage. Although no evidence ever supported this claim, millions of pregnant women took the drug. Meanwhile, millions of Americans were also being exposed to DES through their diet. Beginning in 1947, DES was approved in the United States to promote growth in livestock, first in poultry and then in cattle. At the peak of its use in the 1960s, DES was given to nearly 95 percent of feedlot cattle in the United States, and the estrogenic wastes from feedlots and human sewage made their way into aquatic ecosystems.

In 1939, when the American Food and Drug Administration first deliberated on whether to approve DES for human use, the agency was operating under new regulations that were fundamentally precautionary. The recently passed 1938 Food, Drug, and Cosmetic Act put the burden of proof on the industry to show safety, rather than on the consumer or government to show harm. Moreover, the agency and industry both knew of abundant research studies showing that the chemical was carcinogenic in laboratory animals. Three times, the FDA rejected new uses of DES, arguing that precaution suggested the drug was too risky for a particular use. Three times, however, the FDA quickly retreated from precaution and allowed the drug to make its way into human and livestock bodies, and from there, into broader ecosystems (Langston 2010). Why did these retreats from precaution happen? How did political pressures influence agency decision making?

When the FDA began to study DES in 1939, the Roosevelt administration had just weathered a bitter political battle over precaution and regulation. Industry had strenuously opposed the 1938 Food, Drug, and Cosmetic Act, claiming that it was an unwarranted expansion of government power into business, and industry's efforts delayed the Act's passage for nearly five years. In 1937, a drug manufacturer placed a tainted drug called Elixir Sulfanilamide on the market—legally—and over one hundred people died. Consumer outrage erupted, and that anger helped the Roosevelt administration finally push the Food, Drug, and Cosmetic Act through Congress a year later. DES was the first major test case for the FDA, and regulators felt they had to be extremely careful about how they proceeded with their new and contested authority (Hilts 2003: 72–102).

DES emerged during a larger debate going on in the 1920s and 1930s about the carcinogenic effects of estrogens. Nearly all researchers agreed that natural estrogens were carcinogenic, and that DES had the potential to be at least as carcinogenic, if not more so, because it was more potent at exciting "estrogenic effects." Because of these concerns, and because of research on lab animals, in 1940 the FDA initially denied the drug companies' new drug applications for DES. In rejecting DES, Commissioner Walter Campbell of the FDA argued that regulators must follow what he called the "conservative principle." Given the scientific uncertainty over DES's mechanisms of action and metabolism, and over the applicability of animal studies to women, the FDA refused to approve DES—not because scientists had any proof that the drug would harm women, but because they had no proof the drug would *not* harm women.[3] FDA regulators essentially adopted the precautionary principle sixty years before that term came into common usage.

Within months, however, political pressures on the FDA forced the agency to reverse its decision against DES, and in 1941 the agency approved the use of DES in menopausal women. FDA staff had used scientific uncertainty as a justification for refusing to approve DES, but that strategy was not strong enough to resist court challenges and political pressures. A federal court decision against the American Medical Association (AMA) made the FDA wary of engaging with drug companies over the issues of scientific uncertainty—particularly the applicability of animal models to humans—as justification for stiff regulations on estrogens. In the late 1930s, a company named Hiresta had marketed a breast-enlarging estrogen cream. The AMA had been concerned enough about a possible increase in cancer risk from topical estrogen that it published an editorial decrying the dangers of this cream, and Hiresta sued the Association for defamation. The FDA used animal studies to support the AMA's argument that estrogens were known carcinogens. The federal judge ruled against the AMA, arguing that animal studies failed to prove that estrogen cream would definitely lead to cancer in women—and that clear proof of actual harm to specific women was lacking. This court case led the FDA to abandon its planned campaign to regulate estrogen breast creams and made it wary of continuing to use animal studies in its case against new drug applications for DES.[4]

A similar pattern unfolded several years later when the FDA had to decide whether DES was safe to use during pregnancy. Soon after the initial approval of DES for menopause, drug companies and doctors began petitioning the FDA for approval to treat pregnant women with DES, even though experimental studies on lab animals had shown that DES could cause fetal death and could also harm a woman's future fertility. Starting in 1939, a physician named Dr. Karl John Karnaky of Houston began experimenting with the use of DES in pregnant women, and he soon became an enthusiastic promoter of DES for all pregnancies. As he later recalled: "The drug companies came to Houston, ... fed me and dined me... and I started using it" (Gillam and Bernstein 1987: 67). Research in the early 1940s by the physicians Priscilla White, George

Smith, and Olive Smith encouraged the hope that DES might help to prevent miscarriages. Many physicians convinced themselves that DES was indeed a miracle drug for stopping "accidents of pregnancy." Drug companies lobbied the FDA intensely to approve the drug for pregnancy, sending samples to doctors to create a consumer market for the drug, overwhelming the FDA with short-term data on human effects and ignoring data on animal experiments, and complaining about the safety limits constructed by the FDA.

Initially, FDA staff were quite cautious about DES use during pregnancy, and allowed it to be prescribed only for rare cases of diabetic pregnancies where the mother was almost certain to lose the child otherwise. But this degree of caution quickly faded. Widespread enthusiasm for children in the postwar years, combined with the frustration of the medical community that they had been so powerless to decrease miscarriage rates, helped to persuade much of the medical community that the synthetic estrogen might save babies. In 1947, after the FDA approved limited DES for pregnant women with diabetes, drug companies marketed the drug intensively, urging doctors to prescribe it even for "normal" women "to make a normal pregnancy more normal." The FDA allowed these uses, and soon nearly one-tenth of pregnancies among American women were treated with the synthetic estrogen (Cody 2008: 232–end).

The retreat from precaution occurred a third time with DES, this time with livestock use. While drug manufacturers were promoting DES for pregnancy, the same companies were also looking for new markets in livestock production. During World War II, pharmaceutical companies requested that the FDA approve the use of diethylstilbestrol to treat certain veterinary conditions in livestock, but because of concern about the potential risks to soldiers who might consume estrogen residues, the FDA explicitly forbade DES treatment of livestock that might be eaten. When companies tried to push against wartime FDA restrictions on the use of synthetic hormones in livestock, the FDA insisted on precaution, arguing that the absence of evidence of harm did not prove safety.[5]

As the war came to an end, political pressures once again led the FDA to abandon its position of precaution. The wartime meat rationing had ended in the United States, but food shortages throughout Europe threatened to lead to famine, which many people were afraid might destroy the peace. Grain was being used to feed livestock rather than feed people, threatening shortages, but government officials worried that Americans would be unwilling to voluntarily reduce their meat consumption in order to make more grain available for human food. Rather than reinstitute rationing, the government encouraged research partnerships devoted to learning how to increase meat production while retaining enough grain to prevent famine. The answer appeared to be hormones, which promised more efficient feed utilization. Because animals treated with estrogen fattened up more quickly on less grain, science might allow Americans to eat more meat without guilt (Bentley 1998).

Poultry was the first target, for roosters and turkeys responded readily to DES implants. In February 1946, internal memos within the FDA showed that staff remained skeptical about the use of diethylstilbestrol in poultry. When one company insisted that the pellets were safe because they would be implanted in chicken necks, which middle-class housewives discarded, one staffer scribbled on a memo: "Some people do use the heads of poultry for food!"[6] Throughout 1946, the FDA rejected New Drug Applications submitted for poultry, stating that "No information has been offered to show the amount of diethylstilbestrol remaining in the tissues of treated birds. Until it can be clearly shown that no significant quantity of the drug remains in the tissues which might be capable of producing undesirable effects in human consumers, we will not be disposed to consider any application for the diethylstilbestrol use with this purpose."[7] FDA staff initially insisted on a precautionary approach, telling companies that determining safety was the responsibility of the manufacturer, not the problem of the government.

But in January 1947 the agency reversed course and agreed to allow diethylstilbestrol to be used in poultry implants. None of the problems discussed in the correspondence from the previous several years had yet been fixed. The only research that drug companies offered in support of DES pellet safety actually showed the opposite, indicating that estrogen residues did migrate from the pellets into meat intended for human consumption.[8]

Why did the agency suddenly allow DES implants in livestock, when regulators had resisted for years? Pressures to increase meat production after the war were certainly great, but concerns about the risk of estrogens for men had initially led the FDA to resist these pressures. By 1947, however, DES began to seem much safer to the FDA. Because medical researchers had treated pregnant women with large doses of DES and no deaths had yet resulted, FDA staff began to argue that small doses presented little risk.[9]

Not all scientists or regulators agreed. Immediately after the FDA approved the chicken implants, Canadian regulators wrote to the federal government, urging the FDA to be extremely careful with the use of diethylstilbestrol in animals. A staff member from the Canadian Department of National Health and Welfare wrote: "We have been working on the problem with the poultry division of the Department of Agriculture and our results show that there is a residue of the estrogen in the cockerels, sufficient to change the vaginal smear of the menopausal woman. Of course this is not evidence of any harmful effects but it is possibly an undesirable reaction for some people. We were planning to publish these results and are wondering if any of the results from your division had been published and we had overlooked them." The Canadians, in other words, had data showing that a synthetic estrogen implanted in chicken necks was so powerful that residues left in the meat could change the vaginal smears of the woman who ate that meat.[10]

After the Canadian researchers published their findings in the October 1947 issue of *Endocrinology*, the FDA forbade the use of DES in chicken feed, but continued to allow DES to be used in pellets that were implanted in chicken necks.[11] For decades the FDA insisted on something that made little scientific sense: that although diethylstilbestrol from chicken feed could accumulate in fatty tissues and pose a danger to humans, DES administered in pellets simply wouldn't accumulate. When challenged by members of its own scientific staff, the FDA attempted to explain this logic by arguing that "it is possible to exercise a rigid control over the dosage in the [pellet] process and under these circumstances the estrogen does not accumulate in those portions of the treated bird which are consumed by human beings."[12] A host of assumptions about the possibilities of scientific control are embedded in this statement. First, the statement assumes that technology can offer enough control to sidestep dilemmas posed by pollutants. Second, the statement assumes that people live in an ideal world, one designed by technicians: that no one ever sells a chicken head, that consumers eat what they're supposed to eat, that companies do exactly what they promise to do, that pellets release a specific, measured, infallible dosage that can be carefully controlled. But none of these assumptions were based on empirical evidence. The FDA had never received or examined any data that showed that pellets did release a reliable and controllable dosage, or that this dosage did not accumulate in tissues, even if it were controllable and reliable. Even though the regulatory agencies tried to assure consumers that complete scientific control was possible over hormones, scientists within the agencies agreed that such control was impossible.

Soon after having approved DES for use in livestock, the FDA soon began to receive warnings that the chemical was causing problems with plant workers, farm workers, restaurant workers, and consumers. In 1947, Arapahoe Chemicals of Colorado wrote to the FDA:

> Our Company has recently been approached in regard to manufacturing stilboestrol ... as raw materials for pharmaceutical formulation. We know that these materials are all readily absorbed through the skin and by inhalation. It is our belief that the physiological effect of these materials would constitute a decided industrial hazard. In order to properly evaluate the advantages of undertaking the manufacture of synthetic estrogens, it is necessary that we obtain as much information as possible about them in regard to the seriousness of the health hazard involved, recommended precautions for handling, treatment of affected individuals, cumulative effects, etc. We are particularly concerned over the possibility of carcinogenesis through long continued contact with stilboestrol.[13]

The FDA responded by suggesting the company hire older men who presumably wouldn't mind being made impotent by the chemical: "we have your letter of June 26, 1947 requesting information concerning the health hazard involved and the precautions necessary on the manufacture of stilbestrol. ... It is our understanding that excessive exposure to the substances may cause marked disturbances of the menstrual function in women and have a devirilizing effect in men. For this reason it might be feasible for you to consider the employment of old rather than young men."[14]

A restaurant worker in New York grew breasts after eating the heads of chickens implanted with DES pellets, and his case became immortalized in a medical textbook. Mink farmers began complaining to the FDA that their mink were made sterile by residues from the necks of the implanted chickens. FDA staff discounted these complaints, stating that "a few mink ranchers have alleged that their breeding animals were rendered sterile after having been fed the discarded heads of poultry which were implanted with diethylstilbestrol pellets. As yet we have seen no satisfactory data of a factual or scientifically acceptable nature showing that the offal from birds implanted with these pellets will actually cause sterility in minks or any other animals."[15]

Finally, five months after repeated reports of problems, the FDA checked the chickens, found numerous cases where residue levels violated the law, and seized fifty thousand pounds of chickens. These chickens contained high levels of DES residues, and some birds contained up to four pellets in a single neck.[16] FDA staff had vigorously denied the very possibility this could happen. When empirical data was initially presented to them by mink farmers among others, FDA staff simply denied it. Yet when they went out and collected their own empirical data, they found that their scientific models of what ought to be happening were not supported by actual evidence.[17]

Even as research staff within the FDA and scientific consultants hired to advise the agency were urging the FDA to ban DES for poultry, the FDA approved DES's rapid expansion into cattle feeding. In 1953 the Iowa State University researcher Wise Burroughs published a report showing that "cattle gains could be increased substantially and that feed costs could be reduced materially by placing 5 mg or more of DES in the daily supplemental feed fed to each steer." Burroughs concluded that DES feeding led to 35 percent increases in growth and a decrease in feed cost of 20 percent—astonishing results if they could be reproduced (Burroughs and Culbertson 1954). The FDA almost immediately granted approval on 1 November 1954—just a year after the initial report from the feeding studies. A month later, DES went on the market as Stilbosol. Manufacturers such as Lilly marketed DES feeds intensely, to extension agents, to farmers, and to the farmer's press, and "cattlemen turned to the enhanced feeds in droves" (Marcus 1993: 66; 1994: 22–25). By late 1955, less than a year after DES went on the market, fully half the feedlot cattle in America were receiving DES. Soon, 80 to 95 percent of feedlot cattle received DES. As Marcus points out, the research on, FDA approval process of, and marketing of DES did not happen by accident; it emerged as part of a complex partnership among drug companies, universities, and federal agencies (Marcus 1994).

These partnerships began to unravel with evidence emerging that DES caused cancer in women who had been exposed to the chemical prenatally. In 1971 researchers in Boston reported a cluster of extremely rare vaginal cancers in young women whose mothers had taken DES while they were pregnant. These problems had not been apparent at birth; they emerged only at puberty or young adulthood, sometimes decades after fetal exposure. Mothers and children exposed to DES organized a group called DES Action that lobbied for research and action on the drug, and in 1979 consumers, scientists, and concerned congressional representatives finally forced the American government to ban the chemical for use in livestock.

The full dimensions of the health and environmental disaster that resulted from widespread DES use are only now becoming apparent. By 2002, DES had emerged in toxicological studies as a carcinogen and developmental toxicant so potent that the toxicity of other chemicals is often measured against it. Of the two to five million children who were exposed to DES prenatally, nearly 95 percent of those sampled have experienced reproductive-tract problems, including menstrual irregularities, infertility, and higher risks of a variety of reproductive cancers.

Toxicological Models

Why were regulators unable or unwilling to resist industry pressure? As I argue in detail in *Toxic Bodies,* regulators tended to share certain cultural and conceptual beliefs that industry lobbyists were quick to exploit. Cultural assumptions about gender differences shaped the ways that scientists, regulators, and consumers understood hormones and their effects on the body. Finally, many regulators shared with industry staff a modernist world view that combined faith in scientific expertise with the belief that technological progress could and should control nature. These beliefs often made regulators more skeptical of consumer claims of harm than they were of industry claims of safety. And while individual staff members within the federal agencies worked hard to protect public health, political appointees who headed the agencies often seemed more responsive to industry concerns about profits than to their own staff's concerns about risks (Langston 2010).

Contemporary scientific models of toxicology and development generally did not allow for the possibility that very low levels of synthetic chemicals could influence hormonal actions in the body. Indeed, emerging research that showed the harmful effects of various synthetic chemicals often seemed to violate the standard toxicological paradigms of the era, making it difficult for regulators to interpret scientific results. Even when experimental evidence from laboratory animals seemed to provide compelling proof of harm, uncertainty about the validity of animal studies in assessing risks for people made it difficult for regulators to defend principles of precaution in court.

Ever since endocrine-disrupting chemicals such as DES were first commercially produced in the 1940s, their hormonal mechanisms of action have posed novel challenges for scientists and regulatory agencies seeking to protect public health, because they do not easily fit within traditional risk paradigms. Although the threshold model may be useful for natural toxins such as aspirin, it is rarely relevant for endocrine disruptors. Even at extremely low levels, they can mimic, block, or disrupt the actions of the body's own hormones, thereby altering reproduction and development, often with profound effects later in life. In fact, endocrine-disrupting chemicals can actually have more powerful effects at low doses than at high doses. At low concentrations, hormones normally stimulate receptors, but at high concentrations hormones can saturate receptors, thus inhibiting their pathways. Low doses of endocrine disruptors such as DES might produce adverse impacts, even though higher doses might not. But the idea that a substance can have more powerful effects at low doses than at high doses fundamentally challenged toxicological paradigms (Langston 2010: 5–12).

The effects of estrogenic chemicals such as DES puzzled researchers and regulators because they differed dramatically among individuals, depending on the age of the individual and the timing of the exposure. These findings made little sense when interpreted through a standard toxicological paradigm, but they are less surprising when we consider how the endocrine system functions at different life stages. In adults, hormones mainly regulate ongoing physiological processes such as metabolism. Synthetic chemicals can lead to temporary endocrine changes, but adults are often able to recover from these disturbances. During fetal development, however, hormonal changes can have permanent, irreversible effects. Because a woman accumulates toxic chemicals over an entire lifetime of exposure, she can transfer much of her contaminant burden to her developing fetus during pregnancy, the time of greatest sensitivity.

Risk and Precaution in American Regulation

Questions about risk, profit, and the burden of proof have troubled U.S. regulatory agencies ever since the Food and Drug Administration called for a version of the precautionary principle in the early decades of the twentieth century. Since 1998, they have coalesced around the demand for a precautionary approach placing the burden of proof on those who profit from toxic chemicals. That year, thirty-two scientists and physicians concerned about endocrine disruption published a consensus statement known as the Wingspread Statement on the Precautionary Principle. They wrote: "When an activity raises threats to the environment or human health, precautionary measures should be taken, even if some cause-and-effect relationships are not fully established scientifically. In this context, the proponent of an activity, rather than the public, should bear the burden of proof."[18] Yet the precautionary principle is not easy to implement, for the environmental or health risks of a particular action are usually uncertain and occur in the future, while the costs of averting it are often immediate.

As Sonja Boehmer-Christiansen (1994) argues, a formal precautionary principle evolved out of the German concept of *Vorsorgeprinzip*, which developed in the legal tradition of 1930s democratic socialism. *Vorsorgeprinzip* centered on the concept of good household management, a concept that justified state involvement in planning economic, technological, moral, and social initiatives. Precaution had been adopted well before this in public health efforts, however. When the British physician John Snow recommended removing the handle from the Broad Street water pump in an attempt to stop London's 1854 cholera epidemic, that was a form of precaution. Scientists were still uncertain of the causes of cholera when Snow acted. He had found a correlation between polluted water and cholera five years earlier, but most scientists and physicians rejected his thesis as untenable, believing that airborne contaminants caused cholera. The biological mechanism underlying the link between polluted water and cholera was unknown. Yet even without firm proof, Snow had enough information to judge that the possible costs of inaction would probably be greater than the costs of action (Harremoës et al. 2001: 168).

Snow's vision of protecting the public through precautionary action continued as an important thread in American and European public health. During the first decades of the twentieth century, Harvey Wiley and Walter Campbell argued that the federal government needed to use precaution as the basis of regulation. The Food and Drug Administration, they believed, needed to sift evidence from multiple perspectives, not just the industry standpoint, to find preliminary evidence that might suggest possible links between a compound and an adverse, potentially irreversible, outcome. This preliminary evidence might come from experiments on animals or from structural similarities between a given chemical with unknown effects and one with known effects. Precaution was justified, they believed, when the potential costs were high or irreversible compared to the benefits, when the person who bore the costs did not receive the benefits, and when preliminary evidence suggested a possible link between an action and a harm, even when the exact biological or chemical mechanisms underlying that link were still uncertain. Nevertheless, political pressures in the 1940s and 1950s made it impossible for the young FDA to defend precaution, and each time industry challenged a precautionary decision that the agency had taken against hormone-disrupting chemicals, the agency quickly backed down.

Beginning in the 1970s, scientists and activists made efforts to extend the idea of precaution from the public health arena into broader environmental decision making. German foresters struggled to establish the causes of dying forests and developed a precautionary principle in the 1970s similar to that developed in public health. The 1992 Rio Declaration on Environment and Development was explicitly grounded in precaution, and in 2007 the European Union passed a law mandating that chemical companies demonstrate that their products are safe before they can be placed on the market.

For four decades, American industry has opposed efforts to extend precaution from medical to environmental policy, claiming that such an approach could stall innovation. The Business Roundtable was founded in 1972 to represent two hundred of the nation's largest corporations, and this association has taken an increasingly active role in opposing precautionary regulation. Gerald Markowitz and David Rosner argue that the association's strategy has been to accentuate elements of complexity and uncertainty and then to argue that "economic interests should not be challenged until science has proven danger. Precaution is equated with economic and social stagnancy. ... Progress, as defined by the industrial community, trumps precaution." As consumer concern over environmental pollution placed increasing pressure on industry, industry responded with a "frontal assault on the public health ideals of prevention," hiring product-defense firms, public relations agencies, and scientists who "systematically attacked environmentalists and labor activists as luddites determined to stifle our economy" (Markowitz and Rosner 2002b: 502).

American industry advocates sought to portray precaution as a novel and reckless idea, rather than as a long-held principle at the heart of public health. What was most daring about this campaign was industry's largely successful effort to rewrite history in the public eye, portraying precaution as a new idea and indisputable proof of harm as a historical precedent.

At the core of debates over precaution are questions about the relationship of science and certainty in decision making. When industry argues that no experimental studies have proven endocrine disruptors cause harm to humans, they are correct. Yet they fail to point out that experimental proof of human harm is lacking not because the chemicals are safe, but because those experiments would be both illegal and unethical. In 1947, during the Doctors' Trial in Nuremberg Germany, evidence emerged of Nazi medical experimentation that subjected prisoners to chemical poisons such as mustard gas. The resultant Nuremberg Code forbids any research that might lead to unnecessary pain, suffering, death, or disability. No researcher can design an experiment that subjects a person to a suspected carcinogen to test whether that chemical induces cancer. By definition, then, we will never have firm proof of the links between cancer and pollutants. Scientific uncertainty will *always* remain. Yet refusing to take action because uncertainty exists has continued to be a profitable strategy for industries.

In 2001, a European Union team charged with implementing the precautionary principle examined fourteen case studies of historical hazards. The case studies involved an agent (such as diethylstilbestrol) that most contemporaries had regarded as harmless at prevailing levels of exposure until additional evidence about harmful effects emerged. The goal of the exercise was to identify when the first credible "early warnings" of potential harm emerged, determine how regulatory authorities responded (or failed to respond) to those warnings, and calculate the resulting costs and benefits of that inaction. One critical lesson discussed in the European Union case studies concerns the importance of first recognizing limits to knowledge and then accepting that continued uncertainty is no justification for inaction. As the European Union team writes, "No matter how sophisticated knowledge is, it will always be subject to some degree of ignorance. To be alert to—and humble about—the potential gaps in those bodies of knowledge that are included in our decision-making is fundamental. Surprise is inevitable" (Harremoës et al. 2001: 169). The regulators involved with DES understood that their knowledge about the actions of synthetic hormones was limited, but when it came time to assess risks and make decisions, they seemed to lack the humility that a partial understanding requires.

Ignorance can sometimes be intentional. An industry might prefer not to find out about the potential harm its product might cause, because continued uncertainty means continued profits. Without monitoring of potential hazards, we are almost guaranteed to be more ignorant than we need to be. Yet as DES consumer groups found out, inducing the federal government to monitor industry is difficult, because the political pressures on regulators can be overwhelming.

Several key uncertainties abounded in the DES research, and these foreshadow the uncertainties that haunt today's endocrine-disruptor policies. The significance of laboratory-animal experiments for people, the boundaries between synthetic and natural processes, the risks of low levels of exposure, and the significance of environmental influences on the developing fetus were all uncertain in the 1930s. They remain uncertain today, not because of lack of research effort but because of the complexity of endocrine systems. Using this complexity as a justification for continuing to expose people and environments to synthetic chemicals has proven to be a useful strategy for industry, but it is not one that is likely to protect Americans' health or the environment.

The regulatory agencies' willingness to approve DES was partly derived from the unwillingness of clinicians to pay heed to experimental evidence from laboratory animals. Karl John Karnaky, for example, insisted that the animal studies showing that DES caused fetal harm did not apply to people.

In one publication, Karnaky noted that numerous lab studies had shown that DES was damaging to the fetus. And yet even after summarizing all the reports that DES harmed the fetus or prevented implantation of the fertilized egg, Karnaky went on to state that women are not laboratory animals, and thus there was no reason to believe that DES was harmful to women (Karnaky 1947).

For decades, scientists and regulators debated the possible significance of low levels of exposure to synthetic chemicals. Even when researchers agreed that high levels of estrogens might cause harm, significant disputes remained about what those results might mean at the low levels common in the environment. Traditional toxicological models of risk posited dose response models, where the dose makes the poison; in this model, low levels beneath a given threshold value would not be expected to cause harm. Industry advocates argued that these threshold values were based in sound science, but a careful reading of history reveals that they were often the result of political negotiation.

The boundary between natural and synthetic was also a continuing source of uncertainty. The drug companies argued that because bodies naturally produced estrogens, levels of additional estrogens that were just a fraction of the highest levels of the natural estrogens would not have a toxic effect. When Karnaky argued that DES treatment during pregnancy was safe, he pointed to the fact that a woman's body naturally produced high levels of estrogens during pregnancy, making the additional amounts from DES insignificant. Drug companies promoting DES manipulated the concept of naturalness, with its attendant implications of purity and safety. These same arguments remain potent today in debates over the safety of steroid hormones given to livestock.

Another critical issue focused on the limits of technology and knowledge. If technology did not exist to measure a residue, did that mean the residue did not exist? If an effect could not be measured, was the effect therefore nonexistent? Industry initially argued that only effects and residues that were measurable existed. Scientists consulting with the FDA disputed this, arguing that an inability to detect liver damage from DES, for example, could mean that liver damage did not exist. But it might mean that available tests lacked the sensitivity to show slow, chronic changes. Initially, the FDA regulators agreed with this idea, refusing to assume that an inability to detect a residue or an effect meant the chemical was safe. Yet by 1947, this idea had been discarded, as the FDA joined the industry in arguing that if something could not be measured by available technologies, it did not exist.

Conclusion

Each time regulators reached the limits of their knowledge about the effects of a chemical exposure, they decided to move ahead and allow people to be exposed. Each time they vowed to use that new exposure as an experiment that would be monitored, so that policy makers could learn from the experiment. The toxic chemicals were released with the underlying assumption that "any major problems will emerge in good time for corrective action" (Harremoës et al. 2001: 171). The corollary often cited was that if no major problems emerged, the compound must be safe. Yet when no monitoring is being done, that fundamental assumption is wrong. People may be dying in increased numbers from a particular chemical exposure, yet if their death rates are not being monitored, industry will continue to insist its products are safe. Yet time and again, the federal agencies failed to learn from their own histories—sometimes because they lacked the funding and political power to insist on monitoring, and sometimes because they refused to pay attention to results.

The continuing failure of the FDA to regulate DES and the continuing insistence of physicians on prescribing the drug were closely linked to particular social constructions of diseases and treatments. As the medical historian Robert Bud argues about antibiotics, drugs "came to stand for the technical solution to infection, replacing control through prevention" (Bud 2007: 24). Similarly, DES came to stand for a technical solution to menopause, then to miscarriage, and eventually to grain shortages. Advocates of progress tended to override concerns based in precaution. Rosner and Markowitz show how during Depression-era debates over the safety of lead paint, the lead industry "sought to co-opt the growing public health movement by identifying lead with modernity and health. ... The themes of order, cleanliness, and purity that were hallmarks of the efforts to reform and sanitize American life were quickly incorporated into the promotional materials developed by the industry" (Markowitz and Rosner 2002a; Markowitz and Rosner 2002b: 504). A similar pattern emerged for DES.

Rather than addressing the larger ecological issues of "accidents of pregnancy," DES seemed to promise a technical solution that was cheap and, above all, modern. The pharmaceutical companies played on these themes in their promotions of the synthetic hormone. A crucial lesson from the DES history is that science alone cannot solve our problems with endocrine disruptors. As the history of DES makes clear, the call for "more research!" has often become a way of delaying action, keeping profitable drugs and chemicals on the market as long as possible.

Many environmental advocates currently recommend precaution as a promising new approach to problems with synthetic chemicals that present uncertain risks. Industry lobbyists, on the other hand, often claim that precaution is novel and untested. But as I have argued, precautionary approaches have a long if troubled history in the American regulatory context. The American pharmaceutical industry has thrived since the fundamentally precautionary 1938 Food, Drug, and Cosmetic Act. The drug industry has prospered, not in spite of precautionary regulation, but *because* of it. When doctors and patients began to trust that the risks of new drugs had not been hidden, drug sales rose. Currently, pharmaceuticals are among the most profitable industries in the world, refuting the claim that precaution is a death knell. Yet precaution has been an imperfect tool for protecting public and environmental health. Often, regulators have responded to political pressure by retreating from precaution, putting consumers at heightened risk of harm from chemical exposures. Given the limits of scientific experimentation, precaution is necessary, but without a sustained effort at transparency, disclosure, and public involvement, precaution alone will not protect public or environmental health.

Acknowledgments

This chapter is excerpted from Nancy Langston, *Toxic Bodies: Hormone Disruptors and the Legacy of DES* (New Haven: Yale University Press, 2010). I thank Yale University Press for their kind permission to use the material.

Notes

1. President's Cancer Panel Report: "Reducing Environmental Cancer Risk: What We Can Do Now." Washington, D.C.: NIH National Cancer Institute, May 2010, cover letter to report (unpaginated). http://deainfo.nci.nih.gov/advisory/pcp/pcp08-09rpt/PCP_Report_08-09_508.pdf (last accessed 11 July 2010).
2. See, for example, Elizabeth Whelan of the American Council on Science and Health at http://www.acsh.org/news/newsID.1899/news_detail.asp (last accessed 11 July 2010).
3. Letter, WG Campbell, Commissioner of FDA, to Merck & Co, re NDA 4076, 3 November 1941. In National Archives and Records Administration at College Park, Maryland. RG 88, Records of the Food and Drug Administration [hereafter FDA], A1, Entry 5, General Subject Files, 1938–1974. 1941. Folder 526.1. November–December.
4. FDA, A1, Entry 5, General Subject Files, 1938–1974. 1940. Folder 526.1–.13. Memorandum of Interview. Dr. Robert T. Frank, N.Y., and Dr. Gordon A. Granger, Medical Officer of the FDA. 11 July 1940. In FDA, Folder 526.1–.11. Memorandum 1 October 1941. George Larrick, Acting Chief, New Drug Division, to Walter Campbell, Commissioner. In FDA, Folder 526.1. October.
5. C.W. Crawford, Assistant Commissioner of Food and Drugs, FDA, to Mr. James A. Austin, Jensen-Salsbery Laboratories, Inc., Kansas City, M.O., 8 May 1944, FDA, A1, Entry 5, General Subject Files, 1938–1974 (1944a), Folder 526.1–526.11.
6. Moskey, Chief, Veterinary Medical Section, FDA, to F.B. Hutt, Professor of Animal Genetics, Cornell University, 13 September 1945, FDA, A1, Entry 5, General Subject Files, 1938–1974 (1945a), Folder 526.1.10.
7. P.B. Dunbar, Assistant Commissioner of Food and Drugs, to Dr. David A. Bryce, Lederle Laboratories, Pearl River, N.Y., 13 February 1946, FDA, A1, Entry 5, General Subject Files, 1938–1974 (1946), Folder 526.1.

8. Dunbar to Bryce, 28 January 1947; P.H.E. Moskey, Chief, Veterinary Medical Section, FDA, memorandum of interview with Director C.W. Sondern and various staff of White Laboratories, New Jersey, 30 January 1947, FDA, A1, Entry 5, General Subject Files, 1938–1974 (1947), Folder 526.1.
9. H.E. Moskey, Chief, Veterinary Medical Section, FDA, to E.I. Robertson, Associate Professor, Animal Husbandry, Cornell, 21 July 1947, FDA, A1, Entry 5, General Subject Files, 1938–1974 (1947), Folder 526.1.
10. L.I. Pugsley, Chief, Laboratory Service, Department of National Health and Welfare, Food and Drugs Division, Pharmacology Laboratory, Ottawa, to Dr. Bert J. Vos, Federal Security Agency, Food and Drugs Division, 20 February 1947. FDA, A1, Entry 5, General Subject Files, 1938–1974 (1947), Folder 526.1.
11. Moskey to Austin, 10 September 1947, FDA, A1, Entry 5, General Subject Files, 1938–1974 (1947), Folder 526.1.
12. Crawford to Montgomery, 13 April 1948, FDA, A1, Entry 5, General Subject Files, 1938–1974 (1947), Folder 526.1.
13. Waugh to Stormont, 26 June 1947, FDA, A1, Entry 5, General Subject Files, 1938–1974 (1947), Folder 526.1. The letter was circulated within the FDA for handwritten and initialed comments.
14. Handwritten comments by FDA staff on the letter from Stormont to Waugh, 9 July 47, FDA, A1, Entry 5, General Subject Files, 1938–1974 (1947), Folder 526.1.
15. Collins, "Drugs for Food-Producing Animals and Poultry are a Problem," FDA, A1, Entry 5, General Subject Files, 1938–1974 (1947), Folder 526.1.
16. Collins, "Use and Abuse of Diethylstilbestrol Pellets in Poultry," FDA, A1, Entry 5, General Subject Files, 1938–1974 (1947), Folder 526.1.
17. "Hormones & Chickens," *Time* (December 21, 1959)
18. "Wingspread Statement on the Precautionary Principle." Wingspread Conference, Racine, W.I., January 1998. http://www.gdrc.org/u-gov/precaution-3.html (accessed 14 August 2008).

Bibliography

Bentley, Amy. 1998. *Eating for Victory: Food Rationing and the Politics of Domesticity.* Champaign: University of Illinois Press.

Boehmer-Christiansen, Sonja. 1994. "The Precautionary Principle in Germany—Enabling Government." In *Interpreting the Precautionary Principle,* eds. Tim O'Riordan and James Cameron, 31–61. London: Earthscan.

Bud, Robert. 2007. "Antibiotics: From Germophobia to the Carefree Life and Back Again: The Lifecycle of the Antibiotic Brand." In *Medicating Modern America: Prescription Drugs in History,* ed. Andrea Tone and Elizabeth Siegel Watkins, 17–41. New York: New York University Press.

Burroughs, Wise, and C.C. Culbertson. 1954. "The Effects of Trace Amounts of Diethylstilbestrol in Rations of Fattening Steers." *Science* 120: 66–67.

Cody, Pat. 2008. *DES Voices: From Anger to Action.* New York: Lulu.

Gillam, Richard, and Barton J. Bernstein. 1987. "Doing Harm: The DES Tragedy and Modern American Medicine." *Public Historian* 9: 57–82.

Harremoës, Poul, David Gee, Malcolm MacGarvin, Andy Stirling, Jane Keys, Brian Wynne, and Sofia Guedes Vaz, eds. 2001. *Late Lessons from Early Warnings: The Precautionary Principle, 1896–2000*. European Environment Agency Environmental Issue Report 22. Luxembourg: Office for Official Publications of the European Communities.

Hilts, Philip J. 2003. *Protecting America's Health: The FDA, Business, and 100 Years of Regulation*. New York: Knopf.

"Hormones and Chickens." *Time* (21 December 1959).

Karnaky, Karl John. 1947. "Estrogenic Tolerance in Pregnant Women." *American Journal of Obstetrics and Gynecology* 51: 312–16.

Langston, Nancy. 2010. *Toxic Bodies: Hormone Disruptors and the Legacy of DES*. New Haven, C.T.: Yale University Press.

Marcus, Alan. 1993. "The Newest Knowledge of Nutrition: Wise Burroughs, DES, and Modern Meat." *Agricultural History* 67: 66–85.

———. 1994. *Cancer from Beef: DES, Federal Food Regulation, and Consumer Confidence*. Baltimore: Johns Hopkins University Press.

Markowitz, Gerald, and David Rosner. 2002a. *Deceit and Denial: The Deadly Politics of Industrial Pollution*. Berkeley: University of California Press.

———. 2002b. "Industry Challenges to the Principle of Prevention in Public Health: The Precautionary Principle in Historical Perspective." *Public Health Report* 117: 501–12.

CHAPTER 2

The Political Life of Mutagens
A History of the Ames Test

Angela N.H. Creager

In 1973, Bruce N. Ames, a professor of biochemistry at the University of California, Berkeley, introduced a new assay for use in evaluating carcinogenicity. The test relied on four mutant strains of *Salmonella* that Ames's group had developed, drawing on years of experience using such bacteria in studies of metabolism and mutagenesis. These strains were deficient in their ability to synthesize a particular amino acid, histidine, so they required this supplement in the growth media. Each of the four strains could be used to genetically screen compounds inducing a specific kind of mutation in the DNA sequence. These registered as reverse mutations, or revertants, which compensated for the histidine deficiency. In other words, on test culture plates, cells that grew represented new mutations. The four *Salmonella* strains were further customized with additional genetic changes that made the cells more permeable to large molecules and eliminated some kinds of DNA repair. Ames showed that his test could identify nearly all known chemical carcinogens and he advocated its utilization in assessing the cancer risks posed by new substances. Companies immediately began adopting the Ames test as a way to undertake routine chemical screening; the new method was both quicker and less expensive than traditional animal testing. Facilitating the adoption of his test method, Ames made his strains freely available. Environmental groups were equally enthusiastic about the test, particularly once Ames identified as likely carcinogens a food preservative and a flame retardant being incorporated into children's pajamas.

The value of the Ames test, which was embraced by industry and environmentalists alike in the 1970s, relied on two powerful but vulnerable assumptions. First, as Ames put it, a carcinogen is a mutagen. Human cancer, in this view, could be triggered by exposure to environmental mutagens.[1] Compounds that do not induce mutations were presumed not to cause cancer, either. Second, he assumed that a microbe was a suitable model organism for assaying mutagenicity as it occurred in human cells. Some toxicologists and cancer biologists objected to Ames's simplifying assumptions. But these objections were minor compared to the political controversy that developed around the test in the 1980s, after Ames and others began testing natural substances, such as extracts of vegetables and cooked meats. His assays showed compounds in many foods and beverages to be just as mutagenic as synthetic chemicals. On this basis he began arguing that natural background hazards must be considered in formulating regulatory policy. This point of view led him to oppose some new industrial regulations, even as he was being appointed to government panels to interpret and implement safety standards. Environmentalists felt betrayed.

This paper situates the invention of the Ames test in terms of his experimental trajectory as a biochemist in the broader context of postwar radiation genetics and environmentalism. The reconceptualization in the 1960s of the "somatic" (including cancer-causing) effects of radiation in terms of mutation enabled scientists to directly connect the mutagenicity and carcinogenicity of radiation, and, by extension, synthetic chemicals. Initial uses of the Ames test served to reinforce the emphasis on industrial regulation in the name of public health. Yet, by turning his technique on naturally occurring substances, Ames subverted the environmental presumption that cancer was principally attributable to artificial substances. I also aim to answer the questions of why and how microbial mutations became a key means for visualizing the cancer-causing dangers of environmental substances, and how Ames's attempt to rationalize and rank cancer risks in this way met with opposition from cancer biologists and environmentalists.

Ames's Path to Chemical Mutagens

Bruce Ames's early career was in biochemistry with a special emphasis on biochemical genetics. He earned his Ph.D. at Caltech in 1953, using metabolic mutants of *Neurospora* isolated by his advisor Herschel Mitchell—a former postdoctoral fellow of George Beadle's—to study the biosynthesis of the amino acid histidine.[2] Ames went on to a postdoctoral fellowship at the National Institutes of Health (NIH), in the laboratory of Bernard Horecker, because "I knew I needed to learn enzymology" (Ames 2003: 4370). After one year Ames became a section chief in Gordon Tomkins's unit at the NIH, the Laboratory of Molecular Biology. Ames stayed until 1967.

After arriving in Bethesda, Ames continued to study histidine biosynthesis, but shifted organism to *Salmonella typhimurium*. In this way he could take advantage of an extensive set of histidine-requiring mutants that had been isolated by Philip Hartman at the Carnegie Institution of Washington. Hartman had already isolated and genetically mapped hundreds of histidine mutants in *Salmonella* (eventually it would be thousands), creating a remarkable repertoire of mutants available for work on this biosynthetic pathway. Ames and his coworkers showed that histidine regulated the synthesis of each enzyme in this biosynthetic pathway; they dubbed this kind of regulation "coordinate repression." Hartman had previously shown that these enzymes mapped to the same location of the *Salmonella* chromosome. In fact, the sequence of genes encoding these enzymes on the chromosome was similar to the sequence of enzymatic steps in the biosynthetic pathway. Based on this finding, Ames and his coworkers suggested that histidine regulated its own biosynthesis at the gene level, by repressing "the synthesis of all of the biosynthetic enzymes together" (Ames and Garry 1959: 1459).

In the late 1950s and early 1960s, Ames was among a dozen or so prominent biochemists who were engaged in studies of metabolic responses to physiological or environmental stimuli by measuring changes of enzymatic activities at the cellular level. Papers on "feedback inhibition" of metabolic pathways emerged from members of this loose international network of researchers, which featured in both the 1959 Ciba Foundation Symposium on the Regulation of Cell Metabolism and the 1961 Cold Spring Harbor Symposium on Cellular Regulatory Mechanisms.[3] Ames's research on histidine biosynthesis epitomized this trend, which brought together biochemistry and bacterial genetics and contributed to the vibrancy of molecular biology as it was emerging as a new field during this period. Continuing his affiliation with vanguard institutional niches, Ames took a year-long sabbatical from the NIH in 1961 and split the time between the laboratories of Francis Crick in Cambridge and François Jacob at the Institut Pasteur.

Ames recalls, "Sometime in 1964, I read the list of ingredients on a box of potato chips and began to wonder whether preservatives and other chemicals could cause genetic damage to humans" (Ames 2003: 4371). The thousands of histidine-requiring mutants he had on hand (through collaborator Hartman) provided ready test material for investigating mutagens. The early phase of this work involved classifying mutants with an eye toward studying mutagenesis. At the 1966 Cold Spring Harbor Symposium on Quantitative Biology on the Genetic Code, Ames and laboratory member Harvey Whitfield presented evidence that a group of acridine-like compounds, developed as potential antitumor agents and powerfully mutagenic, added or deleted nucleotides from DNA. They identified a class of mutants in which standard mutagens could not produce reverse mutations, or revertants. (Here, revertants were cells that were no longer histidine-dependent). However, mutants in this class did produce revertants when exposed to one of these acridine-like compounds, ICR 170. The authors inferred that these mutant strains were frameshift mutants. This meant that at the level of DNA, the mutagen caused the addition or deletion of a base, disrupting the reading frame of triplet bases and so changing or ending the sequence of coded amino acids (Ames and Whitfield 1966; Whitfield, Martin, and Ames 1966).

In effect, Ames and Whitfield were able to use current knowledge about mutagens and mutant strains to classify both. For example, the authors used one of the frameshift mutants to test quinacrine, an antimalarial drug. It was a weak mutagen of the strain. As the author commented: "This raises the possibility that the standard antimalarials chloroquine, quinine, and quinacrine, which are known to bind to DNA strongly, are causing frameshift mutations in the human population" (Ames and Whitfield 1966: 225). Ames's eponymous test would build on this practice of using known mutant strains to detect and classify mutagens. After moving to Berkeley in 1967, Ames sought funding for the ongoing project on mutagens. His application to the National Cancer Institute was turned down—as he puts it, "they did not think bacteria could teach us much about cancer"—but supported by the U.S. Atomic Energy Commission (AEC), which had been funding research on radiation genetics and mutations for a decade (Beatty 1991; Ames 2003: 4372).

Ames first presented his mutagen tester strains at the Conference on Evaluating the Mutagenicity of Drugs and Other Chemical Agents, which took place in Washington, D.C., on 4–6 November 1970. According to a report in *Science,* the event was prompted by concern among researchers about the potential hazards of synthetic chemicals, which were becoming ubiquitous (Harris 1971). It was not only scientists who were concerned, of course: Rachel Carson's *Silent Spring,* which appeared in 1962, built on the public fear of radioactive contamination generated by the fallout debates to draw attention to the unseen hazards of pesticides and other synthetic chemicals (Lutts 1985). At this 1970 meeting, six months after the first Earth Day, biologists drew on their familiarity with mutagenesis as a laboratory tool to consider the parallel hazards of ionizing radiation and chemical mutagens in everyday life:

> Using a variety of well-characterized mutagens, scientists have been able to manipulate microorganisms in particular to produce selective mutations in the genes. Their methods are sophisticated enough to produce mutations in the genes governing the synthesis of the macromolecules involved in chromosome duplication (DNA synthesis) and gene expression (RNA and protein synthesis). With these advances has come the realization that similar mutations may be occurring in man by way of less controlled processes, such as radiation damage and alteration of chromosomes by chemicals and drugs. Many workers believe that chemical damage is now a more important problem than radiation hazard. (Harris 1971: 51)

Toxicological and environmental problems could now be understood in molecular biological terms.

Researchers at the meeting presented work on the most up-to-date methods for assessing mutagenicity as well as for detecting the rate of mutations in human populations. In describing his own assay, Bruce Ames acknowledged the limitations of a bacterial test, admitting it is "absurd to extrapolate from bacteria to humans," even as he defended the concept: "But DNA has the same double helical structure and the same four nucleotides in all organisms, and it is logical to believe that mutagens of *Escherichia coli* DNA will also be mutagenic for animal DNA. In general, mutagens for higher organisms are mutagens for bacteria also. More than half of the mutagenic agents for bacteria are carcinogenic for animals" (Harris 1971: 52). Other researchers were attempting to develop laboratory tests using mammalian cells in tissue culture. The 1960s had seen the development of a variety of tissue culture lines, often developed for work with animal viruses, modeled on the investigation of bacteriophages using bacterial cultures. However, the tissue culture systems for scoring carcinogens did not develop as quickly as Ames's bacterial system.[4]

Not all the proposals at this meeting concerned laboratory screening of compounds. James Neel advocated the implementation of mass human screening for mutations, along the lines of the screening of infants for phenylketonuria. By conducting electrophoretic testing of ten proteins in blood samples from 350,000 people per year, such a screening project could detect a "50 percent increase in the human mutation rate" (Harris 1971: 52). The cause of a rise in mutation rate could not be determined by such a screen, but the information gained could serve as a "public health warning system."[5]

The participation at this conference of geneticists like Neel, who had previously participated in the Atomic Bomb Casualty Commission investigations in Japan, carried over directly from earlier studies of radiation.[6] Alexander Hollaender, who had presided over the growth of radiobiology research at Oak Ridge, attended the 1970 meeting and reminded other participants that "the research effort directed toward the investigation of radiation hazards was made possible only by long-range guaranteed support [i.e., of the AEC]" (Harris 1971: 52).[7] In fact, in large part through Hollaender's leadership, the AEC had fostered the research aimed at connecting advances in molecular biology—particularly the growing understanding of DNA replication, DNA transcription, and causes of mutation in bacteria—to studies of human carcinogenesis (Frickel 2004).

The Ames test involved the use of four strains of bacteria (Ames et al. 1972)—for an illustration of what the plates look like, see Figure 2.1. Three of the strains (originally TA1531, TA1532, and TA1534) were designed to detect different kinds of frameshift mutagens. The fourth strain, TA1530, contained a base-pair change, and so it would detect mutations that involve base-pair substitutions. In addition, all four tester strains included a mutation in the *uvrB* gene that disabled DNA excision repair, making them more sensitive to mutagens whose effects would otherwise be corrected by this system. Ames's group soon added two additional features to the system to improve its sensitivity to mutagens. The first was the incorporation of an additional mutation in the

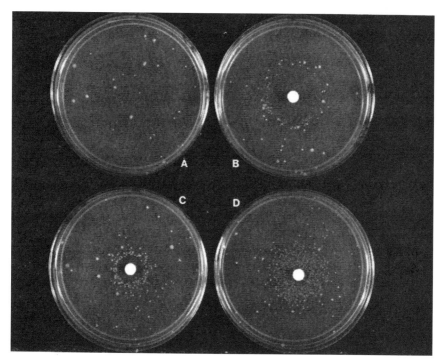

Figure 2.1. Pictures of Ames Tests for (A) Spontaneous Revertants and exposure to (B) Furylfuramide, (C) Aflatoxin B1, and (D) 2-Aminofluorene. The mutagenic compounds in B, C, and D were applied to the 6 mm filter disk in the center of each plate. Each petri plate contains cells of the tester strain in a thin overlay of top agar. (The strain used here is TA98, derived by adding a resistance transfer factor to a *Salmonella* tester strain, mutant *hisD3052*, that scores frameshift mutations.) Plates C and D contain, additionally, a liver microsomal activation system isolated from rats. The spontaneous or compound-induced revertants, each of which reflects a mutational event, appear in a ring as spots around the paper disk. Ames, McCann, and Yamasaki 1975: 358. © Elsevier.

strains that resulted in a deficient lipopolysaccharide (Ames et al. 1973). This compound normally coats the bacterium and poses a barrier to the penetration of large molecules into the cell. The mutation rendered the cells permeable to a wider range of chemical compounds. Second, Ames's group showed that spreading an extract of rat or human liver with the potential carcinogen on the petri dish allowed testing of metabolic derivatives of the compound being tested (Ames et al. 1973). It was known that mammalian microsomal hydroxylase activated many classes of carcinogens and mutagens, including aflatoxin, aromatic amines, and polycyclic hydrocarbons (Ames 1973: 116). Even after these innovations, Ames's group kept improving the strains, so that the "Ames test" was not a fixed assay, but an evolving tool.

As Ames readily admitted, the idea of using microbes to screen compounds had not originated with him. Following Evelyn Witkin's 1947 demonstration that a wide variety of chemical compounds could serve as mutagens in *E. coli*, Milislav Demerec, Giuseppe Bertani, and J. Flint published an article in which they tested a variety of chemicals for mutagenicity in a streptomycin-dependent strain of *E. coli* (Witkin 1947; Demerec, Bertani, and Flint 1951). The system registered mutagenicity by scoring colony growth from back-mutations—cells that grew were mutants. This meant the authors could score mutagenic events at a frequency as low as one per hundred billion (1×10^{-11}), because as many as 500 million bacteria could be screened on a single petri dish (Demerec 1954: 319). Nineteen of the thirty-one compounds tested in this way proved to be mutagenic. It was a chemically diverse group, including boric acid, ammonia, hydrogen peroxide, copper sulfate, acetic acid, formaldehyde, and phenol.

Waclaw Szybalski's laboratory further refined the use of bacterial strains to detect chemical mutagens in the late 1950s, screening over 400 compounds (Iyer and Szybalski 1958; Szybalski 1958).[8] His technique included the paper disk method for screening mutagens, in which the substance was placed on the petri dish on a small circular piece of filter paper, causing revertants to appear in a ring around the substance as it diffused out. Szybalski noted a strong correlation between carcinogenicity and mutagenicity: "these studies demonstrated a close correlation between the carcinogenic effect in mammals and the mutagenic effect on bacteria, stimulating a wide interest in this field" (Iyer and Sybalski 1958: 23). But that was not the principal motivation for his screen, which was aimed at identifying anti-tumor agents (Zeiger 2004). Why did he not see the genetic consequences of these compounds as key to explaining their carcinogenicity, too?

In fact, geneticists tended to view chemical mutagens as inherently different from radiation, which could directly modify genes. Demerec asserted that chemical agents induced mutations in an *indirect* way—that "mutagenic treatment brings about some change in either cytoplasm or nucleus which in turn affects certain physiological processes of the cell, and thereby influences genes" (Demerec 1954: 322). In other words, "treatment with a mutagen does not affect genes directly" (Hemmerly and Demerec 1955: 74). Along similar lines, Joshua Lederberg emphasized that chemical mutagens may interact with genes in indirect, and complicated, ways: "We must be very cautious in interpreting chemical mutagenesis as a direct chemical reaction with the gene. Cells, including bacteria, react in a very complex pattern to treatment with mutagenic agents. The possibility cannot be excluded that some mutations are produced indirectly as a consequence of accidents during recovery or of non-specific and non-localized disturbances of nuclear structure" (Lederberg 1951: 275). Bacterial screens of chemical mutagens, then, were not expected to shed light on the nature of mutation.

Lederberg himself noted that initial research on chemical mutagens might have translated into an earlier engagement with toxicology and public health (Lederberg 1997). Based on his own work on "radiomimetic" chemical mutagens, Lederberg wrote H.J. Muller in 1950 expressing concern that a wide range of common organic reagents might pose a significant genetic hazard to individuals exposed, similar to and even greater than that associated with sources of ionizing radiation. He suggested that the problem be brought to the attention of the National Research Council (NRC). Muller, who certainly did not hesitate to enter into debates about genetics and public safety, felt the evidence was not strong enough to warrant the NRC's involvement. In broaching the topic of who would pay for large-scale investigation, he warned: "It is not right that mutation work should have to be a tail to the cancer kite."[9] Muller's answer reflected a longstanding tendency in radiobiology to differentiate somatic consequences—namely, cancer—from genetic effects. Muller wanted to make sure that work on mutation did not become subordinate.

Any hesitancy to hitch chemical mutagenesis studies to the cancer kite evaporated by the 1970s. Instead, scientists increasingly took the well-established correlation between mutagenicity and carcinogenicity as causal rather than coincidental. And there was no longer any sense of rivalry between a focus on somatic versus a focus on genetic effects—nor even a pronounced distinction, as the somatic mutation theory of cancer gained ground (Jolly 2004: chap. 12). Ames published an overview of his method in *Environmental Health Perspectives* endorsing the theory succinctly: "We postulate that carcinogens cause cancer by somatic mutation" (Ames 1973: 115). Here and in his other publications, Ames drew on recent work both on the genetic code and on the chemical nature of DNA damage, much of which had been funded through the AEC. In contrast to Demerec's perspective that mutation was an indirect consequence of exposure to these agents, biochemists studying DNA damage conceived of *direct* action, such as intercalation between the base pairs of the double helix.

Let me return to a question I posed at the outset—how and why did mutations in bacteria become a key means for visualizing the cancer-causing dangers of chemicals? The answer is not principally technological—microbial screens for chemicals dated to the 1950s—but reflects the convergence of new ideas about genetic damage and cancer with political and institutional developments. Scott Frickel has argued that political engagement and activism on the part of scientists was critical to the founding of the field of genetic toxicology in the late 1960s and early 1970s (Frickel 2004). Exemplifying this trend were the founding of the Environmental Mutagen Society and the establishment of the federal government's National Institute of Environmental Health, both in 1969.

This time period also saw the reorientation of molecular biologists to the challenges of organismal biology, what Michel Morange has termed the "mass migration" of molecular biologists in the 1960s and 1970s from simple microbial systems to eukaryotic organisms (e.g., yeast, flies, mice) to study immunology, development, and, not least, cancer biology (Morange 1997). As Doogab Yi has recently argued, this trend was driven by new political pressures on scientists in the late 1960s and early 1970s to demonstrate that taxpayer-funded research was improving health (Yi 2008a; 2008b). Environmental health and cancer research both became important venues through which experimental biologists could demonstrate the utility of their knowledge. Lastly, the shift in studies of genetic damage from chromosomes to DNA positioned molecular biologists and biochemists to provide new evidence for the mutational theory of cancer. The Ames test registered the confluence of these political, institutional, and disciplinary changes.

By 1976 the Ames test was being used by sixty or seventy major companies. As Gina Kolata observed in *Science*, "This has led to a curious situation in which industries are implicitly endorsing the tests at the same time that scientists and legislators deliberate over whether companies should be forced to use them" (Kolata 1976: 1215). Ames made the tester strains freely available, asking only that recipients request them directly from him rather than from secondary sources (Ames, McCann, and Yamasaki 1975: 350). Companies often contracted with commercial laboratories to conduct their toxicological screening, and Kolata noted that one such outfit, Litton Bionetics in Maryland, had already seen an increase of contracts for screening chemicals. The Ames test was almost always performed first—it cost only $200 per chemical—and if a compound proved mutagenic in the Ames test, other tests could be ordered.

Even as companies eagerly employed the Ames test, its applications identified as worrisome a number of widely used industrial products. By 1975, Ames's laboratory had used its test to demonstrate the mutagenicity of chloroacetaldehyde (a possible metabolic product of vinyl chloride, a commonly used synthetic chemical), cigarette smoke condensate, and hydrogen peroxide–based hair dyes (Kier, Yamasaki, and Ames 1974; Ames, Kammen, and Yamasaki 1975; McCann, Simmon, et al. 1975). The Ames test also revealed the Japanese-developed preservative furylfuramide to be mutagenic; the compound was subsequently banned in Japan (Kolata 1976: 1217). More controversially, Arlene Blum and Ames showed that the most widely used flame retardant for children's pajamas, tris(2,3-dibromopropyl) phosphate, commonly called Tris, was a mutagen (Blum and Ames 1977). In response, toxicologists undertook animal experiments and found that Tris could cause kidney cancer in mice and rats. Other studies showed that Tris could be absorbed through the skin.[10] Soon thereafter, in April 1977, the government banned the sale of Tris-treated garments. The compound was used in clothing for only a few years, having been introduced to meet 1973 federal regulations for decreased flammability in children's sleepwear.[11]

With the newly evident potential of the Ames test to shape government regulation and policy, critics raised doubts about the premise of the test, that cancer should be regarded as a disease induced by mutagens. Harry Rubin, a virologist at Berkeley, voiced his skepticism in a letter to *Science*: "Excessive application of normal steroid hormones causes cancer, as does the simple transplantation of some endocrine organs into the spleen of the same animal. It is difficult to accept mutagenesis as the origin of these cancers.... Acceptance of screening for carcinogenicity by determining mutagenicity lends tacit support to the hypothesis that malignant transformation of cells is caused by somatic mutation. This hypothesis has been tested explicitly in several experiments and has been found wanting in each case" (Rubin 1976: 241). In the assessment of another critic, "The mutation origin of cancers remains an unproven hypothesis, with a substantial body of evidence in support of other mechanisms" (Sivak 1976: 273). Ames's main response was to point to the power of correlation. His laboratory had demonstrated that 90 percent of *Salmonella* mutagens were also rodent carcinogens, and that 89 percent of animal test carcinogens were also bacterial mutagens (McCann, Choi, et al. 1975; McCann and Ames 1976; Zeiger 2004: 364). Of more than 109 "non-carcinogens" tested, none proved to be mutagens. In effect, supporters of the Ames test pointed to the strong correspondence among compounds that screened as mutagenic and those that proved carcinogenic in animal tests as proof enough of the basic principle. As a side benefit, noted a professor of veterinary medicine, wider use of the Ames test could reduce the number of animals consumed in toxicological safety tests (Loew 1976).

Salmonella Strains in Industrial Testing and Government Regulation

In the late 1970s, the Occupational Safety and Health Administration proposed new legislation regulating carcinogens in the workplace. American industry reacted strongly against the threat of regulation, even as they tried to take advantage of less expensive testing methods. Toxicological testing companies routinely used the Ames test alongside several other rapid screens to identify those compounds that warranted further testing.[12] In addition, companies such as DuPont and American Cyanamid used it to test new products before deciding whether to bring them to market. According to Ames, DuPont decided not to produce two Freon propellants because they were found to be mutagenic in *Salmonella* tests (Ames 1979: 593n. 21). The government's own screening program was not extensive enough to provide data on the range of chemicals on the market—the National Cancer Institute screened about 100 compounds a year, out of the 63,000 chemicals being commonly used in the United States (Maugh 1978: 1202).

Environmentalists tended to view the Ames test as allied with the cause of greater industrial regulation. The high-profile identification by Ames and his coworkers of the preservative furylfuramide, hair dyes, and especially Tris as potent mutagens bolstered public concern about the safety of chemicals—and, in the case of both furylfuramide and Tris, led to their ban. However, the apparent alliance between molecular biologists and environmental organizations was already unraveling in the late 1970s over disagreements about the safety of recombinant DNA. As a journalist for *Science* put it in 1978: "Among those who doubt the environmentalists' good faith are National Institutes of Health (NIH) researchers Malcolm Martin, Wallace Rowe, and Maxine Singer—all of whom have been involved in the DNA debate from the outset. Paul Berg of Stanford, Bruce Ames of the University of California at Berkeley, and Norton Zinder of Rockefeller University as well as others not directly involved in the politics of DNA have told the environmentalists that they are flatly wrong in the recombinant DNA case" (Marshall 1978: 1265). The direction that Ames took next in put him further at odds with environmentalists.

Ames became interested in how the risks of somatic mutation from synthetic chemicals compared to those from "natural" sources, particularly prepared food (Ames 1979). Others had already taken his microbial test in this direction. Japanese researcher Takashi Sugimura had first applied the Ames test to screen naturally occurring agents in the mid 1970s, building on a few identified earlier by Ames, such as aflatoxin. Sugimura found that plants were a major source of mutagens (Nagao, Sugimura, and Matsushima 1978).[13] Sugimura also pointed to differences in food preparation to explain the higher incidence of stomach cancer in Japan as compared to the United States. (Abelson 1979). Along similar lines, Barry Commoner's laboratory demonstrated that fried hamburger showed mutagenic activity in the *Salmonella* test (Commoner et al. 1978). Many of these naturally occurring compounds were just as mutagenic as some synthetic chemicals—and animal tests showed some to be just as carcinogenic. While this line of research enlarged the scope of materials that might be tested for mutagenicity in the name of limiting exposure, the attention to the risks of posed by *natural* agents tended to subvert the rationale for increased government regulation. As the editor of *Science*, Philip Abelson, put it in an editorial: "The effort to prove a big role for industrial chemicals diverts attention from what is probably the best hope for reducing cancer incidence—careful study of foods and effects of cooking.... All people ingest the mutagens and carcinogens of food daily" (Abelson 1979: 11).

By the mid 1980s, Ames had become convinced that the greatest danger to human health came from diet and metabolism. Alvin Weinberg was among those scientists who supported Ames's viewpoint: "Cancer is essentially a natural aging process. No matter what we eat, the huge flood of oxygen radicals produced in many metabolic processes overwhelms all but the most heavy external carcinogens, such as tobacco in heavy smokers. To be sure, anticarcinogenic substances are of benefit, but to choose a noncarcinogenic diet would probably be equivalent to starving to death" (Weinberg 1984: 658). Ames was not so defeatist and advocated the ingestion of vitamins and nutrient-rich foods to counteract mutagenicity in foods and chemicals (Ames 1983).

But Ames's touting of nutritional supplements was not what made him controversial. Ames strongly questioned whether cancer rates were increasing in the industrialized world, and was suspicious of putative links between industrial pollutants and cancer incidence, arguing that smoking and poor nutrition could account for most observed cancers. He also pointed out that in both number and amount, we ingested more "natural pesticides and other natural toxic molecules (and traditional mixtures such as cooked food) than we do of manmade substances" (Ames 1984: 758). Thus activists who focused on cancer as a "corporate problem" were misguided, in Ames's view—they should be stopping subsidies to tobacco farmers and improving the diet of ordinary Americans. Pollution and occupational exposure were already sufficiently regulated by the government, given their smaller role in cancer incidence. As he put it, "the preoccupation with tiny amounts of man-made pollution has been blown up out of proportion" (Ames 1984: 668). Needless to say, this viewpoint outraged environmental groups. To add insult to injury, Ames was awarded a major ecology prize in 1985.[14]

In 1987, Ames campaigned against Proposition 65, a "citizens' enforcement law" in California that imposed stringent new regulations on chemical users. The law passed, by a margin of two to one, and Ames was then appointed to a regulatory group to help implement the law. This provoked outrage from the mainline environmentalist groups who had supported the proposition. As Carl Pope of the Sierra Club put it to a writer for *Science*, "I've never seen a clearer fox-in-the-chicken-coop situation" (Marshall 1987: 1459).

In the end, Ames and his coworkers questioned the corporate burden of responsibility for preventing cancer on the grounds of scientific uncertainty:

> In the modern context of being able to measure parts-per-billion and parts-per-trillion levels of substances and the realization that there is universal human exposure to rodent carcinogens of natural origin, it is first important to prioritize among the plethora of possible hazards in order to avoid being distracted from working on the more important problems. The enormous uncertainties in the use of animal data to assess human risk and our lack of knowledge about the mechanisms of carcinogenesis make policy-making especially difficult; however, we do not imply that all problems should be passed over until the last smoker lays down his cigarette. (Ames, Magaw, and Gold 1987: 235)

As Robert Proctor as well as Naomi Oreskes and Erik Conway have shown, scientific uncertainty was increasingly used by politicians and industry representatives (including allied scientists) in the 1970s and 1980s to delay or derail government regulation (Proctor 1995; Proctor 2011; Oreskes and Conway 2010). Intentionally or not, Ames played into this mindset.[15] Industry representatives and libertarian writers cited Ames's skepticism about the role of industrial pollution in causing cancer in their efforts to halt the expansion of government regulation (Efron 1984).[16]

Conclusions

I have emphasized two aspects of the history of the Ames test. One is the way in which its conception of carcinogenicity built on earlier research about the role of radiation in cancer. Here I follow Scott Frickel (2004) in noting how research on chemical carcinogens followed the tracks—conceptual, experimental, and institutional—of research on radiation genetics. The role of the AEC in funding Ames's research, like the establishment of a computer registry for carcinogens at Oak Ridge, attests to the way in which government funding of research on the biological effects of radiation encompassed chemical mutagenesis as well.[17] Indeed, taken together, these two research areas comprised much of what came to be identified under the rubric "genetic toxicology" in the 1960s (Frickel 2004). It is worth considering how this shaped the understanding of chemical agents as mutagens and carcinogens. Charlotte Auerbach has argued that it was hard for researchers to come up with mechanisms that would explain the action of both ionizing radiation and chemical agents, and this initially put the two lines of mutation research in competition. "Sweeping attacks on the target theory were made soon after the discovery of chemical mutagens. The fact that chemicals can produce many of the same effects as X-rays was taken to indicate that X-rays, too, must act by chemical intermediates" (Auerbach 1967: 71). In a sense, the shift from studies of chromosomal damage to the biochemistry of DNA damage provided a substrate, or even boundary object, through which the actions of the two classes of mutagens could be brought into correspondence (Star and Griesemer 1989). Biochemists and molecular biologists such as Ames were also eager to promote the understanding of cancer in terms of DNA damage (though, interestingly, not in terms of particular genes), but this raised the ire of cancer specialists who regarded tumorigenesis as a more complex biological affair.

The second issue concerns how the development and adoption of the Ames test intersected with changing currents in American politics. The Ames test was introduced during a time of popular environmentalism, and Ames himself applied his test to identify dangerous new synthetic chemicals, informing government regulation. However, Ames's subsequent work demonstrating the mutagenicity of natural substances subverted any simple idea of what constituted hazardous exposure, at least from the perspective of DNA damage. Steven Shapin, who happened to work in a government agency using the Ames test, has provided a fascinating firsthand account of how a political sensibility focused on the dangers of drugs and pollutants made it difficult to view a substance like caffeine as legitimately hazardous (Shapin 1995: 264–65). That political climate was changing, however, and opponents of industrial regulation seized on Ames's findings. Moreover, the debates over the safety of recombinant DNA prompted some scientists to sympathize with critics of government regulation. By the mid 1980s environmentalist groups viewed Ames as an adversary as they contested the anti-regulatory movement. They understood that his ongoing work challenged the existence of an inherent difference between the safety of natural and artificial substances, a demarcation crucial to green activism.[18] In the end, the Ames test revealed a world of mutagens, complicating attempts to trace the environmental origins of cancer to industry alone.

Acknowledgments

For their comments on an earlier draft, I thank Bruce Ames, Lindy Baldwin, Soraya Boudia, James Byrne, Luis Campos, Scott Frickel, Michael Gordin, Nathalie Jas, Nancy Langston, Ilana Löwy, Carsten Reinhardt, Hans-Jörg Rheinberger, Jody A. Roberts, Alexander Schwerin, Steve Shapin, Evelyn Witkin, other participants of Making Mutations: Objects, Practices, Contexts at the Max Planck Institute for the History of Science in Berlin on 13–15 January 2009, attendees of Princeton's History of Science Program Seminar on 23 February 2009, and contributors to Carcinogens, Mutagens, Reproductive Toxicants: The Politics of Limit Values and Low Doses in the Twentieth and Twenty-first Centuries, at Misha in Strasbourg on 29–31 March 2010. Work on this chapter was partly supported by a National Library of Medicine Grant for Scholarly Works in Biomedicine and Health from the U.S. National Institutes of Health, award 5G13LM9100.

Notes

1. As Ames has noted (personal communication, 5 August 2009), this did not rule out other causes of cancer, such as hormones, but his test provided a rapid way to evaluate potential environmental carcinogens.
2. Drawing on both biochemical techniques and genetics, Ames used chromatography to separate the precursors of histidine (imidazole intermediates) and made double mutants to determine the order of metabolic steps in the pathway (Ames and Mitchell 1952; Haas et al. 1952; Ames, H. Mitchell, and M. Mitchell 1953).
3. For more on this trend, see Creager and Gaudillière (1996: 6–15).

4. Harris (1971: 52) does not mention any tissue culture researchers by name. The Ames test is compared to some of these mammalian cell culture test systems by the U.S. Interagency Staff Group on Carcinogens (1986: 227).
5. On the development of genetic screening programs, see Lindee (2005).
6. Neel was an architect of the genetics project of the Atomic Bomb Casualty Commission (Beatty 1991; Lindee 1994).
7. For more on Hollaender's role in fostering radiation genetics, see Rader (2006).
8. The list of publications on Professor Szybalski's website is annotated, and the note about these two publications is telling: "Dr. Szybalski and his collaborators studied the mechanism of mutagenesis. First, they developed the 'paper disc mutagenicity test,' which was later adopted in [the] so-called 'Ames test.'" http://mcardle.oncology.wisc.edu/faculty/bio/WSPubl.html (accessed 30 September 2010).
9. H.J. Muller to Joshua Lederberg, 16 March 1950, reprinted in Lederberg (1997: 7).
10. Animal studies before the ban were followed by a study of human absorption published after the ban (Blum et al. 1978).
11. Recently, however, chlorinated Tris has become widely used as a flame retardant in upholstered furniture (Slater 2012).
12. As reported in *Science* magazine, Litton routinely employed four tests: "the Ames test, a test for gene mutation in mouse cells, the SCE test, and an in vitro transformation test" (Maugh 1978: 1204).
13. The importance of Japanese researchers in this arena is attested by an international conference held in Tokyo in 1979, "Naturally Occurring Carcinogens-Mutagens and Modulators of Carcinogenesis," which was attended by a number of American and European researchers (including Bruce Ames) as well as many Japanese scientists.
14. It was the John and Alice Tyler Ecology-Energy Prize, administered by University of Southern California (Dye 1985: 20).
15. Though I would not follow Proctor as far as referring to Ames as "the most powerful anti-environmentalist of the century" (Proctor 1995: 133). Ames has always been adamant about not receiving any money from industry, and argues that he is simply committed to an honest evaluation of the science. On the corporate use of scientific uncertainty, also see Michaels (2008); Oreskes, Conway, and Shindell (2008); and Oreskes and Conway (2010).
16. On the anti-regulatory movement, see Hays (1987).
17. On the Environmental Mutagen Information Center, see Frickel (2004: 58–59).
18. Here, too, the politics of environmental radioactivity (which may be naturally occurring or industrially produced) provides an important precedent. Of course, the persistence of many synthetic chemicals does distinguish them from most natural, biodegradable, substances.

Bibliography

Abelson, Philip H. 1979. "Cancer—Opportunism and Opportunity." *Science* 206: 11.
Ames, Bruce N. 1973. "Carcinogens are Mutagens: Their Detection and Classification." *Environmental Health Perspectives* 6: 115–18.
———. 1979. "Identifying Environmental Chemicals Causing Mutations and Cancer." *Science* 204: 587–93.

———. 1983. "Dietary Carcinogens and Anticarcinogens." *Science* 221: 1256–64.
———. 1984. "Cancer and Diet." (Reply to Letters to Editor) *Science* 224: 668, 670, 757–58, 760.
———. 2003. "An Enthusiasm for Metabolism." *Journal of Biological Chemistry* 278: 4369–80.
Ames, Bruce N., William E. Durston, Edith Yamasaki, and Frank D. Lee. 1973. "Carcinogens are Mutagens: A Simple Test System Combining Liver Homogenates for Activation and Bacteria for Detection." *Proceedings of the National Academy of Sciences, USA* 70: 2281–85.
Ames, Bruce N., and Barbara Garry. 1959. "Coordinate Repression of the Synthesis of Four Histidine Biosynthetic Enzymes by Histidine." *Proceedings of the National Academy of Sciences, USA* 45: 1453–61.
Ames, Bruce N., E.G. Gurney, James A. Miller, and H. Bartsch. 1972. "Carcinogens as Frameshift Mutagens: Metabolites and Derivatives of 2-Acetylaminofluorene and Other Aromatic Amine Carcinogens." *Proceedings of the National Academy of Sciences, USA* 69: 3128–32.
Ames, Bruce N., H.O. Kammen, and Edith Yamasaki. 1975. "Hair Dyes are Mutagenic: Identification of a Variety of Mutagenic Ingredients." *Proceedings of the National Academy of Sciences, USA* 72: 2423–27.
Ames, Bruce N., Frank D. Lee, and William E. Durston. 1973. "An Improved Bacterial Test System for the Detection and Classification of Mutagens and Carcinogens." *Proceedings of the National Academy of Sciences, USA* 70: 782–86.
Ames, Bruce N., Renae Magaw, and Lois S. Gold. 1987. "Risk Assessment." (Response to Letter of Raymond Neutra) *Science* 237: 235.
Ames, Bruce N., Joyce McCann, and Edith Yamasaki. 1975. "Methods for Detecting Carcinogens and Mutagens with the *Salmonella*/Mammalian-Microsomal Mutagenicity Test." *Mutation Research* 31: 347–64.
Ames, Bruce N., and H.K. Mitchell. 1952. "The Paper Chromatography of Imidazoles." *Journal of the American Chemical Society* 74: 252–53.
Ames, Bruce N., Herschel K. Mitchell, and Mary B. Mitchell. 1953. "Some New Naturally Occurring Imidazoles Related to the Biosynthesis of Histidine." *Journal of the American Chemical Society* 75: 1015–18.
Ames, Bruce N., and Harvey J. Whitfield, Jr. 1966. "Frameshift Mutagenesis in *Salmonella*." *Cold Spring Harbor Symposia on Quantitative Biology* 31: 221–25.
Auerbach, Charlotte. 1967. "Changes in the Concept of Mutation and the Aims of Mutation Research." In *Heritage from Mendel: Proceedings of the Mendel Centennial Symposium*, ed. R. Alexander Brink, 67–80. Madison, W.I.: University of Wisconsin Press.
Beatty, John. 1991. "Genetics in the Atomic Age: The Atomic Bomb Casualty Commission, 1947–1956." In *The Expansion of American Biology*, eds. Keith R. Benson, Jane Maienschein, and Ronald Rainger, 284–324. New Brunswick, N.J.: Rutgers University Press.
Blum, Arlene, and Bruce N. Ames. 1977. "Flame-Retardant Additives as Possible Cancer Hazards." *Science* 195: 17–23.
Blum, Arlene, Marian Deborah Gold, Bruce N. Ames, Christine Kenyon, Frank R. Jones, Eva A. Hett, Ralph C. Dougherty, Evan C. Horning, Ismet Dzidic, David I. Carroll, Richard N. Stillwell, and Jean-Paul Thenot. 1978. "Children Absorb Tris-BP Flame

Retardant from Sleepwear: Urine Contains the Mutagenic Metabolite, 2,3-Dibromopropanol." *Science* 201: 1020–23.

Commoner, Barry, Antony J. Vithayathil, Piero Dolara, Subhadra Nair, Prema Madyastha, and Gregory C. Cuca. 1978. "Formation of Mutagens in Beef and Beef Extract During Cooking." *Science* 201: 913–16.

Creager, Angela N.H., and Jean-Paul Gaudillière. 1996. "Meanings in Search of Experiments and Vice-Versa: The Invention of Allosteric Regulation in Paris and Berkeley, 1959–1968." *Historical Studies in the Physical and Biological Sciences* 27: 1–89.

Demerec, Milislav. 1954. "What Makes Genes Mutate?" *Proceedings of the American Philosophical Society* 98: 318–22.

Demerec, Milislav, G. Bertani, and J. Flint. 1951. "A Survey of Chemicals for Mutagenic Action on *E. coli*." *American Naturalist* 85: 119–36.

Dye, Lee. 1985. "Biochemist Will Share Ecology-Science Prize." *Los Angeles Times*, 24 May, 20.

Efron, Edith. 1984. *The Apocalyptics: Cancer and the Big Lie. How Environmental Politics Controls What We Know About Cancer*. New York: Simon and Schuster.

Frickel, Scott. 2004. *Chemical Consequences: Environmental Mutagens, Scientist Activism, and the Rise of Genetic Toxicology*. New Brunswick, N.J.: Rutgers University Press.

Haas, Felix, Mary B. Mitchell, Bruce N. Ames, and Herschel K. Mitchell. 1952. "A Series of Histidineless Mutants of *Neurospora crassa*." *Genetics* 37: 217–26.

Harris, Maureen. 1971. "Mutagenicity of Chemicals and Drugs." *Science* 171: 51–52.

Hays, Samuel P. 1987. *Beauty, Health, and Permanence: Environmental Politics in the United States, 1955–1985*. Cambridge: Cambridge University Press.

Hemmerly, Jean, and Milislav Demerec. 1955. "Tests of Chemicals for Mutagenicity." *Cancer Research*, suppl. 3: 69–75.

Iyer, V.N., and W. Szybalski. 1958. "Two Simple Methods for the Detection of Chemical Mutagens." *Applied Microbiology* 6: 23–29.

Jolly, J. Christopher. 2004. "Thresholds of Uncertainty: Radiation and Responsibility in the Fallout Controversy." Ph.D. diss., Oregon State University.

Kier, Larry D., Edith Yamasaki, and Bruce N. Ames. 1974. "Detection of Mutagenic Activity in Cigarette Smoke Condensates." *Proceedings of the National Academy of Sciences, USA* 71: 4159–63.

Kolata, Gina Bari. 1976. "Chemical Carcinogens: Industry Adopts Controversial 'Quick' Tests." *Science* 192: 1215–17.

Lederberg, Joshua. 1951. "Genetic Studies with Bacteria." In *Genetics in the Twentieth Century*, ed. L.C. Dunn, 263–89. New York: MacMillan Press.

———. 1997. "Some Early Stirrings (1950 ff.) of Concern About Environmental Mutagens." *Environmental and Molecular Mutagenesis* 30: 3–10.

Lindee, M. Susan. 1994. *Suffering Made Real: American Science and the Survivors at Hiroshima*. Chicago: University of Chicago Press.

———. 2005. *Moments of Truth in Genetic Medicine*. Baltimore, M.D.: Johns Hopkins University Press.

Loew, F.M. 1976. "The Ames Assay." (Letter to Editor) *Science* 193: 274.

Lutts, Ralph H. 1985. "Chemical Fallout: Rachel Carson's *Silent Spring*, Radioactive Fallout, and the Environmental Movement." *Environmental Review* 9: 210–25.

Marshall, Eliot. 1978. "Environmental Groups Lose Friends in Effort to Control DNA Research." *Science* 202: 1265-69.

———. 1987. "California's Debate on Carcinogens." *Science* 235: 1459.

Maugh, Thomas H., II. 1978. "Chemical Carcinogens: The Scientific Basis for Regulation." *Science* 201: 1200-1205.

McCann, Joyce, and Bruce N. Ames. 1976. "Detection of Carcinogens as Mutagens in the *Salmonella*/Microsome Test: Assay of 300 Chemicals: Discussion." *Proceedings of the National Academy of Sciences, USA* 73: 950-54.

McCann, Joyce, Edmund Choi, Edith Yamasaki, and Bruce N. Ames. 1975. "Detection of Carcinogens as Mutagens in the *Salmonella*/Microsome Test: Assay of 300 Chemicals." *Proceedings of the National Academy of Sciences, USA* 72: 5135-39.

McCann, Joyce, Vincent Simmon, David Streitwieser, and Bruce N. Ames. 1975. "Mutagenicity of Chloroacetaldehyde, a Possible Metabolic Product of 1,2-dichloroethane (Ethylene Dichloride), Chloroethanol (Ethylene Chlorohydrin), Vinyl Chloride, and Cyclophosphamide." *Proceedings of the National Academy of Sciences, USA* 72: 3190-93.

Michaels, David. 2008. *Doubt is Their Product: How Industry's Assault on Science Threatens Your Health.* Oxford: Oxford University Press.

Morange, Michel. 1997. "The Transformation of Molecular Biology on Contact with Higher Organisms, 1960-1980: From a Molecular Description to a Molecular Explanation." *History and Philosophy of the Life Sciences* 19: 369-93.

Nagao, Minako, Takashi Sugimura, and Taijiro Matsushima. 1978. "Environmental Mutagens and Carcinogens." *Annual Review of Genetics* 12: 117-59.

Oreskes, Naomi, and Erik M. Conway. 2010. *Merchants of Doubt: How a Handful of Scientists Obscured the Truth on Issues from Tobacco Smoke to Global Warming.* New York: Bloomsbury.

Oreskes, Naomi, Erik M. Conway, and Matthew Shindell. 2008. "From Chicken Little to Dr. Pangloss: William Nierenberg, Global Warming, and the Social Deconstruction of Scientific Knowledge." *Historical Studies in the Natural Sciences* 38: 109-52.

Proctor, Robert N. 1995. *Cancer Wars: How Politics Shapes What We Know and Don't Know about Cancer.* New York: Basic Books.

———. 2011. *Golden Holocaust: Origins of the Cigarette Catastrophe and the Case for Abolition.* Berkeley: University of California Press.

Rader, Karen A. 2006. "Alexander *Hollaender's* Postwar Vision for Biology: Oak Ridge and Beyond." *Journal of the History of Biology* 39: 685-706.

Rubin, Harry. 1976. "Carcinogenicity Tests." *Science* 191: 241.

Shapin, Steven. 1995. "Cordelia's Love: Credibility and the Social Studies of Science." *Perspectives on Science* 3: 255-75.

Sivak, Andrew. 1976. "The Ames Assay." (Letter to Editor) *Science* 193: 272-74.

Slater, Dashka. 2012. "Is This the Most Dangerous Thing in Your House?" *The New York Times Magazine,* 12 September, 22-27.

Star, Susan Leigh, and James R. Griesemer. 1989. "Institutional Ecology, 'Translations' and Boundary Objects: Amateurs and Professionals in Berkeley's Museum of Vertebrate Zoology, 1907-39," *Social Studies of Science* 19: 387-402.

Szybalski, W. 1958. "Special Microbial Systems. II. Observations on Chemical Mutagenesis in Microorganisms." *Annals of the New York Academy of Sciences* 76: 475-89.

U.S. Interagency Staff Group on Carcinogens. 1986. "Chemical Carcinogens: A Review of the Science and Its Associated Principles." *Environmental Health Perspectives* 67: 201–82.

Weinberg, Alvin M. 1984. "Cancer and Diet." (Letter to Editor) *Science* 224: 658.

Whitfield, Harvey J., Jr., Robert G. Martin, and Bruce N. Ames. 1966. "Classification of Aminotransferase (C gene) Mutants in the Histidine Operon." *Journal of Molecular Biology* 21: 335–55.

Witkin, Evelyn M. 1947. "Mutations in *Escherichia coli* Induced by Chemical Agents." *Cold Spring Harbor Symposia on Quantitative Biology* 12: 256–69.

Yi, Doogab. 2008a. "The Recombinant University: Genetic Engineering and the Emergence of Biotechnology at Stanford, 1959–1980." Ph.D. diss., Princeton University.

———. 2008b. "Cancer, Viruses, and Mass Migration: Paul Berg's Venture into Eukaryotic Biology and the Advent of Recombinant DNA Research and Technology, 1967–1980." *Journal of the History of Biology* 41: 589–636.

Zeiger, Errol. 2004. "History and Rationale of Genetic Toxicity Testing: An Impersonal, and Sometimes Personal, View." *Environmental and Molecular Mutagenesis* 44: 363–71.

 CHAPTER 3

DES, Cancer, and Endocrine Disruptors
Ways of Regulating, Chemical Risks, and
Public Expertise in the United States

Jean-Paul Gaudillière

On 17 July 1979 *The New York Times* announced that the New York State Supreme Court found the pharmaceutical company Eli Lilly responsible for the vaginal cancer affecting Joyce Bichler and awarded the young woman $500,000 in compensation.[1] The New York state ruling was the first legal decision recognizing that pharmaceutical firms, rather than physicians or regulatory authorities, were liable for the adverse consequences of the medical uses of diethylstilbestrol (DES). This analog of estrogen had been prescribed to millions of pregnant women in the United States as a safety measure against the risk of miscarriage for thirty years, until it was recognized to be the cause of cancers and malformations of the reproductive tract in many exposed fetuses.

A few years earlier, other uses of the same DES had been the target of other media reports and regulatory measures. On 26 April 1973, the U.S. press had announced that the Food and Drug Administration (FDA) was ready to ban DES implants in cattle and other livestock, recalling that one year earlier the same regulatory agency had prohibited the use of DES in animal feeds.[2] The risk of cancer, rather than its occurrence, was the official motive for a path of action whose legitimacy originated in an unusual piece of legislation, the Delaney Clause, which stated that any food additive proved to the carcinogenic in animals or humans should be excluded from food for human consumption.

A major chemical, a substance that had been used for several decades as a potent analog of estrogens in medicine and as a growth enhancer in agriculture, was leaving the scene under the pressure of mobilized consumers, media campaigns, and regulatory interventions. However, the industrial and economic significance of these events were not as dramatic as the pharmaceutical and animal-food industries thought they would be. As the media failed to point out, DES production and prescription did not stop. If pregnancy was contraindicated, other indications like the treatment of prostate cancer remained important. Although the sales of DES for agricultural purposes vanished, other hormones and growth-promoting drugs were soon introduced in animal feed as substitutes.

The importance of the DES affair should not, however, be underestimated. Controversies over its uses had started before 1971, when it was discovered that DES was not only a potential but an actual carcinogen in humans, and they did not stop with the regulatory measures taken in the aftermath. The 1973 ban was for instance immediately contested in court and even though a federal judge finally approved the FDA ban in 1978, illegal uses of DES in agriculture remained a legal issue until the mid 1980s. Similarly, the compensation cases did not end with the few decisions that were actually granted in favor of the "DES daughters" acting as plaintiffs in the late 1970s and early 1980s, but included a wave of cases up to the first years of the twenty-first century, hundreds of them dismissed, many others still being challenged in higher courts.

Controversies originating in the medical as well as the agricultural uses of DES have lasted more than half a century. They have addressed all concerns connected with risk, medical intervention, and industrial practices in the postwar United States. Unsurprisingly, the story of DES in the United States has been investigated in many ways. One group of studies has looked at the unfolding of the medical drama and focused on the lessons to be drawn from the affair, either from the perspective of medical practice or from the feminist viewpoint (Apfel 1984; Meyers 1986; Dutton 1988; Pfeffer 1992). In parallel, historians and sociologists of medicine have investigated DES to integrate its medical uses into the long history of sex hormones and gynecology, while more recent studies on gender have analyzed the role played in the crisis by the then-emerging women's health movement (Bell 1980; Marks 2001; Morgen 2002). Less numerous, analysts of the agricultural controversy have emphasized two different contexts: the rise of industrial agriculture, and the development of the environmental and consumer movements in the 1960s and 1970s (Shell 1984; Marcus 1986; Rifkin 1992). Although they are often alluded to, links between these two developments have rarely been discussed.

From a methodological viewpoint, the literature on DES shares a strong interest in the problem of the "capture," namely, the different ways in which political and economic interests have hinged on or distorted the evaluation of the health risks associated with DES uses. Although the alignments or confrontations between experts and stakeholders, as well as the way they were perceived, are an important element in the DES story, the capture perspective has made the analysis of expertise and risk construction a process and practice of marginal importance since all the important factors, i.e., social interests and alliances, were in place beforehand.

The aim of this paper is therefore not simply to look at the connections between the two types of DES use—agricultural and medical—but to discuss the rise of consumer politics and its impact on the relations between public health and the surveillance of dangerous substances. It is to take the "DES crisis" as a point of entry into a regime of risk management that emerged during the 1960s and 1970s and connected individual and collective consumer actions, stakeholder lobbies, citizens' empowerment, tort legislation, and precautionary action in the legal sphere, as well as the construction of markets.

This politics is still with us and gained widespread importance in handling all sorts of dangers possibly posed by health-threatening substances. As discussed in the last section of this paper, the DES trajectory and the DES affair have in particular played a critical role in the emergence in the 1990s of endocrine disruption as a public health issue and a legitimate research topic. Following this legacy will provide a lens into what may be thought as a new "way of regulating" the industrial uses of life, a mode of regulation rooted in consumer and activist mobilization, in public controversies about risks. Like other ways of regulating—professional, industrial, and administrative—contemporary "public" or "consumer/activist" regulation may be characterized by the values and aims targeted, the main actors involved, the acceptable forms of evidence, and the legitimate means of intervention (Gaudillière and Hess 2012).

Its emergence in the drug/biomedical sector was prompted through two different and often conflicting patterns: on the one hand, by pharmaceutical companies' interest in having more direct access to users, particularly potential consumers of disease-prevention drugs, and on the other, by the critics of paternalistic medicine and the collective empowerment sought by patient groups, inspired for instance by the women's health movement and later the AIDS movement. Emphasis is therefore placed on quality of "service" and on the individual's possibility of making (truly) informed choices. Major attention is given to the risks and potential iatrogenic effects of medical and technical interventions, with observational—eventually social and environmental—epidemiology playing the leading role. Regulatory tools do not only include the systems of postmarketing surveillance organized by administrative or professional bodies like the FDA, but more importantly rely on the precedents and jurisprudence set by court decisions, which may in turn influence regulatory authorities, physicians' prescriptions, and, more generally, users' choices.

The DES story is a good test for this latter hypothesis for at least two reasons. First, it reveals a specific moment in the history of health-related risks and their regulation with a parallel reinforcement of state intervention and social movements. Second, the conjunction of debates about agricultural and medical uses of DES facilitates a broader assessment of regulatory practices beyond the specificities of pharmaceuticals. The argument will be presented in three steps. First, the chapter will recall how DES entered into medical and agricultural practice and how its uses rapidly became contested, leading to the

1970s crisis with its specifically American conjunction between the agricultural controversy and the medical scandal. The second section will explore the form of public expertise that characterized the U.S. debates and their roots in consumer politics, focusing on the evaluation of DES risks conducted by the courts. The last section presents our perspective on the relations between public expertise on DES and the advent of endocrine disruptors.

DES: An Ever-Contested Analog of Sex Hormone

British chemists under the lead of Charles Dodds first synthesized DES in 1938 (Bell 1980; Gaudillière 2003). Working for the Medical Research Council, they did not patent the process, even though the molecule revealed promising properties. Although DES did not present a structural similarity with natural sex steroids, the substance proved a potent analog of estrogens, mimicking most if not all the latter's effects in the animal assays that were then in use for assessing the potency of female hormones. Cheap and easy to produce, DES rapidly became a substitute—and competitor—for industrially produced estrogens, namely, those purified by pharmaceutical firms out of the urine of pregnant women or pregnant mares. Initially, DES was used in gynecology as a therapeutic agent for a variety of indications also handled with estrogens: infertility, menstrual-cycle disorders, uncontrolled bleeding, absence of menses, or problematic menopausal symptoms.

The FDA authorized DES for the U.S. market in 1941. As reported by Susan Bell, the "synthetic estrogen" played an exemplary role in the history of the agency as it was one of the first compound to be approved according to the procedures defined in the 1938 Food, Drug, and Cosmetic Act. The Act partially transferred the burden of proof to industry. A manufacturer seeking authorization was mandated to document the safety of its product, but the new law did not define any type of acceptable evidence or test. Approval was granted for the basic gynecological indications, i.e., amenorrhea, menopausal symptoms, and infertility. However, DES was perceived as a potent estrogen rather than an aromatic compound with an ethylene side chain; "off label" uses rapidly surfaced in the 1950s. The most important of these—in terms of prescription numbers—was for managing the risk of spontaneous abortion during pregnancy. The idea was widely adopted that DES, a quasi-hormone, could replicate the changes in the concentration of sex steroids occurring during pregnancy. This justified the prescription of the drug as "replacement" therapy for a condition attributed to estrogen deficiency (Bell 1980). Despite the fact that clinical trials had produced conflicting results, DES use was gradually extended as a reassurance factor to women with no previous experience of abortion. This widespread use of DES was never considered trivial. Only a small minority of physicians, however, upheld their concern about the fact that a long time prior, already in the 1930s, laboratory experiments with mice had shown that estrogens in general, and DES in particular, could induce tumors in healthy animals.

A decade later, in the 1950s, following pioneering work by animal nutritionists at Iowa State College under Wise Burroughs's lead, DES use was extended to agriculture. Burroughs and his colleagues discovered that DES given in minute amounts accelerated the growth of cattle animals (Marcus 1986: chaps. 1 and 2). The idea of adding DES to industrially prepared premixes was patented by the university and licensed exclusively to the pharmaceutical firm Lilly. The process was a huge success. Lilly sublicensed it to a few dozen companies. As a result, within two years more than 6 million cattle were being fed food containing DES.

This transformation became a matter of concern and an issue of public debate in the 1960s. The trajectory of DES was actually rapidly affected by the emergence of critical voices associated with the consumer movement. Although their origins can be traced back to the 1930s, in the 1960s the heterogeneous organizations targeting "consumer rights" experienced rapid growth and radicalization, often attributed to the favorable climate generated by the economic expansion of the postwar decades (Silber 1983).

The enactment of the so-called Delaney Clause in 1958 was a critical event in the polarization of the debates on the quality of food and the dangers it might pose to public health (Marcus 1986). This amendment to the FDA Act introduced by Congressman James Delaney stated that no food additive that had been found to induce cancer either in animals or in humans could be authorized.[3] The measure was adopted despite the FDA's opposition. The FDA considered this broad ban as both prejudicial to the practice of productive agriculture and impossible to enforce. The approval of the clause was not only a symptom of the mounting influence of consumer activism, but—as testified by the parallel drawn between DES and DDT during the Delaney hearings—was one of the first legislative consequences of an environmental movement that was starting to benefit from a powerful middle-class constituency and no longer strictly targeted conservation issues but also concerns about the dissemination of chemicals in the environment (Dunlap 1981; Gillespie et al. 1984; Brickman et al. 1985; Marcus 1986; Hays 1987; Proctor 1995).

The Delaney Clause crystallized the debates in Congress on industrialized food for ten years after it was enacted. Farmers, industrial-feed producers, and drug companies joined forces in attempts to repeal the clause, arguing that it would ban all sorts of chemicals that could be used for the betterment, preservation, and conservation of food. They claimed that animal tests could not be trusted, upholding that it would always be possible to find models and circumstances under which any chemical could become a carcinogen, and they explored ways of repealing the clause. In 1962, agricultural lobbying succeeded in inserting a short addition to the amended Food and Drug Act, which stated that a substance could be authorized even if proved a carcinogen in the laboratory provided it did not harm the animals and left no residues in the meat for human consumption (Temin 1980; Daemmrich 2004).[4]

Controversies over the dangers of DES grew within the context of the 1960s widespread critique of government, industrial capitalism, and established authorities. The public life of DES was to a large extent determined by the power struggle opposing the alliance for industrial agriculture and the loose front connecting the social movements of the 1960s and their Democratic allies. Best-selling books such as *Silent Spring, The Poisons in Your Food,* and *The Chemical Feast* accompanied the rise of environmental and consumer organizations (Longgood 1960; Carson 1962; Turner 1970).

Most historical work on the relationship between the DES affair and this particular context has been articulated in terms of political interests and distrust of scientific expertise. In a nutshell, the argument focuses on uncertainty and capture. Reassuring discourses about the limited nature of the cancer risk involved in the use of food additives did not convince. Paternalistic attitudes toward the "ignorant lay public" deepened this crisis of confidence: as a consequence of suspicion, the public debate was dominated by wide-ranging generalizations, value-laden evaluations and feelings. Both sides (in favor of and against the use of DES) waged a polemical battle, defending their views and interests in the name of science. From this perspective, experts' statements went far beyond the knowledge of the time, which was remarkably uncertain: they need to be understood above all as pleas in favor of or against a ban on DES. One side upheld that the social movements proved unable to accept the existing facts regarding the limited dangers of DES in meat, while alternatively, the other side argued that big agro-industrial interests produced ignorance and bad science to mislead the public.

It would however be misleading to limit our reading of the DES affair to the making of these political alliances and their confrontation. The production and discussion of expertise on DES was the expression of strong interests, but it was a work in its own right, which in turn redefined attitudes and interests. The production of facts was actually inextricably mixed with public debates, and what characterized the public assessment of DES-risks was the multiplicity and conflicting forms of evidence rather than mere distrust of experts or tainted research. The decisive decade was in this respect not the 1960s but the 1970s, when the links between DES as drug and DES as food were made, bringing in another layer of conflicting and situated scientific discourses.

Medical uses of DES were basically unproblematic until 1971, even though feminist authors occasionally made the analogy to the contraceptive pill and targeted gynecological hormones for their purported carcinogenic properties

(Seaman 1969). The professional "warning" on DES began only in 1970, when gynecologists at Massachusetts General Hospital reported a surprising series of vaginal tumors (Bonah and Gaudillière 2007). Considered very rare, this type of cancer had been diagnosed in rapid succession in—even more unusual—very young women. Following these early observations, Arthur Herbst and his colleagues reinforced their argument that this was a serious public health issue by concentrating on epidemiological evidence. More precisely, they mobilized technologies that had recently gained acceptance among physicians: risk-factor analysis based on retrospective control.

They selected a retrospective control group by looking at the records of women admitted the same day as their patients, matching age and social groups. The statistical comparison of their records revealed one single significant difference: the girls suffering from vaginal cancer were born from women who had been treated with DES during their pregnancy. The correlation was published in April 1971 in the *New England Journal of Medicine*, after Herbst had sent his data to the FDA (Herbst et al. 1971). The report alerted the New York Department of Health and its cancer-control bureau started to look for similar cases. Having found another five, also correlated with "DES mothers," they issued a general warning to the state's physicians (Greenwald et al. 1971). It did not take long for the initiative to find its way into the mainstream media. In less than six months, the professional alert had become a national scandal, relayed by public health authorities and local gynecologists, widely discussed in the news, a topic of congressional hearings, and an ongoing motive for concern within the FDA.

This alert had two effects. First, it gave very concrete meaning to the animal-human translation that had been at the core of all the arguments around the carcinogenic risks of the drug. If the correlation evidenced by Herbst and reinforced by the National Cancer Institute (NCI) survey was accepted, this meant that DES given in therapeutic dosage was inducing tumors in humans, as well as in laboratory animals. As a consequence, the presence of DES residues in meat following its use as growth enhancer could become a significant hazard. Second, the medical crisis reinforced doubts about the policy the FDA had implemented since DES had been put on the market. If—as many observers thought—the agency, blind to the doubts raised by a number of physicians regarding the clinical value of the pregnancy indication, had allied itself with a segment of the gynecological elite and the DES-producing chemical and pharmaceutical firms, a similar scenario might well have taken place within the division of veterinary medicine, leading the FDA to support the claims of nutrition scientists and feed producers. The two crises—on medical DES and on agricultural DES—actually reinforced each other.

Throughout the 1970s, after the FDA had reacted by contraindicating pregnancy, other medical uses of DES were questioned. In parallel with its campaigning on the adverse effects of the contraceptive pill, the women's health movement succeeded in drawing attention to the prescription of DES as a morning-after pill. The pill was then regulated to include mandatory risk-warning leaflets in the packages. A second problematic medical application was the use of DES to inhibit lactation and prevent breast engorgement after childbirth. The FDA plain and simply contraindicated it in the late 1970s.

This conjunction of a medical and an agricultural crisis was all the more powerful as links between the two DES uses were at work at various levels, mobilizing texts, objects, and persons. The most important arenas for this public conjunction were, however, Congress and the courts. The rest of our analysis will concentrate on the latter given their roles (1) in the making of the regulatory knowledge of DES, and (2) in the construction of social responsibility and the reconfiguration of regulatory tools.

The Regulatory Science of Low-Dose Carcinogenesis: The Judicial Construction of the DES Problem

The concept of "regulatory science" is currently being used within regulatory agencies and administrative bodies. It is usually employed to describe their investments in the development of measurement techniques, methodological tools, and decision-making protocols for the standardization, authorization, or control of technological goods. Looking at the regulation of chemicals in the postwar era, analysts of science, technology, and society have used the concept in a broader sense including the entire spectrum of expert activities conducted in collaboration among scientists, state officials, and politicians; this includes laboratory studies for regulatory purposes, testimonies within political fora such as congressional hearings, as well as court rulings on technoscientific cases. Within this perspective, "regulatory science" refers to the production of knowledge for administrative, political, or judicial action, but it also implies the idea that this regulatory perspective "feeds back" to science, leading to the development of specific forms of knowledge. Toxicology and the environmental regulation of chemicals are among the most frequent examples of this dual relationship.

Comparing the fate of ecotoxicology in the United States, the Netherlands, and the United Kingdom, Willem Halffman thus considers regulatory science a form of boundary work with two functions: (1) separating the scientific from the political in order to balance legitimate knowledge and bring specific statements to the status of agreed facts, and (2) facilitating negotiations and closure of conflicts by providing an aura of objectivity to decisions (Halffman 2003). Expert work thus mobilizes boundary entities, i.e., texts, objects, and persons, which operate or circulate among the technical arenas that are linked to or that depend on the regulatory agencies and the more administrative arenas of committee and hearing rooms. An essential aspect of the regulatory knowledge produced for the regulation of chemicals has accordingly been the establishment of tests for effects, protocols for measuring damages or concentrations, model systems, and limit values or exposure thresholds. The 1970s debates on carcinogens in general and DES in particular are no exception to this pattern. The sheer number of congressional investigations on one or both uses of DES (eight between 1971 and 1978) testifies to this and played a special role in intertwining agricultural and medical issues (Gaudillière and Hess 2012).

The second arena where the regulatory science of DES became a matter of public expertise was the judicial system. When the FDA decided to ban the use of implants in all farm animals in 1973, this policy was immediately challenged in court by a coalition of feed producers (the makers of DES followed at a close distance) with the result that the US Court of Appeals for the District of Columbia Circuit invalidated the FDA decision in 1974 as it had been taken without granting the interested parties preliminary hearings. The court thus mandated a hearing on the case. The proceedings started in January 1977 with a final ruling from the judge in September 1978.

Sheila Jasanoff has insisted on the "adversarial" type of expertise that characterizes the mobilization of science in the U.S. judicial system. The DES administrative trial thus provided a new arena for the evaluation of DES risks, since: (1) experts were proffered by and associated with the parties; (2) the corpus of acceptable documents defining the facts to be taken into account was negotiated between the judge and the parties' lawyers before the hearings; (3) each testifying expert was cross-examined by the opposite party with a right to follow up for the party putting forth the expert; (4) one of the aims of the hearings was to include or exclude specific elements or evidence from the body of evidence that the judge would take into account in his final ruling. This procedure led the trial to stage an open controversy on the effects and risks of agricultural DES, which resulted in much greater focus on the experimental and investigation data than had been given during the congressional hearings. The judge, for instance, excluded all questions regarding the economic status of DES on the one hand and its medical benefits on the other, to focus the debate on the arguments for or against a legitimate implementation of the Delaney Clause, for or against a public health threat.

The centrality of this relationship between models, dosage, threshold, and carcinogenic risk echoes Robert Proctor's discussion of "the politics of dose-response curves" (Proctor 1995). Dose response curves, or, more precisely, their purported existence for DES, were mentioned, drawn, cited, commented on, and criticized on many occasions in the DES debate. In his analysis of their role in the "cancer wars," Proctor insists on the fact that the low-dose part of these curves was both more uncertain and politically or administratively more important. Extrapolation was therefore inevitable and contentious, as it was based on very few results and on animal models to be translated into human circumstances. Low-dose effects were for DES, as well as for other putative carcinogens, an obligate point of conflict.

The issue of dose and effect is certainly the best example of the peculiar dynamics of these court hearings. Given the centrality of the dose-effect relationship in pharmacology, toxicology, and nutrition science, the question of dosage was addressed in almost every expert testimony. It was addressed directly, with computations of inoculated or ingested quantities. It was also dealt with in a less direct way when discussing the relationship between DES and natural estrogens. It is not altogether surprising that animal nutritionists argued that DES was an estrogen like any other, thus explaining that its carcinogenic potency was just a manifestation of its ability to mimic the properties of sex hormones. As a consequence of this link, they posited that DES quantities should not be considered in isolation but discussed as a component of a global estrogenic pool that included much higher quantities of normal estrogens, both in the bodies of supplemented cows and in women. As the situation evolved, however, the link became looser. During the last of the hearings, the industry lawyers and experts adopted another position, separating DES from the other estrogens. As the former was generally considered dangerous and was on the verge of being definitively banned, it then become good policy to protect other estrogens—that could possibly replace DES—from an equivalent threat to their use as growth enhancers.

More broadly, three discursive frameworks can be identified in the testimonies of experts who participated in the hearings. The first is an epidemiological discourse focusing on the gynecological alert, its meaning, and the surveys organized to establish the correlation between vaginal cancer and DES. The latter is seen as a risk factor associated with a high probability of adverse events in certain populations or under specific circumstances. The approach is to do everything possible to avoid these adverse effects in order to preserve the health of the population. In other words, every "reasonable precaution" must not only be considered seriously, but also implemented. The second discourse is pharmacological. It focuses on the experiments conducted with mice and other animals, regarding either the induction of tumors or the fate of DES within the body. It mobilizes statistical tools to draw conclusions regarding the best way of defining thresholds, looking for zones of marginal effects (or better, of no effect) and conditions of relatively safe use. It aims at a controlled use of the substance under investigation. The third discourse focuses on animal physiology and nutrition. Using laboratory animals as well, it takes experiments with natural and artificial hormones as evidence of their identical fate and effects in the body. Nature and artifice being equivalent, the latter will not do more harm than the former. DES, the same as many natural hormones, is rapidly eliminated from the body. Moreover, a consequence of the quasi-biological nature of DES action is that potency and carcinogenicity are two sides of the same coin. This physiological discourse is articulated with nutrition studies focusing on the growth-enhancement effect of DES and its value for agricultural productivity. The more general agricultural-engineering perspective combines this veterinary expertise with considerations regarding the detection of residues. This is deemed important because instances of DES in edible meat initially went unnoticed or—according to FDA critics—were not even looked for. The technical answer consisted of seeking a new legal system of chemical measurement that would replace the mouse assay, which was gradually coming to be viewed as not sensitive enough. This was not sufficient, however. If controlled use was to be maintained, sanitary inspection and legal action independent of DES producers and users can be better organized to ensure safety, meaning the absence of DES in meat.

These discourses, although showing a coherence, which makes it not too difficult to link them with the initial statements and responses of experts testifying, respectively, for or against the FDA, did not pass unmodified through the hearings and the court game of examination and cross-examination. The transformation of the low-dose issue during the hearings is the best example of the way in which the legal arrangement reconstructed toxicological evidence.

The hearings repeatedly discussed the results of a study conducted in the early 1960s by G.H. Gass at the University of Illinois with C3H mice, showing unexpectedly high levels of carcinogenesis (when compared with the control group) at low concentrations of DES, thus giving the dose response curve an unusual V- or U-shape. Gass's modeling was judged to be nonconclusive and nonsignificant by the feeders' lawyers and was defended as strongly suggestive evidence by those of the FDA. Whether the curve previously published by Gass was granted the status of fact or artifact depended on many questions, beginning with the possibility of using C3H mice for such a study as these had been selected for a very high genetic susceptibility to mammary tumors. The issue most debated was, however, the nature of the dose response relationship and the low-dose data. Gass's curve presented a statistically significant increase in the number of tumors following exposure at doses as low as 6.5 ppb. The controversial nature of the curve, however, originated in the absence of linearity: higher concentrations did not result in more frequent tumors and it was only for concentrations higher than 20 ppb that a simple relationship could be observed. As explained by most of the feeders' experts, this "abnormality" was a strong enough reason to exclude the data. Hardin Jones, a physiologist of the Atomic Energy Commission testifying against the FDA, explained for instance that Gass had simply botched his experiment.[5]

The experts put on by the FDA countered with three arguments. First, the article was definitely to be taken into account as it had been published in a prestigious, peer-reviewed journal considered a legitimate source of knowledge. Second, a series of mice experiments at the National Center for Toxicological Research commissioned by the FDA potentially (they had not yet been completed) confirmed the Gass data.

The most interesting approach, however, was that of the epidemiologists and cancer specialists of the NCI. Leaving aside the simplistic requisite of linearity, a more sophisticated statistical modeling of the effects could account for the "bizarre" decrease in incidence at 10 ppb. Updated theories of carcinogenesis shared by NCI specialists concurred in holding that "one molecular hit" could induce a mutation that would turn one cell into a cancerous one and induce a tumor, thus rationalizing the idea that there was no "no effect zone" threshold for proven carcinogens like DES. The NCI epidemiologist, Umberto Safiotti thus explained during his oral testimony:

> The experimental evaluation of the carcinogenicity of an environmental chemical by animal bioassays is usually conducted by exposing an animal population to that single test chemical. In contrast the human population is exposed from prenatal life through childhood to adult life to a large variety of environmental carcinogens from any routes of exposure. Relatively little is known about the possible synergism of different carcinogenic compounds, but there are several known case of marked potention. I consider multiple exposure to dozens or possibly hundreds of carcinogens as representing truly "realistic" conditions in relation to human risk. The effect of an individual chemical should be viewed as being added to such a background ... exposure to any amount of carcinogen, however small, will contribute to the total carcinogenic load.[6]

The judge's role was to evaluate the data whose integration in the body of evidence and meanings had been discussed and negotiated during the hearings for his final ruling. Judge Davidson first decided to take the data into account as none of the parties had initially asked for their exclusion. He then defended a balanced, situational approach: the data were not complete proof and could not, taken in isolation, be grounds for a decision—but they were robust enough to suggest that a no-threshold assumption was sound and that low-dose carcinogenesis with DES was a significant risk.

Judge Davidson's final ruling granted the FDA the right to ban all agricultural uses of DES. The conflict of expertise evidenced during the hearings, however, resulted in a new judicial grounding of this choice, which set an administrative DES precedent that completed the idea of collective industrial responsibility. Davidson grounded his evaluation in three elements: (1) the chemical and physiological specificity of DES, which he considered different from natural estrogens; (2) its demonstrated toxicity and carcinogenic potency in humans; and (3) the high probability of low-dose dangerous effects. The legal background for the decision was therefore not the Delaney Clause but the public health emergency. The Delaney Clause was difficult to invoke because of an additional amendment introduced in 1962 stating that an additive could not be banned if it was absent from beef carcasses. During the dispute on DES, the controversy on the detection of data left open the question of the possible disappearance of the artificial hormone within the 120-day quarantine mandated by the FDA. The understanding of a public health alert was easier to define as it was directly rooted in the existence of the medical crisis. At least the no-threshold and low-dose carcinogenesis was accepted.

Beyond the specific fate of DES use in agriculture, the main impact of Judge Davidson's ruling was therefore to transform the problem of low doses and cancer. The debates and the final ruling not only legitimated the idea that traces of DES circulating in the food chain could threaten human health, but it played a central role in stabilizing and reinforcing the peculiar corpus of regulatory science that had surfaced during the hearings and included a combination of ideas and modeling tools regarding: (1) the risk posed by chemicals used in both medicine and agriculture with a special emphasis on "iatrogenic risk," and in that case originating in a form of medical intervention paradoxically targeting health-related risks; (2) the specificity of these chemicals' effects on reproduction with consequences appearing in the second generation only decades after exposure in the mother's womb, thus substantiating the notion of a peculiar "window of sensitivity" during early developmental phases with "lifelong" consequences; (3) the cumulative effects of "multiple exposures" associated with the variety of compounds released in the environment mimicking the effects of sex hormones, which may elicit cancer or affect human reproductive life; and (4) the plausibility of low-dose carcinogenicity as exemplified in the DES U-shaped curve obtained with cancer-prone mouse models.

The Tort Trials: Toward a Collective and Probabilistic Understanding of Responsibility

The iatrogenic risk perspective associated with DES also played a decisive role in the court action against the producing firms that a few of the DES daughters brought in the late 1970s with the support of the patients' association DES Action in order to obtain compensation and—more importantly in the eyes of the collective—a public assessment of responsibility. The choice of suing the industrial actors was both a legal and political choice. Politically, the firms were perceived as the first benefactors of DES use even though physicians could also be viewed as having financial and professional motives for prescribing the drug. Legally, all attempts to mount a case based on medical malpractice would face the fact that DES use in pregnancy had been challenged among academic gynecologists but was widely accepted within broader circles of gynecologists, general practitioners, and their associations, thus making the absence of wrongdoing in regard to "current professional knowledge" easy to document before 1971. One parallel attempt to sue the FDA launched by Nader's Foundation for Taxpayers and Consumer Rights did not meet with success: the case was even not taken to court since the agency had never included pregnancy in the indications of use selected for the marketing permit. Off label prescription was indeed a problem, but not a matter for the law and it escaped the purview of the agency.

The great majority of the legal arguments associated with the tort actions against industrial firms did not challenge the causal relationship between DES and vaginal cancers and/or reproductive malformation as had occurred during the 1977–78 trial. Even the defendants' lawyers accepted that relationship. The epidemiological assessment of causality held in court; legal battles were waged at another level, that of the individual responsibility of each company—in other words, the putative chain of actions linking a given producer of DES and the sick daughter of a treated woman. Arguments thus focused on the strength of the evidence linking the specific product of the company(ies) actually sued and the plaintiff's mother's medical care during her pregnancy. As these trials took place two decades after the event, such evidence was usually missing: prescriptions sheets or pharmacists' books had not been kept, doctors' files could not be traced or the practitioner would not give them, etc. In most instances, this lack of direct evidence was considered important enough to dismiss the case.

The notions that the firms had widely advertised DES prescription to pregnant women and that the (second-generation) innocent victims deserved compensation were, however, strong enough to induce some judges to seek judicial precedents or judicial innovations that would make it possible to condemn the firms in the absence of direct evidence of culpability. The DES jurisprudence thus shifted the classical understanding of industrial responsibility away from the paradigm of the moral person toward an evaluation of parallel and collective action. Three legal notions played a significant role in the early 1980s: concerted action, alternative liability, and market-share liability.

As mentioned in the introduction, in one of the first trials granting compensation, the New York Supreme Court found that Eli Lilly caused the injury to J. Bichler, one of the prominent members of DES Action, on the basis of concerted action. The circumstances necessary for concerted action are that several tortfeasors have acted in concert and injured the plaintiff. In the case of DES, the basis for this concerted action was found in the collective application the DES-producing pharmaceutical firms submitted to the FDA in 1941 in order to get a marketing authorization including all uses of DES as an analog of estrogens, including in the prevention of miscarriages. The problem many lawyers found in this choice was that the FDA had not included such use in the indications it listed, thus leaving open the question of the relationship between the application and the later harmful--but off label—prescription.

An alternative resource for defining liability in the absence of proved personal causation was a famous case in which the plaintiff was injured by a pellet from the gun of one of two hunters who had both fired in his direction. Since both hunters exposed him to the risk of harm, the court had shifted the burden of proof and decided that unless one could prove that the other was responsible for the injury they would be held jointly liable. This precedent was not used directly, but introduced two elements taken up in the formulating the concept of market-share liability: (1) the notion that creating a risk might be enough to start a chain of events leading to injury and (2) the collective responsibility in case of a limited and known number of possible perpetrators.

Market-share liability was introduced by the California Supreme Court in another case originating in the mobilization of the women health associations, *Sindell v. Abbott*. Given the number (300) of DES producers and the unequal structure of the market (six firms produced 90 percent of the drug sold after authorization), the judges considered it unfair to find all DES-producing firms responsible on an equal basis and decided to impose liability according to the producers' respective market shares. In order to reinforce this statistical approach and enhance the probability that the industrialists held liable were those that made the DES causing the plaintiff's injuries, they added that the application of market-share liability was possible if only a group of manufacturers holding a substantial share of the DES market were involved. The *Sindell* case established a form of collective responsibility by requiring that compensation be apportioned among the multiple tortfeasors according to their market shares. Without surprise, the case caused much criticism within the industry and among tort lawyers.

Its most significant dimension was however not to facilitate compensation but to establish a form of consumer-oriented understanding of causation, which focused on exposure and risk. As one legal scholar explained: "the *Sindell* court decision scaled liability to each defendant to the probability that it supplied the product that caused the plaintiff's injury," taking the size of the controlled market as a convenient proxy for such probability, in spite of possible differences in marketing and information patterns. "Fairness in the civil context seems to require only that a defendant's liability be related to his conduct, and liability, where imposed, be roughly proportional to the seriousness of the *risks* he has created" (Robinson 1982: 739).

Within this framework, compensation became a kind of ex post insurance based on an epidemiological definition of health-related risks. The same lawyer thus considered the imagined extension of *Sindell* to the case of Horace Tumor, who had contracted cancer:

> It can be established by substantial evidence that the following events contributed to the risk of his developing cancer: 1) exposure to asbestos (for twenty years Horace installed asbestos insulation, all manufactured by a single firm, Ajax Inc.); 2) exposure to toxic chemical waste (Horace subsequently worked for ten years in a chemical plant operated by Bonanza Inc., where he was exposed to chemical wastes); 3) medication (Horace was treated intensively with a drug manufactured by Consolidate Co.). Although it is not possible to establish precisely the magnitude of the contribution that each of the events made to Horace's cancer, assume that it is possible to estimate their respective contributions as sixty, twenty, and twenty percent. Based on the foregoing, Horace joins Ajax, Bonanza, and Consolidated as codefendants, arguing that each manufactured an unreasonably dangerous product that caused his injury and that each is liable in the amount of the foregoing percentages (Robinson 1982: 750).

Of course, no such case has materialized in the rare instance of market-share liability application but the imagined Horace Tumor case aptly captures the way in which the DES jurisprudence displaced legal notions of causality and responsibility to adapt and reinforce a social and technical package linking the uncertainty of scientific knowledge and the adverse effects of industrial products to redefine the dangers of medical goods as a new category of health risks, namely, the risks of having a medical intervention cause a new disease. The 1970s emergence of iatrogenicity as public issue was not only reflected in the jurisprudence. The timid displacement in the direction of collective guilt associated with the dangers and adverse affects of DES is a marker of a more general change in the legal status of technical and medical expertise from a regime of "personal guilt," within which adverse events could only originated in individual error, lack of training, and insufficient professional control of practitioners' competencies, toward a regime of "risk" within which the mass consumption of technical and medical goods creates unanticipated events whose reality and consequences inevitably become matters of public controversies.

The Legacy of DES: Turning Endocrine Disruptors into a Political Issue and a Scientific Object

During the 1980s, the DES model of iatrogenicity, which had been comforted in the courts, seems to have lived a double public life. First, it circulated in various forms within public arenas. It was for instance adopted by many organizations of the women's health movement, which translated the narrow and technical notions of risk and exposure characteristic of the 1970s expert hearings into a broader social discourse that combined gender and risk.[7] The second life of the DES package echoes the testimony of Umberto Safiotti cited above and needs to be documented in a proper way. It took place within the arenas of regulatory science, among which the epidemiological work conducted at the National Cancer Institute seemed to have been central.[8] DES as reference model in regulatory science, however, experienced a radical change of meaning in the 1990s when it contributed to—and was incorporated in—the rise of "endocrine disruption" as an object of experimentation, a target of expertise, and a political issue.

From the mid 1990s onward, endocrine disruptors have become significant (if not massive) objects in the debates about human health, the environment, and industrial chemicals. This category of substances modifying normal endocrine functions in humans and animals are viewed as increasingly important causes of sterility, reproductive disorders, cancers, and abnormal behaviors. During the past twenty years, they have been the subject of numerous reports, surveys and inquiries, leading to the publication of a few thousand academic articles. As many reviews published since the mid 1990s explain, endocrine disruptors may be a critical, if not the most important, source of ecological and health-related risks in the industrialized world (Sonnenschein and Soto 1998).

Our hypothesis is that endocrine disruptors may be considered a generalized and globalized form of DES. They represent a generalized version of DES since the category includes dozens of substances acting as "xenoestrogens," modifying the entire steroid metabolism and reproductive physiology and having a special responsibility for the rising incidence of hormone-related types of cancer or male sterility. They also represent a generalized version of DES because they link agriculture and medicine and because pesticides and agrochemicals are central among endocrine disruptors both in terms of listing the compounds and of assessing the amounts disseminated in the "environment." Endocrine disruptors finally represent a globalized version of DES because they circulate worldwide and—more importantly—because they are considered to affect entire ecosystems, linking issues of conservation, biodiversity, and human health in an extended version of the 1970s discussion about environmental carcinogenesis.

The hypothesis of a peculiar legacy of DES in the emergence of endocrine disruption as an object of regulatory science requires proper historical investigation. The remarks below must therefore be taken as signposts of the importance DES plays as reference model within the public and scientific discourse about endocrine disruptors. Participants in the field often point to the 1991 Wingspread Conference as the first moment of formal convergence between scientists and activists interested in the three issues incorporated in the endocrine-disruptor perspective: women's health and reproductive cancers, human population dynamics and male sterility, wildlife conservation, pollution, and reproductive abnormalities (Markey et al. 2002). The final declaration of the conference actually gave DES a central meaning.

> A large number of man-made chemicals that have been released into the environment, as well as a few natural ones, have the potential to disrupt the endocrine system of animals, including humans. Among these are the persistent, bioaccumulative, organohalogen compounds that include some pesticides (fungicides, herbicides, and insecticides) and industrial chemicals, other synthetic products, and some metals.
>
> Many wildlife populations are already affected by these compounds. The impacts include thyroid dysfunction in birds and fish; decreased fertility in birds, fish, shellfish, and mammals; decreased hatching success in birds, fish and turtles; gross birth deformities in birds, fish, and turtles; metabolic abnormalities in birds, fish, and mammals; behavioral abnormalities in birds; demasculinization and feminization of male fish, birds and mammals; defeminization and masculinization of female fish and birds; and compromised immune systems in birds and mammals. ...
>
> Humans have been affected by compounds of this nature, too. The effects of DES (diethylstilbestrol), a synthetic therapeutic agent, like many of the compounds mentioned above, are estrogenic. Daughters born to mothers who took DES now suffer increased rates of vaginal clear cell adenocarcinoma, various genital tract abnormalities, abnormal pregnancies, and some changes in immune responses. Both sons and daughters exposed in utero experience congenital anomalies of their reproductive system and reduced fertility. The effects seen in in utero DES-exposed humans parallel those found in contaminated wildlife and laboratory animals, suggesting that humans may be at risk to the same environmental hazards as wildlife.[9]

In order to obtain less anecdotal evidence, our team has started a survey of the literature published on endocrine disruptors between 1966 and 2009 as indexed in Medline, the medical database of the National Institutes of Health. Analysis of correlations between keywords, authors, keywords and authors, or keywords and journals (as well as their changes over time) have been conducted with Réseau-lu, a mapping software.[10] Some results are worth taking into account when discussing the uses of DES in the debates about endocrine disruption.

One way of approaching the emergence of endocrine disruptors is to look at the changes in the keywords associated with the corresponding literature. For that purpose, we have distinguished three periods: (1) 1992–1999, which gathers the first wave of papers on endocrine disruption; it is the period during which, in the United States, endocrine disruptors became not only a label for scientists but an official issue with the first congressional hearing on the topic (in 1995)—which mandated that the Environmental Protection Agency establish a program for specific screening and testing of endocrine disruptors (in 1996) (30 percent of the research corpus has been published during this period); (2) 2000–2005, which is the period of experimentation and massive publication on questions of identification, testing, and screening of chemicals acting as xenohormones—publications that have resulted from this program and the institutionalization of the problem; (3) 2006–2009, which is the period of convergence and generalization of the endocrine-disruptor paradigm both in terms of geography (Europe becoming a major source of publications) and of disciplines with the convergence of ecology and toxicology.

Mapping the most-often-used keywords and their co-occurrence, it appears that DES:

1. is often cited as such during the first phase (1992–1999) (Figure 3.1), associated with cumulative effects on embryos and developmental processes, and occupies a mediating place between clusters focusing on the one hand on epidemiological studies (with keywords focusing on populations) and on the other hand on experimental research on reproduction and the effects of chemicals;
2. appears as a peripheral category during the institutionalization phase (2000–2005) (Figure 3.2) associated with the multiple studies of estrogens and the development of laboratory assays; and
3. disappears from the list of most-cited keywords during the last period (2006–2009), although this last map (Figure 3.3) reveals the mounting importance of reproductive studies and ecotoxicology, which stands in contrast to the marginal status of the epidemiology cluster.

This chronology does not only confirm the importance of the DES reference. It also suggests that it played a peculiar role as boundary-object or mediating package during the first phase of existence of endocrine disruptors, before the establishment of specialized screening programs, thus echoing its status within the sites of public expertise. This temporary mediating function of DES seems to be confirmed with the examination of the relationship between keywords and journals. DES appears only during the first period (Figure 3.4). It is associated with the articles published in the EPA journal *Environmental Health Perspective,* which was (and remains) one of the most important publications in the field. DES is included in a package of words combining epidemiology and reproductive issues with prenatal exposure, estrogens, estrogen receptors, breast neoplasms, embryonic effects, drug effects on reproduction, and preg-

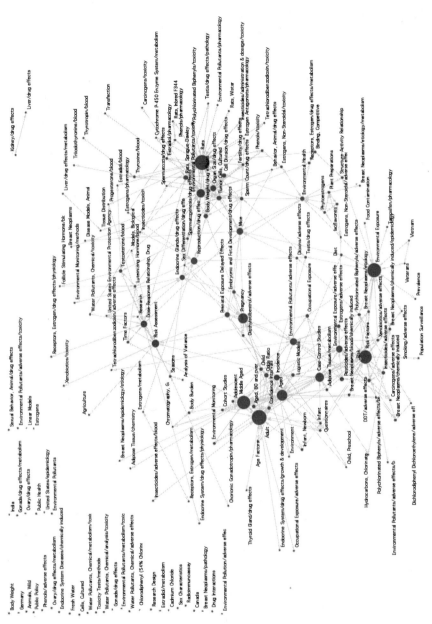

Figure 3.1. Co-occurrence of keywords, 1992–1999.

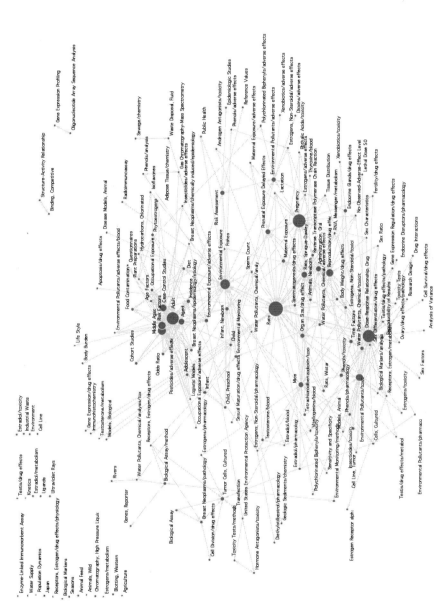

Figure 3.2. Co-occurrence of keywords, 2000–2005.

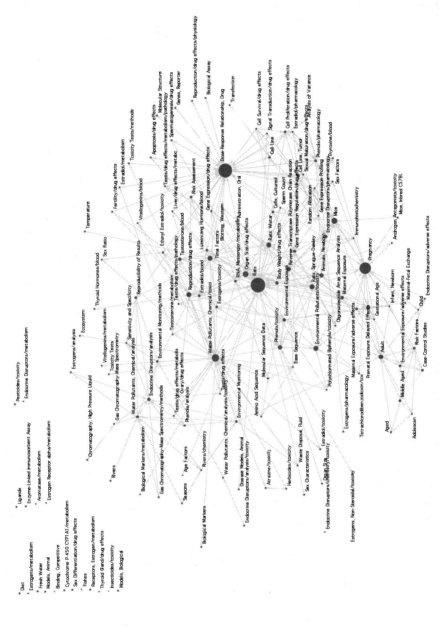

Figure 3.3. Co-occurrence of keywords, 2006–2009.

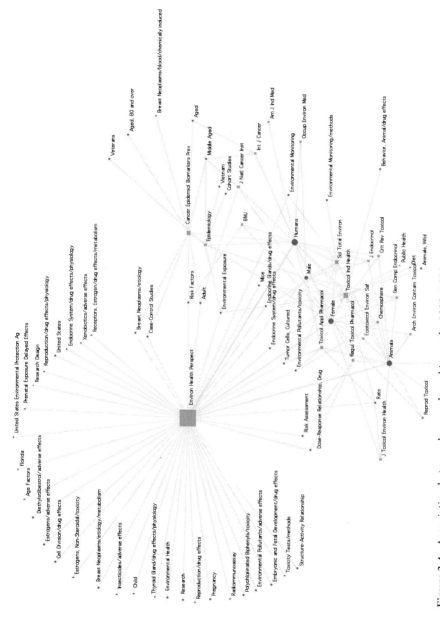

Figure 3.4. Association between keywords and journals, 1992–1999.

nancy. This cluster is however typical of the grounding phase only, disappearing from later maps of EPA keywords and thus signaling the role of the DES reference in the shift in the agency's understanding of environmental carcinogenesis toward a more general questioning of reproductive issues encompassing problems of public health and conservation.

What is the actual role of DES in the new expertise of chemicals? The "mediating" hypothesis suggests that DES has been used as reference to illustrate the general features of endocrine disruption based on a combination of features directly echoing the package of the late 1970s/early 1980s. This can be illustrated with many reviews written in the 1990s. For instance, in 1993, three prominent participants in the Wingspread Conference, the World Wildlife Fund expert Theo Colborn, the toxicologist Frederick S. vom Saal, and the molecular endocrinologist and cancer specialist, Ana Soto published a seminal and programmatic text in *Environmental Health Perspective*. The second section was entitled: "The DES Syndrome: A Model for Exposure to Estrogenic Chemicals in the Environment." It focused on questions of duration, multiple exposure, and low doses, stressing the specificity of prenatal exposure and delayed symptoms as documented in animal models documenting the "detrimental effects of DES exposure during the critical period of organ differentiation" (Colborn et al. 1993). DES was then aligned with a wide range of pesticides and industrial pollutants, beginning with DDT and hypothetically related to the following human health issues: (1) increasing incidence of prostatic and breast cancers; (2) an increase in ectopic pregnancies; (3) an increased incidence of cryptorchidism; and (4) a 50 percent decrease in sperm count. The article added that many of the effects of endocrine disruption linked to DES and estrogenic pollutants have been reported in wildlife species ranging from alligators to fish.

The contemporary value of the DES model is, however, not restricted to DES's status as "precursor." Like all boundary objects, it has also become a local tool with more specific meanings and functions. In contrast to our general mapping indications, DES has, for instance, not disappeared from the research corpus during the period of institutionalization and development of screening assays. In 1997, writing in one of the in-house organs of the EPA, the toxicologist Frederick S. vom Saal thus described what he considered to be the main legacy of the debates on DES toxicity and carcinogenic potency: the U-shaped dose response curve. He pointed to the central role of this type of toxicological evidence in the assessment of chemicals suspected of being "endocrine disruptors" (vom Saal 1997).

The screening program of the EPA has accordingly renewed the experimental uses of DES in two different ways. The first is the most direct one. DES is incorporated in many testing protocols as a positive control of xenoestrogenic potency. It provides a standard point for eliminating false negative results.[11]

The most interesting aspect, however, is the second one, namely, the generalization of the U-shaped curve as means of interpreting low-dose effects. The sociotechnical hierarchy of dose-effect curves mentioned above has accordingly been reordered in reference to DES with the publication of modeling results putatively showing strong effects of endocrine-disrupting chemicals at very low- and high-range doses. Gass's U-shaped curve for carcinogenicity has accordingly been supplemented with a DES-related inverted U-shape concept to describe effects associated with the (distant) consequences of prenatal exposure: low doses increase the response to estrogens and xenoestrogens while high doses reduce the same response (Alworth et al. 2002; Weltje et al. 2005). In the past five years, the inverted U-shape has been associated with several experimental models of developmental disruption, for instance the contrast between the linear response to endocrine disruptors when observed at the molecular level and the nonlinear (U-shape) effect of low doses when analyzed at the level of tissue organization (Vendenberg et al. 2006). Such models remain controversial, but they have gained mounting importance in the context of EPA expertise and, more specifically, of negotiating the screening program. One of the best illustrations of such use of DES and displacement is the present controversy about Bisphenol A (Richter et al. 2007).

Conclusion

The surveillance and control of the dangers associated with the widespread uses of chemicals became major issues in the United States after 1945 when agricultural and medical activities were increasingly based on industrial practices and on the massive use of technoscientific products. Centered on quantification, probabilistic modeling, experimental testing, and controlled use, the ideas of risk, its assessment, and its management thus emerged as central features in the handling of old dangers as well as of the specific social and political tensions engendered by this industrialization. The spread of risk in all domains of health and sanitary policy during the period between 1945 and 1975 has been accompanied by the institution of a large regulatory apparatus comprising a number of state agencies somehow modeled on the experience of public health institutions. Historians as well as law and political science specialists have tried to understand the mounting importance of risk regulation by focusing their analysis on the status and practices of bodies such as the FDA or the EPA. The DES controversy of the 1970s provides a lens for investigating other aspects of postwar risk regulation, which takes into account the relationships among the main stakeholders, the types of evidence they favor, and the regulatory tools they rely on. The controversies that led to the final ban of agricultural DES are in effect a privileged entry point into a specific way of regulating that combined the politics of consumption and a form of regulatory science conducted in political and judicial arenas.

The conjunction of the medical and agricultural DES is in this respect an essential dimension of the debates in the United States. It was not repeated in European countries where affairs related to the medical uses of DES emerged. It radically changed the meanings and the issues at stake, both scientifically and politically. As mentioned above, this convergence played a critical role in turning low-dose carcinogenesis into a major risk with environmental (pesticides and agrochemicals) and medical (hormone prescriptions) dimensions. The relations between the agricultural and the medical DES crisis thus reinforced the management of health with risk in general, and regulation through the politics of consumption in particular.

As other observers have noted, the medical crisis reinforced an emerging women's health movement rooted in 1970s feminism with the institution of powerful DES-daughters organizations that played the card of empowerment, adopted a critical perspective on professional medicine, and sought "alternative" expertise conducted by and for women's collectives. This approach of regulatory science as "situated" rather than "independent" led to singling out DES as another example of medical practice gone wrong for the sake of industry (if not the profession), of medicine that turned risk or women's biology into pathologies to be treated chemically, and of medicine leading to iatrogenic disorders. Agricultural DES comprises another context, which focused on the consumer rather than on the victim and the activist. Even though "radical" collectives like the Nader campaign exerted significant influence on the crisis, their role has been to question agro-industrial practices in the name of quality and informed customers.

This latter developments highlight the links between this management of health with risk and new forms of governmentality. Debates on risk management are very often framed in terms of "risk-benefit" balancing, resource optimization, and consumer choice, giving empowerment an individual meaning. Within this perspective, consumers need to organize in order to balance the power of industrial monopolies and gain influence in the regulatory arenas, but the focus of their action remains the consumer and his or her ability to make informed choices from the market. Fair trade is a key concept here, while practical investments are in tests, information campaigns, and judicial tort actions making claims for compensation. It would be misleading to consider this consumer regulation to be simply the outcome and reflection of consumer organizations' postwar developments. The DES controversy shows that other actors in the regulatory system, from administrative agencies to industrial firms, have adopted various aspects of this very same way of regulating, beginning with the emphasis on certified quality, information, risk assessment protocols, and cost-benefit analysis. The latter proved a powerful instrument for bringing together heterogeneous social, economic, and professional interests under the umbrella of numerical reasoning. The consumer way of regulating is in this respect closely related to forms of political management that rely on the lobbying capabilities and organized participation of stakeholders.

This peculiar way of regulating associated with risk assessment and risk management has strongly reinforced the role of public expertise (Table 3.1). Associated with open controversies, the assessment of the threats to health posed by agricultural and pharmaceutical chemicals has not been conducted in "closed" experimental settings, whether academic or depending on the administration. As revealed by the dynamics of DES expertise, court hearings were even more important than directly political arenas (like congressional hearings) for the production of legitimate evidence regarding the nature of risks, their magnitude, and the measures necessary to control them. The final ban of DES is accordingly less the result of social activism than the direct product of a judicial assessment that set both legal and scientific precedent.

Public expertise in general and the DES administrative trial in particular have legitimized a "precautionary" approach to cancer risks, which did not only call for attention to the effects of low doses, but constructed a reference package for further expertise. The DES model thus included a combination of ideas and modeling tools regarding: (1) the risk posed by chemicals used in both medicine and agriculture with a special emphasis on "iatrogenic risk"

Table 3.1. Four Ways of Regulating Health-Threatening Food and Drugs

Way of regulating	Professional	Administrative	Industrial	Consumerist
Aims, values	Compliance, competency	Public health, efficacy, access	Productivity, profit, quality	Individual choices, quality of life
Forms of evidence	Pharmacology, animal models, dosage, indications	Statistical, controlled trials	Animal testing, market research, cost-benefit analysis	Observational epidemiology, risk-benefit analysis
Main actors	Corporations, scientific associations	Agencies, governmental committees	Firms, business associations	Patient or consumer groups, the media
Regulatory tools	Pharmacopoeia, Codex Alimentarius, prescription guidelines	Marketing permits, public statements, labeling	Quality control, scientific publicity, package inserts	Postmarketing surveillance, court decisions

and in that case originating in a form of medical intervention paradoxically targeting health-related risks; (2) the specificity of these chemicals' effects on reproduction with consequences appearing in the second generation only decades after exposure in the mother's womb, thus substantiating the notion of a peculiar "window of sensitivity" during early developmental phases with "life-long" consequences; (3) the cumulative effects of "multiple exposures" associated with the variety of compounds released in the environment mimicking the effects of sex hormones, which may elicit cancer or affect human reproductive life; and (4) the plausibility of low-dose carcinogenicity as exemplified in the DES U-shaped curve obtained with cancer-prone mouse models. As argued in the last section of this paper, the main legacy of both this model and the practices of public expertise that grounded it was the rise of endocrine disruption as object of regulatory science in the 1990s.

Notes

1. "Woman wins suit in DES case. Earlier suits dismissed." *New York Times*, 17 July 1979.
2. "DES Livestock Implants Are Prohibited by FDA Because of Cancer Link." *Wall Street Journal*, 26 April 1973.
3. "Bill on Food Additives Gains." *New York Times*, 14 August 1958; "An Act to Protect the Public Health by Amending the Federal Food, Drug, and Cosmetic Act to Prohibit the Use in Food of Additives Which Have Not Been Adequately Tested to Establish Their Safety." *US Statutes*, 1958, 1784–89.
4. On the amended clause, see *Congressional Record*, 87th Congress, II, 12713.
5. FDA Archives, 76N002, 2 November 1977.
6. FDA archives, 76N002, Minutes of hearings, 26 October 1977.
7. See for instance National Women's Health Network, *DES Resource Guide*, 1980.
8. A good test of this hypothesis would be an analysis of the epidemiological and biostatistics branches of the NCI between 1975 and 1995.
9. Statement from the Work Session on Chemically-Induced Alterations in Sexual Development: The Wildlife/Human Connection. Wingspread Conference Center, Racine, W.I. July 1991.
10. Slightly more than 3,200 references published during the period 1966–2004 were identified in the Medline database using a reference including "endocrine disrupt" as the main term. This collection was screened manually for pertinence and cross-checked with the Web of Science database for information regarding citation patterns, leaving a reference set of 2,670 articles, which were analyzed with the Réseau-Lu software for coauthorship, co-occurrence of keywords, author-keyword associations, and journal-keyword associations. This work was conducted within the framework of a research project on the history of endocrine disruptors supported by the French Ministry of Ecology and Sustainable Development and coordinated by Nathalie Jas and me.
11. Endocrine Disruptors Screening and Testing Advisory Committee, Final report, EPA, August 1998.

Bibliography

Alworth, L.C., K.L. Howdeshell, R.L. Ruhlen, J.K. Day, D.B. Lubahn, T.H. Huang, C.L. Besch-Williford, and Frederick S. vom Saal. 2002. "Uterine Responsiveness to Estradiol and DNA Methylation are Altered by Fetal Exposure to Diethylstilbestrol and Methoxychlor in CD-1 Mice: Effects of Low versus High Doses." *Toxicology and Applied Pharmacolology* 183: 10–22.

Apfel, Roberta J., and Susan M. Fisher. 1984. *To Do No Harm: DES and the Dilemmas of Modern Medicine.* New Haven: Yale University Press.

Bell, Susan. 1980. "The Synthetic Compound DES, 1938-1941: The Social Construction of a Medical Treatment" Ph.D. diss, Brandeis University.

Bonah, Christian, and Jean-Paul Gaudillière. 2007. "Faute, accident ou risque iatrogène? La régulation des évènements indésirables du médicament à l'aune des affaires Stalinon et Distilbène." *Revue Française des Affaires Sociales* 3–4: 123–51.

Brickman, Ronald, Sheila Jasanoff, and Thomas Ilgen. 1985. *Controlling Chemicals: The Politics of Regulation in Europe and in the United States.* Ithaca: Cornell University Press.

Carson, Rachel. 1962. *Silent Spring.* Boston: Houghton Mifflin.

Colborn, Theo, Frederick S. vom Saal, and Ana M. Soto. 1993. "Developmental Effects of Endocrine-Disrupting Chemicals in Wildlife and Humans." *Environmental Health Perspective* 101: 379–84.

Daemmrich, Arthur A. 2004. *Pharmacopolitics: Drug Regulation in the United States and Germany.* Chapel Hill: University of North Carolina Press.

Dunlap, Thomas R. 1981. *DDT: Scientists, Citizens, and Public Policy.* Princeton: Princeton University Press.

Dutton, Diana. 1988. *Worse Than Disease: Pitfalls of Medical Progress.* Cambridge: Cambridge University Press.

Gaudillière, Jean-Paul. 2003. "Hormones at Risk: Cancer and the Medical Uses of Industrially Produced Sex Steroids in Germany, 1930-1960." In *Risk and Safety in Medical Innovation,* ed. T. Schlicht and U. Ströhler, 148–69. London: Routledge.

Gaudillière, Jean-Paul, and Volker Hess, ed. 2012. *Ways of Regulating Drugs in the Nineteenth and the Twentieth Centuries.* Palgrave Macmillan, London.

Gillespie, Brendan, Dave Eva, and Ron Johnson. 1984. "Carcinogenic Risk Assessment in the United States and Great Britain: The Case of Aldrin/Dieldrin." *Social Studies of Science* 14: 265–301.

Greenwald, Peter, Joseph J. Barlow, and Philip C. Nasca. 1971. "Vaginal Cancer after Maternal Treatment with Synthetic Estrogens." *New England Journal of Medicine* 285: 390–92.

Halffman, Willem. 2003. *Boundaries of Regulatory Science. Eco/toxicology and Aquatic Hazards of Chemicals in the US, England and the Netherlands, 1970-1995.* Ph.D. diss, University of Amsterdam.

Hays, Samuel P. 1987. *Beauty, Health and Permanence: Environmental Politics in the United States, 1955-1985.* Cambridge: Cambridge University Press.

Herbst Arthur L., Howard Ulefelder, and David C. Poskanzer. 1971. "Adenocarcinoma of the Vagina: Association of Maternal Stilbestrol Therapy with Tumor Appearance in Young Women." *New England Journal of Medicine* 284: 878–81.

Longgood, William. 1960. *The Poisons in Your Food.* New York: Simon and Schuster.

Marcus, Alan I. 1986. *Cancer from Beef. DES, Federal Food Regulation and Consumer Confidence*. Baltimore: Johns Hopkins University Press.

Markey, Caroline M., Beverly S. Rubin, Ana M. Soto, and Carlos Sonnenschein. 2002. "Endocrine Disruptors: From Wingspread to Environmental Developmental Biology." *Journal of Steroid Biochemistry and Molecular Biology* 83(1–5): 235–44.

Marks, Lara. 2001. *Sexual Chemistry. A History of the Contraceptive Pill*. New Haven: Yale University Press.

Meyers, Robert. 1986. *DES, The Bitter Pill*. New York: Putnam.

Morgen, Sandra, 2002. *In Our Own Hands: The Women's Health Movement in the United States*. New Brunswick: Rutgers University Press.

National Women's Health Network. 1980. *DES Resource Guide*. Washington: NWHN.

Pfeffer, Naomi. 1992. "Lessons from History: The Salutary Tale of Stilboestrol." In *Consent to Health Treatment and Research: Differing Perspectives*, ed. Peter Alderson, Report of the Social Science Research Unit Conference.

Proctor, Robert. 1995. *Cancer Wars. How Politics Shape What We Know and Don't Know About Cancer*. New York: Basic Books.

Richter, Catherine A., Linda S. Birnbaum, Francesca Farabollini, Retha R. Newbold, Beverly S. Rubin, Chris E. Talsness, John G. Vandenbergh, Debby R. Walser-Kuntz, and Frederick S. vom Saal. 2007. "In Vivo Effects of Bisphenol A in Laboratory Rodent Studies." *Reproductive Toxicology* 24: 199–224.

Rifkin, Jeremy. 1992. *Beyond Beef: The Rise and Fall of the Cattle Culture*. New York: Dutton.

Robinson, Glen O. 1982. "Multiple Causation in Tort Law: Reflections on the DES Cases." *Virginia Law Review* 68: 713–69.

Seaman, Barbara. 1969. *The Doctors' Case against the Pill*. New York: P.H. Wyden.

Shell, Orville. 1984. *Modern Meat*. New York: Random House.

Silber, Norman. 1983. *Test and Protest: The Influence of Consumers Union*. New York: Holmes & Meier.

Sonnenschein, Carlos, and Ana M. Soto. 1998. "An Updated Review of Environmental Estrogen and Androgen Mimics and Antagonists." *Journal of Steroid Biochemistry and Molecular Biology* 65: 143–50.

Temin, Peter. 1980. *Taking Your Medicine: Drug Regulation in the United States*. Cambridge: Harvard University Press.

Turner, James S. 1970. *The Chemical Feast: Ralph Nader's Study Group Report on the Food and Drug Administration*. New York: Grossman Publishers.

Vendenberg, Laura N., Perinaaz R. Wadia, Cheryl M. Schaeberle, Beverly S. Rubin, Carlos Sonnenschein, and Ana M. Soto. 2006. "The Mammary-Gland Response to Estradiol: Monotonic at the Cellular Level, Non-monotonic at the Tissue-Level of Organization?" *Journal of Steroid Biochemistry and Molecular Biology* 101: 263–74.

vom Saal, Frederick, et al. 1997. "Prostate Enlargement in Mice Due to Fetal Exposure to Low Doses of Estradiol or Diethylstilbestrol and Opposite Effects at High Doses. *Proceedings of the National Academy of Sciences* 94:2056–61.

Weltje, Lennart, Frederick S. vom Saal, and Jörg Oehlmann. 2005. "Reproductive Stimulation by Low Doses of Xenoestrogens Contrasts with the view of Hormesis as an Adaptive Response." *Human Experimental Toxicolology* 24: 431–37.

 CHAPTER 4

Managing Scientific and Political Uncertainty
Environmental Risk Assessment in a Historical Perspective

Soraya Boudia

On the evening of Monday, 28 February 1983, a sumptuous and meticulously organized dinner took place in Washington, D.C., on the initiative of the National Research Council (NRC). The list of 129 handpicked guests was drawn up with the collaboration of Edwin Behrens of the American Industrial Health Council. This list included twenty-seven key figures from the U.S. Congress, including several senators, and twenty-four industrial personalities, including Richard Leet, chairman of Amoco Chemicals Corporation; Barclay Morlay, chairman and executive director of Stauffer Chemical Company; and Hunter Henry, chairman of Dow Chemical Company.[1] With this event, the NRC inaugurated a new event-driven policy for the promotion of expert reports that are useful for science-related public policies. The report at the origin of this initial dinner was entitled "Committee on Institutional Means for Assessment of Risks to Public Health," better known as the "Red Book" of risk assessment. It was one of the NRC's most influential reports in the United States and at the international scale. It provided a general framework in the management and regulation of health and environmental hazards, including those of carcinogenic, mutagenic, and reprotoxic substances. The Red Book marked an important stage in the extension of environmental risk assessment.

This chapter examines the origins of this Red Book and thereby establishes a genealogy of risk as a category of analysis and management of environmental pollutants and related health issues. For the last thirty years, risk has been pervasive in the management of scientific, technological, and medical innovation (Beck 1992; Giddens 1990 Lupton 1999; O'Malley 2004; Rothstein, Huber, and Gaskell 2006; Power 2007). As several researchers in governmentality studies have stressed, the effectiveness of the notion stems from the fact that it relates not to any particular category of events but to the way in which they are considered and to how one attempts to objectify, assess, and process them (Ewald 1986; 1991).

From the end of the 1960s on, there were increasing numbers of attempts to use risk tools to examine and process—on a transversal basis—a set of environmental issues, especially those dealing with hazards. In this article, I will attempt to identify and map the places, actors, issues, institutional framings, and dynamics that led to the identification, in terms of risk, of a heterogeneous set of problems, and to the management of a large number of health-environmental activities by risk assessment. The rapid development and proliferation of risk techniques is not simply a process of generalization; it goes hand in hand with the creation of new concepts, tools, and devices that are broadening the significance of risk and changing the political rationality in play. By exploring several avenues and following the debates in those areas, I will attempt to account for the effervescence found in various institutional spaces through the 1970s and which, in the early 1980s, led to the publication of the "Red Book" for environmental risk assessment.

Elaborating and formalizing an environmental risk assessment was a major concern in several industries that produced, used, and managed substances and activities known to entail health and environmental hazards. The problem of toxic substances in the environment and their health effects had been growing in importance since the 1950s. Several specialists were becoming alarmed by the increasing number of cancer cases being attributed to environmental factors (WHO 1964), which reinforced the idea that humans had contributed to making their environment toxic for their own health (Hays 1989; Scheffer 1991). After the wave of controversies on radioactive fallout effects in the 1950s (Wittner 1993; 1997; Boyer 1994; Boudia 2007) and the publication of Rachel Carson's book, *Silent Spring* (1962) on the effects of chemical pollutants, the number of health-environmental issues on the political agenda multiplied. DDT, water pollution, and low-dose radiation were some of the health and environmental hazards that became major public policy issues. During the 1970s, with the rapid increase in health and environmental concerns, carcinogens became the subject of a twin intention: to define a methodology with which to characterize these substances, and to construct a homogeneous framework for their management and regulation (Brickman et al. 1985; Proctor 1995; Cranor 1997; Epstein 1998). Two international initiatives were determining factors in achieving this. The first was the creation in 1967 of the carcinogen unit within the International Agency for Research on Cancer (IARC), affiliated with the World Health Organization, and the purpose of which was to produce scientific monographs on carcinogen risk assessment. Responsibility for this unit was given to Laurenzo Tomatis, who had worked at the University of Chicago under Philipp Shubik, one of the leading international experts on carcinogen substances. The second initiative took place after the Stockholm conference in 1972, with the creation of the International Programme on Chemical Safety, which was tasked with establishing the criteria for defining and testing toxic substances. This program was financed by the United Nations Environment Programme (UNEP) and supervised by the WHO, in partnership with several institutions such as the International Labour Organization (ILO) and the Organisation for Economic Co-operation and Development (OECD).

At the start of the 1970s, two regulatory strategies were set up to manage the dangers of these substances through their prohibition and the setting of exposure levels. Regulation through prohibition was promoted by several environmental activists. In the United States, it was only considered for food additives with the Delaney Clause in 1958. In other sectors the regulation of carcinogenic and mutagenic substances was largely dominated by the setting of exposure limits below which pathogenic effects were not supposed to be produced. Yet during the 1970s, several scientific studies converged to show that various chemical substances had a certain effect on cancers at low levels, well below what had been suggested by the more conservative estimates. The idea that numerous substances had mutagenic and carcinogenic effects with no thresholds was gaining ground in scientific circles. The issue of low-dose pollutants and their effects led to a major change in philosophy regarding protection against the toxicant hazards in the expert community and regulatory institutions (Boudia 2009). The idea that there was no safe level of exposure was born. This meant that whatever the dose, there was always a potential risk, and a probability of the occurrence of health injuries. Despite this, many scientists felt that the notion of banishing such substances from the market remained unthinkable and unrealistic.

This bias was strongly defended—either directly or through the intermediary of paid experts—by a powerful chemical industry that had never lacked energy or financial resources when it came to promoting its image and defending its interests. Many scientists put forward two ideas: on the one hand the extent of uncertainty relating to the problem of low-dose exposure and the importance of cumulating more data using properly defined scientific methodologies, and on the other hand the difficulties in basing all regulatory decisions on scientific data alone. From their point of view, protection against toxic substances could not mean aiming for zero risk and banning a substance or an activity; it meant making a decision that took different parameters into account. Gilbert Omenn, Professor of Medicine at the University of Washington and Associate Director at the Office of Science and Technology Policy (OSTP) summarized the two regulatory options for toxicants as follows: "There was quite a struggle between those who insisted on 'zero risk' and those who proposed methods of risk assessment to identify what Lowrance (1976) called 'acceptable risk' and most of the rest of us preferred to call 'negligible risk', realizing that this conclusion was in the eye of the beholder" (Omen 2003: 2). Several experts felt that it was a matter of urgency to define some rules while taking into account the fact that "sound decisions require technical, economic and sociological considerations of a complex nature" (Omen 2003: 2). Defining a clear and accountable methodology in risk assessment and decision became a matter of concern for several scientific and regulatory institutions.

How Safe Is Safe Enough? The Rise of Risk Assessment in the 1960s and 1970s

To tackle the big issue of risk evaluation and decision making under uncertainty, collective reflection was organized discreetly, within the confines of the National Academy of Sciences (NAS), the National Research Council (NRC), and the National Academy of Engineering (NAE). The NAE's Committee on Public Engineering Policy (COPEP) established a work group called the Benefit-Risk Decision Making Subcommittee (COPEP 1972). The members of this committee were from a wide range of disciplines: engineering, psychiatry, economics, medicine, law, political science, and decision theory. Their work was to lead to the development of methods for comparing risks and benefits in decision-making processes with regard to public acceptance of potentially dangerous products and technologies. To that end, they defined priority domains such as health, air transport, and nuclear energy, in which research would serve to develop and improve methods of analysis.

A key actor in the COPEP's work was Chauncey Starr, now seen as a founder of risk as an approach to health and environmental problems. Starr was a nuclear engineer and dean of the School of Engineering and Applied Science at the University of California Los Angeles (UCLA). In the late 1960s, he was also working for the Electric Power Research Institute of Palo Alto in California, of which he was a founder. His article entitled "Social Benefit versus Technological Risk," published in *Science* in 1969 (Starr 1969), aroused considerable interest in several scientific domains and provided a general framework, which would channel debates on the issue. Starr argued that his approach was an empirical one and justified it by the absence of any economic or sociological theory that might offer systematic assessment of risks and benefits. Two elements structured his approach. The first was the systematic comparison of the different types of risk, based on one central element: morbidity. The second element was an estimation of the risk acceptability or refusal, based on an examination of previous cases. On this basis he confirmed that the public was far more prepared to accept voluntary risks than risks it had not chosen. Starr recognized the limits of his approach, mainly "the uncertainty inherent in the quantitative approach," but in spite of that he was certain that its application "to other areas of public responsibility is self-evident" and that it provided a method with which to answer the question "how safe is safe enough?" (Starr 1969: 1237). While Starr's conceptual framework attracted many authors who readily embarked on this type of research, others remained skeptical about the possibility of comparing risks of such varying natures. Some of them wanted to demonstrate how Starr's ideas included many political and social values (Otway and Cohen 1975).

A second committee of the NAS-NRC, the Committee on Science and Public Policy (COSPUP), was deeply involved in risk issues. The COSPUP was one of the most powerful committees of the NRC-NAS. Several of its reports had a strong influence on federal science and technology policy (Cochrane 1978) and it played an active role in the creation of the Office of Technology Assessment in 1972, for example (Bimber 1996). Its chairman, Harvey Brooks, believed that one of the most interesting avenues for quantifying risk, costs, and benefits was that suggested by the group run by British economist Chris Freeman.[2] Freeman had founded and was in charge of the Science Policy Research Unit at the University of Sussex, and was developing a program backed by the Social Science Research Council in order to "try to identify and quantify both benefits and hazards of different classes of industrial activities" (Sinclair 1969: 120). The kingpin of this program was Craig Sinclair, whose previous work related to safety issues for nuclear reactors. The question behind the group's work was "how much should we be prepared to pay for safety measures?" (Sinclair 1972; Sinclair et al. 1972). It therefore offered an approach based on the assessment of the cost effectiveness of reducing risks compared to the various costs engendered by those risks. Sinclair thus assessed the cost of risk prevention in various sectors of British industry, particularly in the steel and pharmaceutical industries. In order to establish these comparisons, he used different values generally borrowed from the insurance sector: loss in productivity, cost of sick leave, cost of treatments, etc. His idea was to try to attribute an absolute value to life in order to carry out a cost effectiveness analysis. Brooks felt the idea to be "fairly primitive, but quite do-able." The main results were used in a book that marked an important moment in thinking about risk issues, that of William Lowrance, who had taken part in the COSPUP works. His central issue was that which was preoccupying the majority of decision makers: what is an acceptable risk? Lowrance put forward the notion that an acceptable risk is an accepted risk, which in turn led to his definition of "safety": "A thing is safe if its risks are judged to be acceptable" (Lowrance 1976: 8).

In parallel to the work being done by the NAS and the NRC, these issues were also receiving attention at international level. This was particularly true within the new United Nations Environment Programme (UNEP), which was created after the Stockholm summit in 1972. Reflection within UNEP was run by the Scientific Committee on Problems of the Environment (SCOPE). As early as 1973, in partnership with the Electric Power Research Institute founded by Starr and funded by industrial electricity companies, SCOPE carried out a study on environmental risk assessment. The study was run by Robert Kates, a geographer from Clark University who was a specialist in natural disasters and a participant in NAS-NRC reflections on risk. The group's work gave rise to a series of workshops on an international scale and led to the proposition for a methodology for risk assessment in several steps (Kates 1978; Whyte and Burton 1980). Another institution promoted several studies on risk: the International Institute for Applied Systems Analysis (IIASA), founded in 1972 with a view to developing partnerships between scientists from the West and scientists from the East. Based in Vienna and run by economist and statistician Howard Raiffa, IIASA was a driving force behind energy and ecology issues. Within this framework, several risk-related aspects were examined and regular reports were written.[3]

In these works from the first half of the 1970s, it would appear that in the deployment of risk technologies, cost effectiveness, and risk-benefit analyses were given a central position. This process is reflected in a "monetization" of various activities in different sectors. From an historical point of view (Porter 1986; Espeland and Mitchell 1998) the long-term processes which tend to monetize various activities and aspects of social life continued into new fields such as the environment and health (Benamouzig 2005). Many experts maintain that the economic approach—translating health and environmental damage into financial damages, especially the costs generated by risks or by improvements to protection—is the only operational methodology (Otway 1975; Linnerooth 1975). One place in which this type of work flourished was the OECD. Its subcommittee of economists worked on preparing guidelines to provide a conceptual framework and a set of operative definitions for the development and use of financial estimates (OECD 1972; Mäler and Wyzga 1976).

But this type of approach was still not satisfactory. Numerous experts, including certain economists, were quick to point out the limits of a purely insurance-based approach to health and environmental risks. Indeed, these risks posed a range of new problems for the experts tasked with defining an overall methodology. One of their characteristics was the expansion of the spatial and temporal scales on which these problems occurred. Pollution is not just local, but can affect the entire planet; it does not affect just health, but also the entire ecosystem; its consequences are not just immediate, but can have an impact over several generations. One of the cruxes of criticism of a monetary approach is that some damage is irreversible. This irreversibility could be included in the economic calculations but those calculations would correspond to the preferences of the present generation, whereas it is future generations who will suffer from the irreversibility. Because of the multiplicity and latency of the effects, to this must be added an extension in time of the causality, which requires one to imagine retroactive responsibility. These dangers, which affect life's capacity to reproduce and to exist are henceforth deployed from the infinitely small scale, with exposure to low doses of pollution that are responsible for tens of thousands of cancer victims, to the infinitely large scale, with major technological risks such as the Seveso accident. These risks, the full consequences of which are unknown, and which are capable of endless proliferation, are a priori difficult to define and even harder to quantify. Their amplitude, irrevocable consequences, latency, and potential deployment over several generations make it difficult to evaluate their impact and almost impossible to assess their insurance value as no institution has the financial capacity to fully cover them over time and space.

In these debates there is another recurrent issue, that of risk acceptability (Douglas 1985). In the final analysis, the idea is that the problem is not so much whether there is a potential danger as whether it is accepted by a given population. This supports the idea that for institutions, risk is created by the dispute that a given activity might provoke. With the question of no-threshold effects, a major shift takes place: risk is no longer a random event that escapes human control, but is inherent to a given activity when functioning normally. The problem cannot therefore be approached in a binary manner, because the activity is no longer considered to be "dangerous" or "not dangerous," "safe" or "unsafe." For the community it becomes a question of "balance" between different parameters, and of "judgment." In such a configuration, the central issue is ultimately the decision of whether or not to maintain an activity that may give rise to a given level of risk. Yet by its very political nature, such a situation will cause disagreement. Such risks divide opinions about what is in the community's best interests, which means that the issue is no longer a matter of determining a threshold below which there is no danger, but rather of determining one that minimizes the level of public dispute. Risk technology should help to define and implement decision-making methods that allow decisions to be (or to seem to be) collective. In this way responsibility can be of a collective nature and no longer be borne solely by those involved in the activity or by policy makers. This would help to defuse the criticism to which they are subjected. While the avenue of research into collective decision-making procedures has been marked out, it would appear to be difficult to come up with an overall solution.

Despite this intense activity in expert committees and regulatory agencies around risk evaluation and cost-benefit calculations, the results in terms of risk management were somewhat limited. The National Science Foundation (NSF) consequently launched a program devoted to the definition of a risk-evaluation methodology. Its financial and institutional support enhanced the professionalization of a body of experts specialized in risk analysis and laid the foundations for contemporary practices in this domain. In 1979 the House Committee on Science and Technology asked the NSF to come up with a systematic research strategy to improve risk methodologies, taking into account a broad range of domains, such as energy, materials, environmental quality, foods, and drugs. The NSF's arrival on the scene led to an increase in the amount of research on the issue and helped to structure a professional community dedicated to risk analysis and management (Golding 1992).

Starting in August 1979, after consulting the Academy of Sciences, the NSF set up a program entrusted to its Policy Research Analysis (PRA) division. A new group called Technology Assessment and Risk Analysis (TARA) was created under Joshua Menkes, who had headed the NSF's technology assessment activities until then. Vincent Covello, who had joined Menkes's group in 1978, was appointed as program manager, responsible for various activities. TARA was assigned the mission of promoting research, outside the NSF, on risk analysis and its application to regulatory policy making. As its name suggests, the group was expected to highlight the complementarity between technology assessment and risk analysis. Its job was to develop research and to perform analyses that could be used to help decision makers in foreseeing and planning the risks and impacts associated with new technologies.

The various initiatives driven and funded by TARA helped to widen the circle of scientists and professionals concerned with environmental risk assessment. At the end of the 1970s, they were in sufficient number to be able to envisage the foundation of a journal dedicated to risk analysis. This project was managed by Robert B. Cumming, from the Biology Division of Oak Ridge Laboratory and member of the Environmental Mutagen Society (Frickel 2004). Cumming went on a tour of Europe to discuss the issue with a few of his colleagues. During his stay in Stockholm, he had in-depth discussions with the biochemist and geneticist Lars Ehrenberg. They agreed to promote the creation of a journal and to develop international collaboration. To achieve this, with the help of Robert Tardiff from the Board on Toxicology and Environmental Health Hazards at the NRC, Cumming brought together in October 1979 a committee made up of Vincent Covello, Gary Flamm, Allen Newell, Tim O'Riordan, and Joseph Rodricks. They decided to create a scientific society, the Society for Risk Analysis, and a related journal, *Risk Analysis*, the first issue of which appeared in March 1981. As its success has shown, this initiative met a need. As early as 1981 it had 300 members, climbing to 1,500 six years later (Golding 1992; Thompson, Deisler, and Schwing 2005).

Despite the difficulties encountered by the various experts working on the issue, TARA did not abandon the idea of helping to develop a general decision-making methodology for risk situations and controversies. It once again contacted the NAS in order to carry out a new study, which would this time be entrusted to the NAS-NRC's Assembly of Behavioral and Social Sciences (ABSS). The Committee on Risk and Decision-Making (CORADM) was created with, as chairman, Howard Raiffa, who was a specialist in decision-making theory, a professor at Harvard, and a former chairman of IIASA, and, as secretary, William Ruckelshaus, the former Environmental Protection Agency's first administrator, currently Senior Vice-President of Legal Affairs of Weyerhaeuser. This committee illustrated the shifts that had taken place within the field, one of which was to cease focusing on the determination of benefits. The president of the Academy commented to David Goslin, executive director of the ABSS, that "what seems remarkable is the absence of the word 'benefit' in your brief write-up. Granting the validity of all that is said, risks are assumed only to secure stated ends. And it is the fact that the risks and benefits are incommensurable and assigned to different population groups that makes government so difficult at this time."[4] In this committee, there were several people who believed in the theory of rational choice and whose works had been carried out in part at RAND—people such as Raiffa, the sociologist James Coleman, and the psychologist Amos Tversky. At a later date, Raiffa was to clearly consider this work to be a failure: "I found my task daunting and in the end completely frustrating, although along the way the experience was absolutely fascinating. Besides my disappointment with the discontinuance of the Ph.D. program in the decision sciences at Harvard, my CORADM experience was my only other major disappointment in my professional life" (Raiffa 2002).

After a decade of collective reflection and abundant production, the elaboration of a cross-cutting methodology for risk analysis was still no more than a project. In several sectors, however, including health environmental hazards, concepts and approaches developed within the different expert committees started to be adopted at the end of the 1970s.

Working Toward the Red Book of Environmental Risk Assessment

The issue of risk assessment of toxic substances was highly political and quite controversial in the public arena in the 1970s (Patterson 1987; Jasanoff 1990; Proctor 1995). One of the major actors involved in the debates was the chemical industry, in particular through the American Industrial Health Council, an advocacy group created by several major chemical companies in 1977. This group brought together 130 chemical corporations, headed by Shell, Proctor and Gamble, and Monsanto, which combined produced a wide range of basic chemical and metallic substances, drugs, and consumer goods. Officially, this council was formed with the aim of studying occupational exposure and drawing up a list of proposals for the Occupational Safety and Health Administration (OSHA) in charge of occupational health policy in the United States, and for improving federal policy on the identification and regulation of carcinogens.[5] The AIHC argued that it was necessary to find the means for clearly separating scientific aspects—identification and quantification of risk—from political aspects, whose regulation took into account social and economic judgments. It had the support of the chairman of the Office of Science and Technology Policy (OSTP)[6] and engaged in actively lobbying members of the U.S. House of Representatives and the Senate to promote their project.[7] The council suggested to the senator from Missouri, Thomas Eagleton—who presided over the Subcommittee on Agriculture, Rural Development, and the related Agencies, that Congress devote $500,000 to a study by the Academy of Sciences on alternative means to elaborate scientific judgments on the quantitative aspects of human risk, especially in the case of chronic diseases.[8]

The AIHC also had many discussions with the NAS-NRC. Two figures played an important part in the interaction between the two institutions: Paul Sitton, chairperson of the Assembly of Life Sciences, and Edwin Behrens of Proctor and Gamble. On 29 November 1979 they organized a meeting between Philip Handler, chairperson of the Academy of Sciences, Miner Joe Sloan, in charge of "regulatory affairs, health, safety and environment" at Shell, and William J. McCarville, director of environmental affairs at Monsanto. The main objective of the meeting was to discuss AIHC proposals for the use of science in regulatory policy making.[9] This first meeting was followed by regular contact. On 13 March 1980 a new meeting was devoted to in-depth discussions on the basis of a document produced by the AIHC. Handler proposed several amendments to a new version of the document. The AIHC set up two task forces. The first, headed by Paul F. Deisler, had the mission of clarifying the concept of a "science panel," while the second, headed by Monte Throdahl from Monsanto, was to reflect on its implementation.[10]

On 24 November 1980, the U.S. Senate approved the allocation of $500,000 for the risk assessment study, to examine "alternative programs and institutional means to insure that federal regulatory policies with respect to carcinogens and other public health hazards of particular national significance are developed on the basis of reliable scientific assessments."[11] The required objectives of the study were to assess the merits of an institutional separation of scientific, political, and social functions, of considering the feasibility of unifying the different aspects of risk analysis, and of looking at the possibilities of developing a coherent methodology for risk analysis to be taken up by all regulatory agencies. Given the origin of the funding (agriculture), it was the FDA that contacted the NAS-NRC to make the order.

Before the FDA had initiated anything at all, internal tensions appeared within the NAS and the NRC with regard to the localization of the study. At the time, the expert committee chaired by Raiffa was still working on the production of a report on risk and decision processes for the NSF. As soon as the Senate granted $500,000 to the new study, David Goslin, chairman of the ABSS, with which Raiffa's group was affiliated, proposed that this committee be entrusted with drawing up the new report. He deemed it necessary to complete the expert committee with specialists in biomedical science and to undertake the study in collaboration with the Assembly of Life Sciences. Sitton, one of the people to whom the letter was addressed, reacted immediately, pointing out his role and therefore that of the Assembly of Life Sciences in the adoption of the idea and the financing of such a study. He wrote to the director of the Academy: "You will recall that a number of meetings and other contacts between the American Industrial Health Council and our staff (including a meeting which you attended) was the basic driving force behind the legislative action on this study. Al Lazen, on behalf of ALS, worked closely with me on the formulation of the language and in carrying out other parts of the staff work" (Mirer 2003: 1136).

Sitton affirmed that the study ought to be focused on health issues, justifying his point of view to the chairman of the Academy who agreed that "a precise subject limited to risk assessment of health hazards probably is manageable from the viewpoint of looking at institutional considerations. If it is broader we may not be able to accomplish much."[12] He furthermore noted that such a choice would make it possible to carry out the study relatively quickly. Handler also wanted a study that could be undertaken without delay. He granted great importance to the results of this work, pointing out that the aim was to determine how the United States could regulate the problem of chemical substances in the environment—a vast issue concerning several agencies. He also affirmed that the proposals that emerged should serve as a model for other countries.

As the perimeter of the study was limited, in early May 1981 discussions were engaged on the composition of the expert committee to which the study would be entrusted. In the end, the committee, called the Committee on Institutional Means for Assessment of Risks to Public Health, consisted of experts from various disciplines. It was chaired by an epidemiologist specializing in cardiovascular risk factors, Reuel A. Stallones, dean of the School of Public Health, University of Texas, and member of the Board on Toxicology and Environmental Health Hazards at the NAS-NRC. Its fourteen members looked at the state of the art of risk analysis for chemical and physical substances in order to feed reflection on the advantages and disadvantages of a uniform approach to risk analysis.

First, the members of the Committee on Institutional Means for Assessment of Risks ruled on the proposal from the AIHC's science panel. The members of the committee told Frank Press, the president of the NAS, that: "We believe strongly that it would be inappropriate to remove such an essential and growing analytical function from the responsible agencies or to duplicate the agency activities. Therefore, we recommend that this Board make its contributions through open discussion of contending scientific positions, through guidelines, and through advancement of the field."[13] Another argument that had been put forward related to the illusion of using science to end the controversies. The committee highlighted the fact that the regulations for the twenty carcinogenic substances suggested by OSHA were based on the scientific works of the WHO's IARC, which was similar to the science panel. Yet this had in no way reduced the controversies or the attacks on OSHA policy.

Second, the Committee on Institutional Means for Assessment of Risks mainly worked on the formalization of risk analysis. Its kingpin was Gilbert Omenn, who had worked on the question when he was with the OSTP (Omenn and Ball 1979). The committee used what had been learned from numerous other studies in the field of carcinogens by attempting to generalize a staggered approach to risk management. It suggested, within the framework of public health issues, that a risk analysis be carried out in four stages: (1) hazard identification, (2) dose response assessment, (3) exposure assessment, and (4) risk characterization. This report, even with amendments, marked a clear distinction between "facts" and "values," between "science" and "politics." It proposed a two-step approach: first the implementation of an initial process of objective scientific analysis of risks, and then a second process in which economic, ethical, and other considerations would be taken into account for policy-making purposes.

The heads of the National Academy of Sciences and the National Research Council were particularly attentive to the process of publicizing the report. In addition to the traditional press communiqué, they devised several initiatives, including the dinner symposium organized by Sitton and Lazen on 28 February 1983.

The report was very well received. Even the AIHC, whose suggestion had not been accepted, stated that it "is comprehensive and offers constructive suggestions for improving the risk assessment processes of government regulatory agencies. In many respects it agrees with the position expressed by AIHC that we have supported, but differs on the role of a central board of scientists. The NAS approach appears more practical and offers a better fit with current procedures of the regulatory agencies."[14] During the following years, even though the debates continued, the report became the real "Red Book" for the risk assessment of health and environmental hazards. It led to the development of several guidelines for the different agencies in charge of assessing and regulating health and environmental problems on an international scale. The proposals set out in the report were not fundamentally new. They were the fruit of the convergence and synthesis of a set of ideas. The Red Book was thus one step in the synthesis and closure of a certain number of debates that took place in an attempt to reach a consensus. Its success was also due to the important institutional and political work developed after its publication by the Society of Risk Analysis (SRA), NAS, EPA, OSTP, and UNEP. It made it possible to adapt and transfer its recommendations to other fields. This was possible because the Red Book provided decision makers with at least two sets of advantages.

First, this work of assessment helped to make risk analysis both more sophisticated and more understandable and operational. It provided precise elements with which to construct procedure, while at the same time setting out a framework, a way of working that was sufficiently general to allow it to be adapted to different situations. One of the central ideas that it helped to promote was that the analytical and decision-making methodology must go through different stages. In this sense the Red Book borrowed from the intellectual logics of decision sciences and systems that promoted the importance of a formal analysis that used scientific knowledge and statistics, in particular. There are at least two advantages to such a bias. First of all, it reinforces the "impersonal" and "abstract" nature of the forthcoming decision. The methodology proposed by the Red Book also targets the implementation of a process, the temporality of which is lengthy and concerted. With problems relating to technoscientific dangers, time is an important element. Giving oneself time allows one to examine the different aspects of the problem. Yet this is probably not its prime objective. One of the criticisms of the Red Book was expressed later by one of the members of the committee who had helped write it, Franklin E. Mirer, director of the Health and Safety Department of the United Automobile, Aerospace & Agricultural Implement Workers of America International Union (UAW). He saw the Red Book as an instrument of paralysis, a stalling mechanism that separates the time at which a regulation is developed from that of its implementation: "The primary purpose of the study was to judge early proposals for adding 'procedural botox' to paralysis-by-analysis then slowing public health intervention against chemicals." He also pointed out that "since 1983, evidence for carcinogenicity of many exposures in the occupational environment has greatly proliferated, but public health protection has stagnated" (Mirer 2003: 1136). And indeed, since then and despite decades of programs and thousands of meetings and scientific and political actions, the issue of toxic substances has still not been resolved. Their growing numbers and their large-scale circulation, in often poorly controlled proportions, have merely increased the range of their effects. The problems caused by their toxicity are not the only ones not to have been solved by the Red Book. Political issues also remained very open, as their regular return to the public arena and the large amount of scientific work and public policy relating to them show.

Conclusion: Risk and Government Paradigms

The "Red Book" general framework is still the dominant paradigm, even if it has been facing several critics and questions concerning its limits in the assessment and regulation of health and environmental hazards (NRC 1994). Its success story means in fact the expansion of risk technology and its establishment as a central category in the conception, formulation, and management of a set of issues. Indeed, risk would appear to be a polysemic concept and tool, capable of comprising different expectations and challenges. If we were to retain just one of its functions, I would highlight the fact that it comes across as a central strategy for managing technical and scientific (of course) but also social and political uncertainty. The question of uncertainty appeared in the specifications set down by the committee of experts behind the Red Book. It was expressed as follows: "The Committee will perform an in-depth study of the nature and the extent of the uncertainties that result from deficiencies in scientific knowledge concerning health risk assessment [and] describe how and the extent to which these uncertainties are accommodated by current practices in regulatory agencies."[15]

Uncertainty covers several aspects. Behind this term we find a number of questions which from a scientific standpoint are formidable: the multifactorial nature of pathologies—cancers in particular, their latency period, their causality, the status and nature of proof in these pathologies, individual variability, and the type of link between clinical research and laboratory research. These different aspects are clearly set out, discussed, and analyzed within the framework of formalizing risk assessment. On the other hand, what is far less clearly expressed but nevertheless very present is the question of political uncertainties. First and foremost, these concern the behaviors of the various players within the "public" category, including environmental and activist associations. They also concern—and this is related—institutional and political players, the results of whose actions during periods of crisis can never be fully controlled. Ultimately, the challenge is to gain robustness. This does not mean the end of uncertainty, but rather an increase in the capacities of administrative and political institutions to better resist public disputes and critics.

The successes of risk and of a conception like a "risk society" contribute to the idea of a world where antagonisms must be set aside in order for people to think together about common solutions. They convey the notion of a world that is pacified and reconciled through the absorption of conflict and the eradication of insurmountable antagonisms. Yet while risk has contributed to the construction of the scientific and political robustness of various technoscientific options, it has not succeeded in fully containing criticism. In certain cases it has even become a lever for dispute. This means that governing through risk has proved to be insufficient. It was therefore adapted and changed throughout the 1990s and 2000s. One of the main changes was a broadening of the decision-making base, with the inclusion of different groups of actors and stakeholders, including industrial companies and NGOs. As early as the 1970s there was a call for public participation in the decision-making process, with institutional reflection taking place within both the Office of Technology Assessment in the United States and the OECD at the international level. Since then, the idea of including stakeholders at various stages of the decision-making process has been developed and theorized within the framework of participative governance. The result is a superposition of governmental technologies that policy makers can mobilize and adapt to different situations. However, the health-environmental issues related to toxic substances are far from being resolved.

Notes

1. Memorandum of Paul Sitton to Frank Press, 10 February 1983, in NAS-NRC Archives, Assembly on Life Science, Committee on Institutional Means for Assessment of Risks to Public Health.
2. Letter from Harvey Brooks to Philip Handler, 9 November 1970, in NAS-NRC Archives, Medical Division, Committee Advisory to FRC.
3. http://webarchive.iiasa.ac.at/docs/history.html (last accessed 5 November 2012).
4. Letter from Philip Handler to David Goslin, 7 May 1979, in NAS-NRC Archives, Behavioral and Social Sciences Assembly. Committee on Risk and Decision Making, 1979.
5. AIHC, "AIHC recommended Framework for Identifying Carcinogens and Regulating Them in Manufacturing Situations," 11 October 1979, in NAS-NRC Archives, NAS-NRC Executive Offices Organization. Projects proposed: Risk Assessment and Federal Regulation Policies 1980.
6. Letter from AIHC's director to Senator Eagleton, n.d. (March–April 1980), in NAS-NRC Archives, NAS-NRC Executive Offices Organization. Projects proposed: Risk Assessment and Federal Regulation Policies 1980.
7. AIHC, "AIHC recommended Framework for Identifying Carcinogens and Regulating Them in Manufacturing Situations," 11 October 1979, in NAS-NRC Archives, NAS-NRC Executive Offices Organization. Projects proposed: Risk Assessment and Federal Regulation Policies 1980, 1.
8. Letter from AIHC director to the Senator Eagleton, n.d. (March–April 1980), in NAS-NRC Archives, NAS-NRC Executive Offices Organization. Projects proposed: Risk Assessment and Federal Regulation Policies 1980, 1.
9. Letter from Miner Joe Sloan to Philip Handler, 22 February 1980, in NAS-NRC Archives, NAS-NRC Executive Offices Organization. Projects proposed: Risk Assessment and Federal Regulation Policies 1980.
10. Letter from Paul Deisler to Philip Handle, 17 March 1980, in NAS-NRC Archives, NAS-NRC Executive Offices Organization. Projects proposed: Risk Assessment and Federal Regulation Policies 1980.
11. Letter from Edwin Behrens, 4 December 1980, in NAS-NRC Archives, NAS-NRC Executive Offices Organization. Projects proposed: Risk Assessment and Federal Regulation Policies.
12. Letter from Paul Sitton to Philip Handle, 16 April 1981, in NAS-NRC Archives, NAS-NRC Executive Offices Organization. Projects proposed: Risk Assessment and Federal Regulation Policies.
13. Letter Alvin Lazen to Frank Press, 7 September 1982, in NAS-NRC Archives, Assembly of Life Sciences; Committee on Institutional Means for Assessment of Risks to Public Health.

14. Letter from Joe E. Penick to Frank Press, Fred Robins, Larry McCray, and Paul Siton, 31 March 1983, in NAS-NRC Archives, Assembly of Life Sciences; Committee on Institutional Means for Assessment of Risks to Public Health. General.
15. FDA contract 282-81-8251, 18 September 1981, in NAS-NRC Archives, Assembly of Life Sciences; Committee on Institutional Means for Assessment of Risks to Public Health.

Bibliography

Beck, Ulrich. 1992. *Risk Society: Towards a New Modernity.* London: Sage.
Benamouzig, Daniel. 2005. *La santé au miroir de l'économie.* Paris: Presses Universitaires de France.
Bimber, Bruce. 1996. *The Politics of Expertise. The Rise and Fall of the Office of Technology Assessment.* Albany, N.Y.: The State University of New York Press.
Boudia, Soraya. 2007. "Global Regulation: Controlling and Accepting Radioactivity Risks." *History and Technology* 23, 4: 389–406.
———. 2009. "Les problèmes de santé publique de longue durée. Les effets des faibles doses de radioactivité." In *La définition des problèmes de santé publiques,* ed. Claude Guilbert and Emmanuel Henry, 35–53. Paris: Editions de la découverte.
Boyer, Paul. 1994. *By the Bomb's Early Light: American Thought and Culture at the Dawn of the Atomic Age.* Chapel Hill: The University of North Carolina Press.
Brickman, Ronald, Sheila Jasanoff, and Thomas Ilgen. 1985. *Controlling Chemicals: The Politics of Regulation in Europe and the United States.* Ithaca: Cornell University Press.
Carson, Rachel. 1962. *Silent Spring.* Boston: Houghton Mifflin.
Cochrane, Rexmond Canning. 1978. *The National Academy of Sciences: The First Hundred Years, 1863-1963.* Washington, D.C.: National Academy of Sciences.
COPEP. 1972. *Perspective on Benefit Risk Decision Making.* Washington: Committee on Public Engineering Policy, National Academy of Engineering.
Cranor, Carl F. 1997. *Regulating Toxic Substances: A Philosophy of Science and the Law.* Oxford: Oxford University Press.
Douglas, Mary. 1985. *Acceptability According to the Social Sciences.* New York: Russell Sage Foundation.
Epstein, Samuel S. 1998. *The Politics of Cancer Revisited.* New York: East Ridge Press.
Espeland, Wendy, and Mitchell Stevens. 1998. "Commensuration as a Social Process." *Annual Review of Sociology,* 24: 313–43.
Ewald, François. 1986. *L'État-Providence.* Paris: Grasset.
———. 1991. "Insurance and risk." In *The Foucault Effect: Studies in Governmentality,* ed. Graham Burchell, Colin Gordon, and Peter Miller, 197–210. Chicago: Chicago University Press.
Frickel, Scott. 2004. *Chemical Consequences: Environmental Mutagens, Scientist Activism and the Rise of Genetic Toxicology.* New Brunswick, N.J.: Rutgers University Press.
Giddens, Anthony. 1990. *The Consequences of Modernity.* Cambridge: Polity Press.
Golding, Dominic. 1992. "A Social and Programmatic History of Risk Research." In *Social Theories of Risk,* ed. Sheldon Krimsky and Dominic Golding, 23–52. Westport: Praeger.
Hays, Samuel P., 1989. *Beauty, Health and Permanence: Environmental Politics in the United States.* Cambridge: Cambridge University Press.

Jasanoff, Sheila. 1990. *The Fifth Branch: Science Advisers as Policymakers.* Cambridge, M.A.: Harvard University Press.

Kates, Robert W. 1978. *Risk Assessment of Environmental Hazard, Scientific Committee on Problems of the Environment, SCOPE 8.* New York: Wiley.

Linnerooth, Joanne. 1975. *Theoretical Modelling to Determine Life-Values: Some Comments.* Laxenburg: International Institute for Applied Systems Analysis.

Lowrance, William 1976. *Of Acceptable Risk: Science and the Determination of Safety.* Los Altos, C.A.: W. Kaufmann.

Lupton, Deborah. 1999. *Risk.* Londres: Routledge.

Mäler, Karl Göran, and Ronald E. Wyzga. 1976. *Economic Measurement of Environmental Damage.* Paris: OECD.

Mirer, Franklin. 2003. "Distortions of the 'Mis-Read' Book: Adding Procedural Botox to Paralysis by Analysis." *Human and Ecological Risk,* 9: 1129–43.

NRC. 1983. *Risk Assessment in the Federal Government: Managing the Process.* Washington, D.C.: National Academy Press.

———. 1994. *Science and Judgment in Risk.* Washington, D.C.: National Academy Press.

OECD. 1972. *Recommendation of the Council on Guiding Principles Concerning International Economic Aspects of Environmental Policies [C(72)128].* Paris: Organisation for Economic Co-operation and Development.

O'Malley, Pat. 2004. *Risk, Uncertainty and Government.* London: Cavendish Press/ Glasshouse.

Omenn, Gilbert. 2003. "On the Significance of 'The Red Book' in the Evolution of Risk Assessment and Risk Management." *Human and Ecological Risk Assessment,* 9: 1155–67.

Omenn, Gilbert, and John Ball. 1979. "The Role of Health Technology Evaluation: A Policy Perspective." In *Health Care Technology Evaluation,* ed. J. Goldman, 5–32. New York: Springer-Verlag.

Otway, Harry J. 1975. *Risk Assessment and Societal Choices.* Laxenberg: International Institute for Applied Systems Analysis.

Otway, Harry J., and Jerry J. Cohen, 1975. *Revealed Preferences: Comments on the Starr Benefit/Risk Relationships.* Laxenburg: International Institute for Applied Systems Analysis.

Patterson, James T. 1987. *The Dread Disease: Cancer and Modern American Culture.* Cambridge, M.A.: Harvard University Press.

Power, Micheal. 2007. *Organized Uncertainty: Designing a World of Risk Management.* Oxford, Oxford University Press.

Porter, M. Theodor. 1986. *The Rise of Statistical Thinking, 1820–1900.* Princeton: Princeton University Press.

Proctor, Robert N. 1995. *Cancer Wars: How Politics Shapes What We Know and Don't Know about Cancer.* New York: Basic Books.

Raiffa, Howard. 2002. "Decision analysis: a personal account of how it got started and evolved." *Operations Research* 50, 1: 179–85.

Rothstein, Henry, Michael Huber, and George Gaskell. 2006. "A Theory of Risk Colonization: the Spiralling Regulatory Logics of Societal and Institutional Risk." *Economy and Society* 35, 1: 91–112.

Scheffer, Victor. 1991. *The Shaping of Environmentalism in America.* Seattle: University of Washington Press.

Sinclair, Craig. 1969. "Costing the Hazards of Technology." *New Scientist* 16 October, 20–122.

———. 1972. *A Cost-Effectiveness Approach to Industrial Safety*. London: Stationery Office.

Sinclair, Craig, Pauline Marstrand, and Pamela Newick. 1972. *Innovation and Human Risk: The Evaluation of Human Life and Safety in Relation to Technical Change*. London: Centre for the Study of Industrial Innovation, University of Sussex, Science Policy Research Unit.

Starr, Chauncey. 1969. "Social Benefits versus Technological Risks." *Science* 165 (3899): 1232–38.

Thompson, Kimberly M., Paul F. Deisler, and Richard C. Schwing. 2005. "Interdisciplinary Vision: The First 25 Years of the Society for Risk Analysis (SRA), 1980–2005." *Risk Analysis* 25, 6: 1333–86.

Wittner, Lawrence S. 1993. *The Struggle against the Bomb*. Volume 1. *One World or None: A History of the World Nuclear Disarmament Movement through 1953*. Stanford: Stanford University Press.

———. 1997. *The Struggle against the Bomb*. Volume 2. *Resisting the Bomb: A History of the World Nuclear Disarmament Movement, 1954-1970*. Stanford: Stanford University Press.

Whyte, Anne V., and Ian Burton, eds. 1980. *Environmental Risk Assessment, Scientific Committee on Problems of the Environment, SCOPE 15*. New York: Wiley.

WHO. 1964. *Prevention of Cancer*. Report no. 276. Geneva: World Health Organization.

 PART II

Activism and Nonactivism
Alternative Uses of Knowledge

 CHAPTER 5

Work, Bodies, Militancy
The "Class Ecology" Debate in 1970s Italy

Stefania Barca

During the two and a half centuries since the industrial revolution, health risks in the factory have not been eliminated, or even radically reduced, compared to the nineteenth century: they have simply changed.[1] Older pathologies have been replaced by newer ones mostly derived from the large-scale spread of organic chemistry, especially in the petrochemical sector, and the marketing of an impressive quantity of products with high content of CMR substances. Workers' bodies have thus become sites of social struggles that have, on occasion, led to legislative reform in the broader field of environmental policy (Elling 1986; Rosner and Markowitz 1986; Berlinguer 1991; Sellers 1997; Carnevale and Baldasseroni 1999; Johnston and McIvor 2000; Bartrip 2001; Markowitz and Rosner 2002).

Diseases induced by the petrochemical industry, however, were less easily recognizable as occupational diseases because the number of synthetic substances produced in chemical laboratories increased at very high rate every year, and it was hence virtually impossible for medical science to keep track of them and ascertain their dangerousness preventively. Thus workers often found themselves playing the role of human guinea pigs until the environmental toxicity of some widely used material or substance was clearly identified.[2]

Although much ecological criticism of contemporary society is founded on the exposing of the environmental damage caused by modern industry (Massard-Guilbaud and Scott 2002; Allen 2004; Platt 2005; Santiago 2006), environmental history has not yet dealt with this subject systematically. It seems that environmental historians so far have had trouble seeing the factory as something lying within their sphere of interest (McEvoy 1995; Meisner Rosen and Sellers 1999). The story told in this chapter, however, shows how the workplace and workers' bodies lay at the core of the new environmental consciousness of the 1970s.

Italy's Labor Environmentalism (1961–1978)

In Italy, scientific expertise and political regulation on CMR substances have been forged out of the experience of what I call "labor environmentalism," i.e., the coalition between workers' organizations and "militant" scientists in the struggle for the recognition and regulation of industrial hazards, eventually producing important social reforms such as the Labour Statute (1970) and the Public Health System (1978) (Barca 2006; Barca 2012). Focusing on the work environment, that peculiar type of environmentalism was based on the recognition of the centrality of the industrial manipulation of nature in determining the deterioration of both occupational and public health (von Hardenberg and Pelizzari 2008). Such new ecological consciousness arose from the totally new conditions of production and reproduction that were formed in the country's tumultuous economic boom of the late 1950s, during which Italians experienced such a rapid and massive industrialization that all aspects of social life were revolutionized. From 1951 to 1971 the agriculture sector expelled almost five million people, 2.3 million of whom entered the factory gates; in the same period, industrial employment in different sectors grew from 40 to 55 percent of the total workforce. The core of this cycle of expansion was the crucial five-year period of 1958–1963, the "economic miracle" during which the GNP doubled and industry surpassed agriculture as a source of income for the first time in the Italian history (Signorelli 1995; Crepas 1998; Musso 1998).

In the aftermath of economic boom, the country experienced the epidemiological shift typical of advanced industrial economies, namely, from infectious to degenerative diseases. Yet, a clear vision of the new risk factors was hardly produced within medical science and public health institutions. Among the occupational diseases recognized by the Workers' Compensation Authority (INAIL) there was a gradual shift from silicosis and lead poisoning to pathologies related to the manipulation of mercury and benzene hydrocarbons. Nevertheless, national statistics severely underestimated cases because often workers did not disclose their illnesses for fear of being fired. Compensation, however, was the very obstacle to the prevention of hazards: the law, in fact, still sanctioned the total non-liability of employers in the matter of industrial accidents and health hazards (Berlinguer 1991; Calavita 1986; Carnevale and Baldasseroni 2009: 138–39).

Spurring from the "economic miracle," the Italian experience of "labor environmentalism" was generated in the cultural context of the 1960s and 1970s, marked by a strong cultural hegemony of the left parties and the labor movement, but also by student protests and new political movements pressing for radical changes in the organization of social life. This new Italian environmentalism was also crucially influenced by the spread of a new international environmental movement (Luzzi 2009: 95–114), much less devoted to conservation than in the past and more concerned with the toxicity of industrial production, especially of petrochemicals (Gottlieb 1993; Rome 2003). What marked the Italian experience, however, was the much stricter link existing between the new environmentalists and the labor movement, unions in particular, which makes it appropriate to speak of a very "labor environmentalism." This had begun to take shape in the early 1960s, when a group of sociologists at the University of Turin formulated what was to become the new methodology of research on occupational health. Soon renamed the "environmental club," this group categorized the four main factors of work-related risk: unspecific risk (noise, microclimate, radiations, vibrations, etc.), risk specific to the work environment (exposure to toxic or explosive substances), risk related to fatigue (physical effort and posture), and psychological risk (linked to labor relations within the workplace). In addition, the group theorized a new methodology of research, based on the direct production of knowledge on the part of workers. Having been successfully tested in 1961 at the plant of Farmitalia, a consociate of the powerful petrochemical group Montedison, those theories were accepted by the Italian labor movement and became the core principles of labor environmentalism. Courses and lectures on the ecology of the work environment were organized throughout the country by the Trade Union Confederation. In 1970, with the passing of the new Labour Statute, the principle of workers' direct control over the work environment became law. A golden age for labor environmentalism had started (Calavita 1986; Tonelli 2006; Tonelli 2007).

The Italian experience was also connected to that of other affluent societies in the same period, especially from a cultural point of view. The translations of Rachel Carson's *Silent Spring* (1962) and of Barry Commoner's *The Closing Circle* (1971) were instrumental in the making of a new cultural scenario, demanding more attentive consideration of the social costs of economic growth, and especially of oil-related production.[3] In this context, the relationships among industrial pollution, ecology, public health, and politics were conceptualized by the Italian Left for the first time in the country's history. The debate involved individual scientists and politicians, but it also required some effort in reorienting the strategy of well-structured organizations such as the Communist Party, the confederation of unions, a number of university labor clinics, and the Association of Industrial Hygiene.

Due to the rapid industrialization experienced in the preceding decade, the 1970s were also a time of significantly increased CMR risk in Italy, affecting not only the workforce, but the Italian population at large, through widespread and largely uncontrolled pollution. Given the favorable trend for Italian organic chemistry and oil-related productions, petrochemicals—and the Montedison company in particular—came to occupy a top position among the new polluting industries. The Bormida river valley in Lombardy, the Tyrrhenian coast near Scarlino in Tuscany, Porto Marghera in the Venice lagoon, the Sicilian coast in the area of Gela, the area of Sarroch in Sardinia, and the area of Manfredonia in Apulia were only some of the places where Italians started to become familiar with petrochemical contamination during the 1970s. A public health disaster was openly recognized in 1972 in Cirié, Piedmont, where forty-one workers of a dye factory were stricken by cancer of the bladder and the river Stura got seriously contaminated with sulphuric acid and other chemical residues. In contrast, no such recognition was granted to the area of Manfredonia, Apulia, when an accident occurred at the ANIC petrochemical plant causing some 32 tons of arsenious dioxide to fall upon a population of fifty thousand, also seriously compromising local agriculture and fisheries (Di Luzio 2003; Tomaiuolo 2006; Luzzi 2009: 152–55; Barca 2012).

In the rising awareness of chemical risk as the dark side of economic growth, falling upon both workers and the environment, a turning point was the accident that occurred at the ICMESA chemical plant near Seveso, in Lombardy. On 10 July 1976, the explosion of a chemical reactor caused a cloud of dioxin to rise over the town and its rural hinterland, directly affecting a population of ten thousand (Centemeri 2010). Among all industrial disasters, the one occurring in Seveso no doubt spurred the greatest attention on the part of the Italian government and the media, national and international. Urging collaboration among labor physicians, professional ecologists, public health agencies, and elected representatives from the local to the national level, the ICMESA disaster turned out to be a remarkable experiment in the interaction of science and politics in the country. It also played a crucial role in the birth of a new ecological consciousness in the Italian Left (Centemeri 2006; Luzzi: 2009: 140–55).

Laura Conti: A Working-Class Ecologist in Seveso

In the convulsive post-disaster scenario that fell upon Seveso between July 1976 and April 1977, a scientist and regional councilor for the Communist Party, Laura Conti, found herself at the forefront of the battle for citizens' right to know and participative science that characterized the political relevance of the accident. As a participant observant with a dual identity of scientist and politician, Conti clearly exposed government's pro-corporate policies, systematically excluding citizens from participation in knowledge formation and the management of risk. The whole point of Conti's political activity, in Seveso and beyond, was exactly that of struggling against "deceit and denial" politics (Markowitz and Rosner 2002) played by corporate as well as government agencies. This was not an easy task, considering that Conti was a communist representative in an area of solid Catholic traditions and politically dominated by the Christian Democrats, also the strongest government party in the country (Zigliioli 2010).[4]

More than anything, however, it was the "politics of low doses and limit values"—as defined by Soraya Boudia and Nathalie Jas—clearly appearing in the public arena for the first time in the country's history, that became a central concern for the communist councilor. Dioxin, Conti observed, seemed to have "*all* the characteristics of the most terrible poisons that modern chemistry spreads over the planet": stability, the tendency to accumulate in organisms, extreme toxicity (such that no micro quantity can be considered innocuous), embryo-toxicity, mutagenicity on bacteria (implying the possibility that it be mutagenic and carcinogenic in humans), and immuno-depressivity. Moreover, its effects can manifest over long time periods. "These aspects, outlined before my eyes in the first few days, made up to the most typical ecological catastrophe that can be imagined," Conti (1977a: 20–21) wrote in her journal. Uncertainty, which the government claimed as the single most important reason for underplaying the risks, was not a case in point: what was uncertain, Conti remarked, was not the dangerousness of dioxin, but the extent to which the environment and the people of Seveso had been contaminated.[5]

Measuring the presence of tetrachlorodibenzodioxin (TCDD) in the soil and the vegetation of the affected area and, on the other hand, establishing a Maximum Allowable Concentration (MAC) for dioxin became, in fact, the most important political tasks in the following weeks. How local and national authorities arrived at establishing such limit values, affecting the definition of different zones of dangerousness, and consequently the lives of thousands of people and future generations, is the topic of the fascinating story that Conti narrates in the book she published roughly a year later, reporting on the decision-making process at the local and regional level. Here I choose to concentrate on one particular aspect of that story, which exemplifies the crucial link existing between working-class history and the history of the environment: the fact that, in explaining *how* the MAC of dioxin in Seveso had been decided, government officials claimed to have relied on "US standards for farm work" (Conti 1977a: 56).

As a labor physician by training, and as a communist representative, Laura Conti could not help but develop an immediate interest in getting as much information as she could concerning the MAC of dioxin in American farming, and she insistently pressed the regional council to reveal the source of their knowledge on the matter. Answers were vague and elusive, referring to a book on which someone had orally reported, but whose title and author(s) never materialized. To complicate things, Conti heard from Barry Commoner, who was in Seveso in September following the disaster, that no such standards existed in the United States. In any case, and whatever the source, the scientific information to which government officials referred appeared reasonably dubious to Conti. First, she observed, why the need to establish a maximum concentration of dioxin in the soil—a volatile standard, difficult to measure, and subject to local variations—being much easier to do it for the pesticide? Second, a document released by NATO officials in Italy advised a MAC of 50 micrograms per acre, that is, a much lower dose than that established by the Lombardia regional government on the basis of "US farm-work standards." Why should the American military authorities suggest standards so different from those accepted for farm workers in their own homeland? Conti asked. It soon became clear that the "standards" were nothing more than a pseudoscientific justification for decisions made in obedience to political considerations and organizational issues: in particular, the decision to circumscribe a "zone B," from which evacuation was not necessary.

The American farm-work standard, however, was soon appropriated by Italian labor physicians, who reinterpreted it as a starting point for further negotiations: having known that the techniques for measuring dioxin in the soil had improved up to the point of being able to detect 1 part per 70 billion, they obtained that the MAC within workplaces be lowered to 0.75 ppm for the ground and to 0.01 micrograms per square meter for indoor walls and equipment. "Good job!" Conti (1977a: 61) commented, "Now, we must extend that to the whole population…" She took on the work of the Medicine and Epidemiology Commission of the Lombardia regional council to advance the idea that, on the day on which cleanup of the area would start, workers' MACs become the general accepted standard for backyards, roads, public parks, playgrounds, and all open spaces, especially those frequented by children, as well as for indoor spaces, public and private.

Conti's connections with Italian "militant" medicine were instrumental for her understanding of dioxin contamination and for her political activity. Colleagues of the "communist cell" within the Istituto Superiore di Sanità—the country's higher scientific authority for public health—informed Conti that the official MACs adopted by the regional government, advised by two academic toxicologists, were based on incorrect calculations. From the scientist Nora Frontali, who directed the industrial hygiene lab of the same institute, Conti obtained precious information about the MAC of dioxin in humans. Those values were incomparably lower that those accepted in Seveso: in fact, they were counted in picograms, a measurement that is one-millionth of a microgram. However, "militant," and woman-led, science was not granted the authority of official medicine: the report that Dr. Frontali and her team had sent to the Lombardia regional government in March of 1977 had been ignored, with the pretext that it was not an official document and it only represented the opinion of one group of scientists.

In relying on occupational medicine to establish a safety standard for the whole population, Laura Conti was applying an approach quite common to environmental health science, which had developed internationally since the times of Alice Hamilton (Sellers 1997). But she was doing it with a particular emphasis: that of a "militant" scientist, committed to the working-class political cause and to the articulation of a working-class ecology. In other words, she was also applying a Gramscian vision of the hegemony of the working class over Italian society and following the Communist Party's strategic view of "progressive democracy," that is, the coincidence between working-class interests and needs and those of the nation. Conti's crucial contribution to the development of a new environmental consciousness in Italy was the clear perception of how working-class needs and interests crucially included environmental health.[6]

Born in Udine in 1921, Laura Conti had actively participated in the anti-Nazis resistance and, at age 23, was interned in a camp near Bolzano. That experience inspired her first novel, *La condizione sperimentale* (Conti 1965), and alimented a writing vocation that she cultivated throughout her life.[7] After the war she graduated in medicine and started working as a traumatologist at the Workers' Compensation Authority and as children orthopedist in the public schools of the Milan district. At the same time, she enrolled in the Italian communist party (PCI), where she started her long political career. She was an elected councilor of the Milan district between 1960 and 1970, then of the Lombardia regional government between 1970 and 1980, and a deputy in the national parliament from 1987 to 1992, where she worked at the Agriculture Commission.

During all her public life, Conti was, at the same time, a politician and an engaged scientist. Not having a family, she devoted most of her uncommon energy endowment to her two main interests: (1) the popularization of ecology as a science of political and social relevance and (2) the inclusion of citizens and ordinary people in scientific decision making, especially as regarded public and environmental health. Probably the most significant example of her commitment to social inclusion is her direct involvement into the post-crisis management of the Seveso disaster. Conti's action/research investigation into the politics of industrial hazard in Seveso was a result of the reflections and experimentations conducted within the Italian labor environmentalism in the previous decade; nevertheless, her own reflections also constituted the beginning of a new ecological consciousness, reaching out from the factory into the larger web of the country's ecological relationships and political-economy scenario.

In the very same year of the accident, Conti was completing her first ecology book, which was to become a seminal reading in Italian environmentalism: with the title *Che cos'è l'ecologia. Capitale, lavoro e ambiente* (Conti 1977b), the book represented a first comprehensive account of relationships between ecology and politics in Italy. From the first page, the author posits organic chemistry and CMR risk at the center stage of her clear, vivid explanation of what ecology is. The book started with the image of a petrochemical plant, which—during the production of artificial fibers—released polluting substances that damaged the health of workers first, then of nearby residents. This first level of ecological relations, from the factory to the body through work, was then intrinsically connected to a broader level, that of bio-geo-chemical cycles: from the factory to the living and nonliving world, and eventually to humans, through water and the food chain. She continued: "As living organisms have similar physiology and biochemistry features the polluting substances produced in the making of artificial fibers enter the watercourses which irrigate pastures, damaging livestock that feeds on those pastures; when gathering into a river they damage fish, and in so doing they eventually damage a source of proteins indispensable to man" (Conti 1977b: 7). The third level of Conti's vision of ecology was the one concerning the limitedness of resources and the non-renewability of mineral matter—the entropy vision. Once consumed in the production of petrochemicals, oil was not available anymore for other human needs; furthermore, the increasing replacement of cotton, linen, flax, and mulberry with artificial fibers would eventually lead to a significant reduction of biodiversity and the loss of age-old human abilities to cultivate and process natural fibers.

After this brief introduction, the author structured her explanation of ecology into four chapters: (1) water, (2) the cycle of matter and the flow of energy, (3) agriculture, food, and population, and 4) ecology and politics. CMR substances and organic chemistry were core topics throughout the chapters. Organic chemistry was vividly described as the science that—like nature itself (which Conti called "life")—could link carbon, hydrogen, and oxygen into an infinite variety of different structures. Unlike nature, the author pointed out, organic chemistry produces totally new molecules without producing enzymes that can degrade them; thus, these new molecules can be unnaturally stable. Conti insisted this was a fundamental break with the laws of evolution: if only one molecule existed that could escape degradation, the world today would be full of it; similarly, the human body functions on the equilibrium between hormones and enzymes. Organic chemistry, in sum, acted as an endocrine disruptor in the environment just as in the human body (Conti 1977b: 32–39).

The major successes of organic chemistry, Conti remarked, were also its greatest hazards. Among those, chlorinated hydrocarbons took the lead: PCB, PVC, and DDT were all highly toxic for humans. One of them, trichlorophenol, when brought to high temperatures released another chlorinated hydrocarbon, dioxin. Toxic substances, Conti explained, acted on the organism according to quantities, and their effect varied from molecule to molecule and also according to the age and general health condition of the organism. Mutagenic substances were a different matter, for there was no threshold under which contact may be innocuous.

That said, the point in Conti's book was to understand by which political system CMR substances were allowed to make their way into human and environmental health. To do so, she chose DDT to exemplify "how the mechanism of profit exploits the mechanisms of nature" (Conti 1977b: 39). The paradoxical aspect of DDT, Conti explained, was how, by killing birds who ate great quantities of poisoned insects, it had indirectly caused the increase of the number of insects themselves. In the meanwhile, insects easily developed resistance to DDT (but not birds, which were far more complex organisms). While ecological reasoning would suggest stopping this vicious circle and restore that of natural predation, the existing structure of political-economic opportunities in capitalist countries encouraged chemical industries to invest in the marketing of newer and newer poisons. In short, by eliminating birds, industry created a virtually endless market for insecticides. In this way capitalism made profits out of the manipulation/destruction of life.

Things being this way, chemical industry had already completely pervaded agriculture, a problem dramatically felt in Italy, where DDT content in human tissues, Conti reported, was 20 ppm, the highest among industrialized countries (Conti 1977b: 38).[8] The result, was that "Water is poisoned, fish die, frogs have almost disappeared, birds are disappearing, man gets intoxicated, children get mercury in the womb and suck DDT with breast milk. Insects, instead, are thriving, and so is chemical industry" (Conti 1977b: 42). In 1976, to limit the poisoning of Italy's rivers by organophosphates, the parliament had passed a "clean water" bill—the so-called *legge Merli*. Conti was highly disappointed with it, as the law clearly exemplified the paradoxes of the political economy of low doses: while it established a table of maximum concentrations of pollutants in industrial effluents, it did not pose any limit to the quantity of total discharge from each plant. In practice, pollutants had to be diluted, but they could be released into the environment in any amount by an ever-increasing number of plants. Moreover, in order to comply with the limits imposed by the law, industrialists diluted not only the non-filterable pollutants, but also those that were filterable, mixing all effluents in the same drainage. As a consequence, filtering and purification processes would become more costly. A chemical plant near Milan, for example, released yearly 120 kg of mercury mixed with other pollutants, making the purification of its effluents very difficult. The European Community was aware of such paradoxes, Conti observed, and in fact it had adopted the criterion of "quantity of pollutant per unit of product," albeit equally unsatisfactory—for, if industry can produce as much as it wishes, then it can also pollute as much—at least this "polluter payer" principle spurred industrialists to invest in cleaner technologies (Conti 1977b: 43–44).

The *legge Merli* treated the environment as the ultimate, unlimited sink where Italian industry flushed away its poisons. However, Conti remarked, the environment (the sea in this case), did not have its own "environment": it couldn't get rid of toxins. It would become filled with them. By passing a bill on industrial effluents based on the concentration principle, the Italian legislator had acted like a physician who instructs her patient to dilute a bit of salt in each glass of water, without considering that the patient has diabetes—thus drinks a lot—and does not have kidneys (Conti 1977b: 44–45).

Eventually, by the very functioning of natural cycles, poisons would return to society in the form of mercury accumulated in fish, or eutrophication—which caused tourists and swimmers to avoid popular recreational sites along the Adriatic Coast in the summers of 1975 and 1976. An effect of discharging the excess of human and animal waste into surface water, eutrophication was of course exponentially increased by the discharge into runoff waters of chemical fertilizers used in agriculture. As such, Conti considered it an indirect effect of organic chemistry. Moreover, since chemical fertilizers had replaced animal excrement in agriculture, the latter had become "waste" to be discharged into the sea. When agriculture and raising livestock are organically connected and use the same soil, no water pollution occurs, she emphasized; once separated, each becomes a polluting activity (Conti 1977b: 96–101).

Such a complex web of interrelationships between natural and social mechanisms needed a good dose of environmental planning. The book's final chapter, "Ecology and Politics," contained Conti's proposed measures to counteract the environmental crisis that was occurring in the country. Taken as a whole, her proposals made no eco-technocracy; rather, they were based on a philosophical-Marxist view of social relationships as intrinsically and organically ecological. The struggle against those who damage nature, "the life of our and other species," Conti wrote in the conclusion, must have society as a protagonist, and specifically one social class: the one that opposes capital, that is, the working class. In defending not only its own interests, but those of humanity itself as belonging to the sphere of nature, the working class would find substantial solidarities and coalitions in society—or at least so Laura Conti believed.

As this overview of the book reveals, Conti's ecology was profoundly human-centered. At the core of all ecological relations lay the manipulation of nature by human work and the human body. The human body was also a recurrent metaphor through which the author—a medical doctor by training—evoked and explained the environment itself in physiological terms. Focusing on CMR risk, but also enlarging the view to society, Conti's ecology was very similar to that of another woman scientist who had convincingly argued that petrochemicals posed a terrible menace to all living creatures including humans: the American biologist Rachel Carson. Unlike Carson, however, Conti was also a politician. Her idea of ecology must be linked to her political militancy as a communist. As her numerous publications testify, her engagement on environmental issues was never disentangled from her political engagement, the two linked in a unique vision of the relationships between society and nature that might be described as radical, or political, ecology (Merchant 2005). In fact, Conti explained, the science of ecology was much broader than the three levels laid out in her first chapter. It was the science of interrelationships among all living and nonliving matter, independent of human interactions. Only part of this vast science was relevant to economic activities, and thus to political choices. Preserving environmental and human health from toxic contamination, saving water not only for industry and agriculture, but for recreation and enjoyment as well, and conserving nature for future generations were matters concerning the sphere of political action. Politics was, to Conti, the realm of "will," counterbalancing the impersonal "mechanism" of economic laws. "A blind mechanism is all is needed to degrade the environment," she concluded. "In order to rebuild it, will is needed. A will based on science and finding expression into well coordinated political action" (Conti 1977b: 10).

As Laura Centemeri (2006: 120) remarks, the Seveso experience added to Conti's vision of ecology a sense of the role of culture and symbolic meaning into the shaping of human-nature relationships: places and people's connection to them must find their way into the science of ecology. Such a vision was probably what led Conti to join the effort that others were making in those same years to build a new environmental movement in Italy. In 1979 she participated in the creation of the *Lega per l'Ambiente*, today a highly established environmental organization; born as a subsection of the Communist Party's cultural/recreational activities, the organization was mainly concerned with the problems originating from industrialization—from energy to pollution and food contamination, from the impact of automobiles to waste management (Della Seta 2000: 46). The novelty of this organization, in respect to other preceding experiences of Italian conservationism, was its being a "popular" environmentalism, initially much connected to the politics of the Left. Conti was not the only militant scientist to participate in the making of this new organization: she was joined by the chemist and communist deputy Giorgio Nebbia, the urban ecologist Virginio Bettini and the public prosecutor Gianfranco Amendola (both of whom later become Green deputies), and the American biologist Barry Commoner, who played a key role in the formation of an environmental consciousness in the Italian Left. Probably the most authoritative among the founders of Legambiente (also for generational reasons), Laura Conti was also the "organic" intellectual of the movement. Her numerous publications, and especially *Il dominio della materia* (Conti 1973) and *Questo planeta* (Conti 1983), were the basic readings of a generation of Italian environmentalists. With a series of articles published in *l'Unitá* and *Rinascita* (respectively, the newspaper and cultural magazine of the Communist Party),

Conti articulated the environmentalist reasons against nuclear energy and for a stricter regulation of game hunting, as well as those for sexual education in schools, for public health reform, for the pro-abortion law. Various prizes, a number of Legambiente's territorial sections, a laboratory of environmental education of the University of Milan, and a school of environmental journalism are now dedicated to her. Her personal papers are conserved at the Fondazione Micheletti in Brescia.

"Class" vs. "Power": A Tale of Two Ecologies

In delineating her political ecology vision, Conti's sources of inspiration were Marx and Engels themselves, but also a few seminal works published in those same years.[9] In fact, Conti was not alone in her search for ecological Marxism: in the fall of 1971, at its yearly cadres' school in Frattocchie, the Italian Communist Party had held its first national meeting on the theme "Man, nature, society." Opening the conference, physician and party executive Giovanni Berlinguer admitted the need to update Marxist orthodoxy in order to take into account the concept of natural limits; he also highlighted how toxicity had become the existential condition of global capital. Berlinguer, along with other top-ranking cadres and "organic intellectuals," compared ecology to socialist planning and emphasized the need for the party to consider the environment a working-class priority (Luzzi 2009: 100–01; von Hardenberg and Pelizzari 2008). A landmark in the making of an ecological consciousness among a generation of militants, the conference had an enormous symbolic meaning—certainly greater than the sum of its speeches—for it implied the possibility of developing a totally new line of critique of capitalist society, *and* a new kind of environmentalism. In a sense, the whole experience of labor environmentalism in Italy can be considered a product of that meeting, which had encouraged communist activists to link ecology and class struggle. In 1972, one year after Frattocchie, a national conference of the confederation of unions was held in Rimini on the theme "Industry and Health." Many other signals throughout the 1970s testify to both intellectual and activist ferment in linking Marxism and ecology. The publisher Gian Giacomo Feltrinelli, for example (he also being one of the most prominent leftist intellectuals of the period) initiated a book series dedicated to "Medicine and Power," collecting books on health risks in industrial societies. Even more radical was the position of another leftist intellectual, the journalist Dario Paccino, author of *L'imbroglio ecologico* (Paccino 1972) which exposed nature conservation as an elitist concern and put workers' bodies firmly at the center stage of a true environmentalism.

Among Conti's references, there was a collective volume in the philosophy of science called *L'ape e l'archietto. Paradigmi scientifici e materialismo storico,* published by Feltrinelli; edited by the physicist Marcello Cini, and destined to become a landmark contribution to the dialogue between the social and the natural sciences in Italy, the book posited the Marxist critique of science as a search for the imprint of class relationships within the very methods and contents of scientific practice (Cini 1976). Conti commented that a thorough contestation of capitalism's use of science could only come "from that global outlook over the world which is ecology." In fact, *political* ecology, that is, "the study of how social relationships within the human species influence the natural world and other species," seemed to Conti even more relevant as a critique of capitalism itself (Conti 1977b: 135–36).

The most relevant novelties in the field of occupational/environmental health consciousness in Italy, however, had taken place in the couple of years immediately before the Seveso disaster (1974–1976), with the birth of the grassroots organization Medicina Democratica (MD), whose founder and inspirer, Giulio Maccacaro, was also directing the major Italian scientific magazine *Sapere*.[10] MD was destined to have a key role in several judicial inquiries concerning Italian industrial plants in the following decades, including that in Porto Marghera, the biggest petrochemical area in Italy, located in the Venice lagoon. The articles published in *Sapere* during the 1970s—some of which were written by Barry Commoner—testify to the remarkable level of political-ecological consciousness within the country's new generation of militant scientists, and also to the hegemonic capacity that the Italian Left exercised in the realm of scientific culture (if not at the governmental level).[11]

The Seveso experience also inspired another seminal book of the Italian radical ecology, significantly entitled *Ecologia e lotte sociali. Ambiente, popolazione, unquinamento*, also published by Feltrinelli in 1976.[12] Coauthored by Virginio Bettini and Barry Commoner, the book linked environmental hazard to a Marxist analysis of the capitalist economy, highlighting the toxicity of most industrial productions and the need to democratize the management of risk. In his introduction, Bettini theorized a distinction between "power" and "class" ecology: the first was represented by company experts and government agencies, the second by the "popular scientific committees" organized in Seveso by the Communist Party, coalescing working-class people and militant scientists. These committees were an advanced experiment in working-class ecology in the sense that they practiced a participated and emancipatory form of knowledge production (Terracini 1977). Their point of reference was the methodology practiced in those same years by the Servizi di Medicina per gli Ambienti di Lavoro (SMALs), the Medical Services for the Work Environment, where material evidence and bodily experience of toxicity were actively recorded by the workforce and elaborated with the help of militant experts into officially recognized "science," of practical relevance in the public arena (CGIL-CISL-UIL Federazione Provinciale di Milano 1976, Calavita 1986,

Barca 2012). In Bettini's view of ecology, industrial pollution represented the most compelling and politically relevant aspect—in contrast to those who approached problems of environmental contamination as if these were not borne and paid primarily by subaltern social groups. In his view, "the debt towards nature is a debt towards the working class" (Bettini and Commoner 1976: 6).

It is not clear, however, how much the working class, and even the workers of the ICMESA plant, actively participated in Seveso's "popular scientific committee," or whether this only comprised a number of "militant experts," including university researchers, SMAL personnel, and organic intellectuals.[13] Despite their generous efforts at helping local people to struggle for their rights (and not only for monetary compensation), communist activists in Seveso met with diffidence and even open resistance, which was also significantly related to their pro-abortion stance. A political battle of great significance, the passage of women's right to abortion was being fought over at the national level during those very years by the government and the left oppositions. Seveso became one crucial terrain of that battle, a place of enormous symbolic power—and local people did not like that. Furthermore, there was the issue of evacuation: accepting safer MACs, like those proposed by the Left, would imply that the authorities would revise the zoning, and that the thousands of residents of zone B must leave their homes forever, a price that Sevesians were not willing to pay (Centemeri 2006).

The problem with the strategy of working-class ecology was that, however ideally correct, it met with a dual challenge: it had to overcome political-economic constraints, corporate/governmental resistance, and power-science coalitions, but it also met the inevitable noncompliance of real working-class people, who struggled for things different, and also thought differently, from what was expected. As Laura Conti wrote in an illuminating passage of her *Visto da Seveso*: "People had never been put in the condition to understand that, to have a healthy environment, it is necessary to sacrifice something: everything has always been done to get more salary, more cars, more highways, even—in the best cases—more hospitals and schools, but almost nothing to get cleaner air, cleaner water, safer food. At this point, why expect that all of a sudden the *Brianzoli* recognize that living in a healthy land is worth a mass exodus?"[14] (Conti 1977a: 54).

On this point—an issue of enormous relevance such as the formation of ecological consciousness, and, implicitly, its relationship with class consciousness—Conti's critique was directed against her own party, which had never taken a real stance toward the protection of nature. She found it outraging that only the people of Seveso were stigmatized as "immature" or "stubborn," and concluded, "none of us has the right to criticize the *Brianzoli*" (Conti 1977a: 54).

Conclusions

The chapter has shown the existence of a working-class ecology in the making in 1970s Italy. This radical political ecology was an intellectual project that heavily rested on the organizational support of the Communist Party and was also partially constrained by ideology. It nevertheless introduced into the Italian environmental debate and political scenario a perception of ecology as something having to do with the human body and its situatedness within the configuration of power relationships, both inside the factory and in the local space. Consciousness of the political link between occupational, environmental, and public health was not a philosophical speculation for a few militant scientists; in fact, it was largely shared throughout the Left and in the confederation of unions, and led to a period of intense struggle for the recognition of workers' control over the work environment, eventually leading to the creation of the National Public Health System in 1978.

The conceptual and political link between anti-capitalist struggles on the shop-floor and outside the factory gates also led many to think in terms of working-class ecology: a political project that did not survive the harsh economic recession of the late seventies, nor the contemporary recrudescence of political conflict in the country, including terrorism. Moreover, by the end of the decade, the political-economy scenario began to change: factory work, especially that employed in big high-tech industry, represented less and less of the Italian workforce, while the political and symbolic power of blue-collar workers started to erode and entered an irreversible crisis by the end of the 1980s.

All considered, however, the radical ecology project did have a durable legacy. Numerous anti-toxic struggles, involving more or less grassroots organizations especially at the local level, have concerned petrochemical sites throughout the country in the last fifty years. The time has come perhaps to tell the story of these struggles, tracing their material and ideal connections with each other and with the story of class ecology in Italy.

Notes

1. According to the International Labor Organization (data from 2010), "every year more than 2 million people die from occupational accidents or work-related diseases. By conservative estimates, there are 270 million occupational accidents and 160 million cases of occupational disease." See http://www.ilo.org/global/Themes/Safety_and_Health_at_Work/lang—en/index.htm.
2. This problem was already highlighted by Barry Commoner in his seminal *The Closing Circle* (Commoner 1971), and has increased exponentially since, as most of these chapters clearly show.

3. Commoner visited Italy frequently between 1975 and 1976, giving talks and publishing articles, and established a direct and durable relationship with Italian environmentalists (Commoner 1975).
4. The battle for citizens' right to know remained a constant in Conti's environmentalist activity: in 1979, for example, commenting on the Public Health Reform approved by the Italian parliament, she opposed article 20 of the law, establishing that industries be compelled to disclose to local authorities the list of substances they manipulated, for the article still granted the "protection of industrial secrets" against public disclosure (Conti 1979).
5. Conti's experience in Seveso also inspired her to write the novel *Una lepre con la faccia da bambina* (*A Hare With the Face of a Girl*, 1978). For an eco-critical reading of that novel, see Iovino, forthcoming. A series of annotations of a more technical and legislative nature are now conserved at the Fondazione Micheletti, Brescia: Fondo Laura Conti. See http://www.fondazionemicheletti.it/public/Scheda_Fondo_Conti.pdf.
6. Such perception was also present in the experience of U.S. environmentalism of the 1960s and 1970s (Gottlieb 1993; Rome 2003; Montrie 2008: 106–12).
7. See: http://scienzaa2voci.unibo.it/scheda.asp?scheda_id=914.
8. That was an average value: in some areas of intensive monocrop cultivation, like the highly mechanized Po Plain, values reached 40 ppm. The average was 11 ppm in the United States, 10 in Israel, and only 2 in the United Kingdom; by contrast, it was 31 ppm in India. This pattern seems to follow the relevance of agriculture in each national context. It is not clear what Conti's source was for these data, but likely enough it was Rachel Carson's *Silent Spring* (New York 1962); thus, they may have been fifteen years old.
9. She quoted Marx's passage in *Capital* on capitalist production as a fundamental break in social metabolism, and Engels's remarks on nature's revenge on human domination in the *Dialectics of Nature*. But she also abundantly relied on Barry Commoner's work and on a collective volume on socialism and the environment published by Feltrinelli a couple of years before.
10. Maccacaro, who died prematurely in January 1977, is considered a father of biometrics in Italy, and was a founder of *Epidemiologia e Prevenzione*, the most important Italian epidemiology journal.
11. The list of articles published by the magazine on the topic of industrial hazards would be long: some examples are articles on the Minamata disaster (K. Myamoto, "Il progresso avvelenato," April 1976, 2–12), on titanium dioxide and the contamination of the Tuscan coast with "red mud" (Gruppo Prevenzione Montedison di Castellanza, "Eliminazione dei fanghi rossi," July–August 1978, 45–46); on air pollution in the petrochemical site of Porto Marghera (G. Mastrangelo and G. Moriani, "Porto Marghera: per la salute contro l'inquinamento," July 1976, 14–17); on asbestos hazard in Trieste (P.M. Biava et al., "Cancro da lavoro a Trieste: il mesotelioma della pleura," August 1976, 41–45); on industrial pollution in the Po Plain (S. Bernardi, F. Mandelli, and L. Mussio, "Inversione termica e nocività ambientale," August 1976, 36–40); and on PCBs (A. Fraser, "I PCB, un'altra Seveso?" December 1977, 29–34). A special issue was entirely devoted to the accident in Seveso ("Seveso, un crimine di pace," November–December 1976), plus various other articles in the following years (for example, the forum "Seveso due anni dopo," July–August 1978, 2–27).

12. The book also included the text of a number of lectures that Commoner had given at the Istituto Superiore di Sanità in 1976 (Bettini and Commoner 1976: 5–6).
13. This is the impression given by the list of members reported by Bettini on page 8 (Bettini and Commoner 1976: 8).
14. See Conti (1977a: 54). *Brianzoli* is the term defining the people of Brianza, a sub-area of Lombardy of which Seveso is part.

Bibliography

Adorno, Salvatore, and Simone Neri Serneri, eds. 2009. *Industria, ambiente e territorio. Per una storia ambientale delle aree industriali in Italia*. Bologna: Il Mulino

Allen, Barbara L. 2004. *Uneasy Alchemy: Citizens and Experts in Lousiana's Chemical Corridor Disputes*. Cambridge, M.A.: MIT Press

Barca, Stefania, ed. 2005. "Massimo Menegozzo: salute e lavoro in Italia." *I Frutti di Demetra* 5: 63–70.

Barca, Stefania. 2006. "Health, Labor and Social Justice: The Environmental Costs of Italian Economic Growth." Paper presented at the Agrarian Studies Colloquium: http://www.ces.uc.pt/myces/UserFiles/livros/271_stefania_barca_yale.pdf..

———. 2012. "Bread and Poison: The Story of Labor Environmentalism in Italy." In *Dangerous Trade: Histories of Industrial Hazards Across a Globalizing World*, ed. Christopher Sellers and Joseph Melling, 126–39. Philadelphia: Temple University Press.

Bartrip, Peter W.J. 2001. *The Way from Dusty Death: Turner and Newall and the Regulation of Occupational Health in the British Asbestos Industry, 1890-1970*. London and New York: The Athlone Press.

Berlinguer, Giovanni. 1991. *Storia e politica della salute*. Milano: Franco Angeli.

Bettini, Virginio, and Barry Commoner. 1976. *Ecologia e lotte sociali. Ambiente, popolazione, inquinamento*. Milano: Feltrinelli.

Calavita, Kitty. 1986. "Worker Safety, Law and Social Change: The Italian Case." *Law & Society Review* 2: 189–228.

Carson, Rachel. 1962. *Silent Spring*. Boston: Houghton Mifflin.

Carnevale, Francesco, and Davide Baldasseroni. 1999. *Mal da lavoro. Storia della salute dei lavoratori*. Roma-Bari: Laterza.

Centemeri, Laura. 2006. *Ritorno a Seveso. Il danno ambientale, il suo riconoscimento, la sua riparazione*. Milano: Mondadori.

———. 2010. "The Seveso Disaster's Legacy." In *Nature and History in Modern Italy*, ed. Marco Armiero and Marcus Hall, 251–73. Athens, O.H.: Ohio University Press & Swallow Press.

CGIL-CISL-UIL Federazione Provinciale di Milano. 1976. *Salute e Ambiente di Lavoro. L'esperienza degli SMAL*. Milano: Mazzotta.

Cini, Marcello, ed. 1976. *L'ape e l'architetto. Paradigmi scientifici e materialismo storico*. Milano: Feltrinelli.

Commoner, Barry. 1971. *The Closing Circle: Nature, Man and Technology*. New York: Alfred Knopf.

———. 1975. "Le fabbriche del veleno." *Sapere* (April–May): 3–15.

Conti, Laura. 1965. *La condizione sperimentale*. Milano: Mondadori.

———. 1973. *Il dominio sulla materia*. Milano: Mondadori.

———. 1977a. *Visto da Seveso. L'evento straordinario e l'ordinaria amministrazione*. Milano: Feltrinelli.
———. 1977b. *Che cos'è l'ecologia. Capitale, lavoro, ambiente*. Milano: Mazzotta.
———. 1979. "Il segreto industriale." *Sapere* (January): 58–59.
———. 1983. *Questo pianeta*. Roma: Editori Riuniti.
Crepas, Nicola. 1998. "Industria." In *Guida all'Italia contemporanea, 1861-1997*, ed. Marcello Firpo, Nicola Tranfaglia, and Pier Giorgio Zunino. Milano: Garzanti.
Della Seta, Roberto. 2000. *La difesa dell'ambiente in Italia*. Milano: Franco Angeli.
De Luna, Giovanni. 2008. *Le ragioni di un decennio. 1969-1979. Militanza, violenza, sconfitta, memoria*. Milano: Feltrinelli.
Di Luzio, Giulio. 2003. *I Fantasmi dell'Enichem*. Milano: Baldini Castoldi Dalai.
Elling, Ray H. 1986. *The Struggle for Workers' Health: A Study of Six Industrialized Countries*. New York: Baywood.
Gottlieb, Robert. 1993. *Forcing the Spring: The Transformation of the American Environmental Movement*. Washington: Island Press.
Iovino, Serenella. Forthcoming. "Toxic Epiphanies: Dioxin, Power and Gendered Bodies in Laura Conti's Narratives on Seveso." In *International Perspectives in Feminist Ecocriticism*, ed. Greta Gaard, Simon Estok, and Serpil Oppermann. London: Routledge.
Johnston, Ronald, and Arthur McIvor. 2000. *Lethal Work: A History of the Asbestos Tragedy in Scotland*. Tuckwell: The Mill House 2000.
Luzzi, Saverio. 2009. *Il virus del benessere. Ambiente, salute e sviluppo nell'Italia repubblicana*. Roma-Bari: Laterza.
Markowitz, Gerald, and David Rosner. 2002. *Deceit and Denial: The Deadly Politics of Industrial Pollution*. Berkeley: University of California Press.
Massard-Guilbaud, Genevieve, and Dietrich Scott, eds. 2002. *Le Démon Moderne. La Pollution dans les Sociétés Urbaines et Industrielles d'Europe (The Modern Demon. Pollution in Urban and Industrial European Societies)*. Clermont Ferrand: Presses Universitaires Blaise-Pascal.
McEvoy, Arthur. 1995. "*Working Environments: An Ecological Approach to Industrial Health and Safety*." *Technology and Culture* 36: S145–73.
Meisner Rosen, Christine, and Christopher Sellers. 1999. "The Nature of the Firm: Towards an Ecocultural History of Business." *Business History Review* 73: 577–600.
Merchant, Carolyn. 2005. *Radical Ecology: The Search for a Livable World*. London: Routledge.
Montrie, Chad. 2008. *Making a Living: Work and Environment in the United States*. Chapel Hill: University of North Carolina Press.
Musso, Stefano. 1998. "Lavoro e occupazione." In *Guida all'Italia contemporanea, 1861-1997*, ed. Marcello Firpo, Nicola Tranfaglia, and Pier Giorgio Zunino. Milano: Garzanti.
Paccino, Dario. 1972. *L'imbroglio ecologico*. Torino: Einaudi.
Platt, Harold. 2005. *Shock Cities: The Environmental Transformation and Reform of Manchester and Chicago*, Chicago: University of Chicago Press.
Rome, Adam. 2003. "'Give Earth a Chance': The Environmental Movement and the Sixties." *Journal of American History* (September): 525–54.
Rosner, David, and Gerald Markowitz. 1986. *Dying for Work: Workers' Safety and Health in Twentieth-Century America*. Bloomington: Indiana University Press.

Ruzzenenti, Marino. 2001. *Un secolo di cloro e... PCB. Storia delle industria Caffaro di Brescia*. Milano: Jaca Book.

Santiago, Myrna. 2006. *The Ecology of Oil: Environment, Labor, and the Mexican Revolution, 1900-1938*. Cambridge and New York: Cambridge University Press.

Sellers, Christopher. 1997. *Hazards of the Job: From Industrial Hygiene to Environmental Health Science*. Chapel Hill and London: University of North Carolina Press.

Signorelli, Amalia. 1995. "Movimenti di popolazione e trasformazioni culturali." In *Storia dell'Italia repubblicana*, vol. 4. *La trasformazione dell'Italia: sviluppo e squilibri*. Torino: Einaudi.

Terracini, Benedetto. 1977. "Le risposte non date alle popolazioni colpite sulle conseguenze del TCDD." *Sapere* (October–November): 36–43.

Tomaiuolo, Francesco. 2006. "1976-2006: Trent'anni di arsenico all'Enichem di Manfredonia." *I Frutti di Demetra* 12: 33–41.

Tonelli, Fabrizio. 2006. "Salute e lavoro." In *Il '900: alcune istruzioni per l'uso*, ed. Luigi Falossi. Firenze: La Giuntina.

———. 2007. "La salute non si vende. Ambiente di lavoro e lotte di fabbrica tra anni '60 e '70." In *I due bienni rossi del '900. 1919-1920 e 1968-1969. Studi e interpretazioni a confronto*, ed. Luigi Falossi and Fabrizio Loreto. Roma: Ediesse.

von Hardenberg, Wilko Graf and Paolo Pelizzari. 2008. "The Environmental Question, Employment and Development in Italy's left." *Left History* 2: 77–104.

Ziglioli, Bruno. 2010. *La mina vagante. Il disastro di Seveso e la solidarietà nazionale*. Milano: Franco Angeli.

 CHAPTER 6

What Kind of Knowledge is Needed about Toxicant-Related Health Issues?
Some Lessons Drawn from the Seveso Dioxin Case

Laura Centemeri

Dioxins, a class of chemical contaminants produced in both natural and industrial processes, were discovered in the late 1950s and have been extensively studied since the early 1970s. The majority of studies have focused on the most toxic congener, 2,3,7,8-TCDD, simply called dioxin,[1] with much toxicology, biochemistry, and epidemiology research having been aimed at determining its effects on humans, in particular its carcinogenic effects. Nevertheless, despite thirty years of intensive research, exactly how dangerous dioxin is remains a controversial issue. In 1997 the International Agency for Research on Cancer (IARC) classified TCDD as a group 1 carcinogen based on limited evidence on humans, sufficient evidence on animals, and extensive mechanistic information. This classification has stirred controversy, in particular concerning the use of mechanistic data to interpret cancer risk in humans (Cole et al. 2003; Steenland et al. 2004). In 2009, IARC confirmed the inclusion of TCDD in group 1, citing sufficient epidemiological evidence for all cancers combined (Baan et al. 2009).

All the direct evidence on acute dioxin effects on human health comes from epidemiological studies of human populations exposed accidentally or occupationally to elevated dioxin levels. One of the cases most studied in the dioxin carcinogenicity literature concerns the population living in the area of Seveso, Italy. In 1976 an industrial accident in the chemical factory ICMESA (owned by the Swiss multinational corporation Roche) exposed the residents of the surrounding area—in particular the inhabitants of Seveso, Meda, Cesano Maderno, and Desio—to the highest exposure to TCDD known to have occurred in humans (Eskenazi et al. 2001). To quote an epidemiologist involved in the follow-up studies investigating the health consequences for the population affected: "The accident was a tragedy, for sure, but for us scientists, I must admit, it has been a rare chance to have a sort of laboratory situation, so to explore how dioxin works on human beings."[2] This chapter focuses on the paradox of this "laboratory population" that is playing such a crucial role in the controversy concerning dioxin carcinogenicity.

The paradox is as follows: the vast scientific output concerning dioxin effects in Seveso is having no impact in terms of public health measures implemented in the area affected, in particular as far as prevention is concerned. This scientific output is oriented exclusively around the problems and discussions that have emerged over the uncertainties surrounding dioxin toxicity and the problem of its regulation.

At the same time, the population affected has not engaged in collective action to seek full disclosure of the impact of dioxin contamination or of epidemiological studies concerned with local environmental health and prevention. In the ICMESA disaster area, environmental health—and most specifically long-term dioxin health effects—are not questions of public concern and mobilization, but they are rather the source mostly of "rumors" circulating in the community, or of the personal troubles of those directly touched by diseases that might be linked to dioxin exposure. Using a dichotomy introduced by Charles Wright Mills (1959), dioxin in Seveso is not a public "issue" but a matter of personal "troubles."[3]

In this paper, I discuss how this double framing of dioxin's long-term health effects—either as a pure scientific problem or as purely personal troubles—has emerged. The hypothesis I advance is that this double framing has affected the kind of scientific knowledge produced on local effects of dioxin contamination. Moreover, the Seveso case shows how the global regulation of toxicants relies on a very specific kind of knowledge, focused on the issue of carcinogenicity and employing a mono-causal explicative model of the onset of cancers. There is a gap between this kind of knowledge output meant for regulation and the knowledge relevant for the implementation of local prevention policies to assure environmental health.

In order to develop this argument, I will first analyze the responses to the Seveso disaster, in particular the choices made by the public authorities (at the regional and national levels) to manage the crisis. The role of the public authorities in responding to the dioxin contamination emerges as crucial when trying to give an account of the specific way in which dioxin was interpreted as a collective threat by the population affected. I will then focus on the local pressure groups and the conflicting interpretations of dioxin risk they supported. Through investigating the dynamics of these organizations and their interplay with the public authorities' crisis management approach, I show how the dioxin risk was never framed as a problem of public health in the area affected. I also show how an interpretation of dioxin as a cultural threat came to prevail among the population exposed to the contamination. This prevailing interpretation of dioxin seems to have acted as an obstacle against any popular movement toward asking the public authorities to respond to environmental health problems in the contaminated area. These problems are related not only to the dioxin contamination but also more generally to the chronic chemical pollution caused by chemical factories located in the area since the 1950s.

In the final section, I discuss the lessons from the Seveso case about the construction of environmental risks as public problems, and in particular the role played by participation, which is to be understood here as meaning dialogue among scientists, citizens, activists, and public authorities. This dialogue seems to be necessary to prompt scientific research to address health issues not in terms of individual problems but as part of the condition of local populations. This dialogue is also necessary to ensure that issues of uncertainty, which are ubiquitous in the study of toxicants, are not dealt with only within restricted circles of scientists but in public arenas in which priorities—for research and action—can be defined in a more democratic way, that is, in a more inclusive plural way. When this dialogue fails, research on environmental health issues becomes more easily detachable from its geographical dimension, and tends to focus exclusively on a laboratory approach that only partially responds to public health concerns. Moreover, this approach obscures under a veil of objectivity the political dimensions of making regulatory choices in situations involving uncertain scientific knowledge.

Seveso: The Disaster and the Response of the Public Authorities

It is always difficult to give a concise description of a disaster and its consequences when addressing the problem from a sociological point of view. The official "toll of the tragedy" is often the object of endless controversy and, moreover, it tells nothing of the long-term impact of an event on the community affected. A variety of sociopolitical processes, including framing processes, shape disasters, making them generative of social change. These processes take place at different times and in different, but intertwined, public arenas: local and global political arenas, and expert arenas, in particular legal and scientific ones (Jasanoff 1994).

The main feature of the Seveso disaster is that it was the first major accident in the chemical industry at the European level. It contributed to the definition of the European directive (Directive 82/501/EEC, the "Seveso Directive") on the major-accident hazards of certain industrial activities (De Marchi, Funtowicz, and Ravetz 1996; De Marchi 1997; De Marchi, Funtowicz, and Guimares Pereira 2001). Another important feature is that "there were no fatalities following the accident," as Stavros Dimas, European Commissioner for the Environment, stated when commemorating its thirtieth anniversary in 2006.[4] In fact, at the European level, the Seveso disaster is considered an "information disaster" (van Eijndhoven 1994). It helped to highlight the fact that a lack of information about hazardous industrial processes is a major source of vulnerability in our highly industrialized societies.

To quote Dimas again: "The reason for this particular accident becoming such a symbol is because it exposed the serious flaws in the response to industrial accidents." The absence of "fatalities," connected to the recovery of the contaminated area (Ramondetta and Repossi 1998), also explains why ecoskeptic books often cite the case of Seveso as an example of "unjustified alarmism" (Kohler 2002).

This emphasis on the event itself and its consequences has completely concealed the reality of a community exposed to chemical pollution since 1945. This reality has never been seriously investigated in terms of its human and environmental costs. The harmfulness of ICMESA, even though known to the local community, became a public concern only with the accident of July 1976 and merely in terms of the specific consequences of dioxin contamination. Moreover, although there has not been the health catastrophe expected by some back in 1976, dioxin contamination has affected people's health with various degrees of gravity. A 25-year follow-up study of mortality in the population exposed shows excesses of lymphatic and hematopoietic tissue neoplasms, diabetes mellitus, and chronic obstructive pulmonary disease (Consonni et al. 2008). A more recent study examining the relation of serum TCDD with cancer incidence in 981 women from the most contaminated areas—and part of the wider project Seveso Women's Health Study run by researchers of the School of Public Health, University of California, Berkeley—shows a significantly positive all-cancer incidence in this cohort, thirty years after the accident (Werner et al. 2011).

In spite of the disaster, its direct effects, and the reality of previous chronic pollution that it brought to light, the issue of environmental health has never been a cause for public concern or activism in the communities affected, thus contributing to the absence of this issue in local public debate. At the same time, the Seveso case has been extensively studied by epidemiologists within the frame of research on the toxic effects of dioxin on human beings. This scientific output has had no impact on the area directly concerned, either in terms of local public health policies or victims' mobilization. How can this paradox be explained?

In order to develop our analysis, we first need to introduce some context. Seveso is a town of twenty thousand inhabitants, located 15 km north of Milan, the regional capital of Lombardy, in the geographical area known as *Brianza Milanese*. *Brianza* is a "district area" (Bagnasco 1977) with a strong Catholic cultural tradition, specializing in furniture production and design by small, family-owned firms. After World War II, chemical companies began to install plants in the area because of the rich water resources and the good transport infrastructure. Thus, two different models of production organization and integration came to coexist in the area. The accident at the origin of the Seveso disaster occurred in the ICMESA chemical plant (located at Meda, near Seveso), which had 170 workers and had been owned since 1963 by the Swiss multinational corporation Roche through its subsidiary Givaudan. It produced intermediate compounds for the cosmetics and pharmaceutical industry among which, since 1969 and more intensively in the 1970s, was 2,4,5-trichlorophenol (TCP), an inflammable toxic compound used for the chemical synthesis of herbicides.[5]

On Saturday, 10 July 1976 at around 12:30 A.M., the reactor where trichlorophenol was produced released a toxic cloud of dioxin and other pollutants because of a sudden exothermic reaction caused by the breakdown of a safety valve.[6] The hazardous gas produced by the twenty-minute emission settled on a large area of about 1,810 hectares in the municipalities of Seveso, Meda, Desio, Cesano Maderno, and, although to a less serious extent, seven other municipalities in the province of Milan.

The 2,3,7,8-tetrachlorodibenzo-p-dioxin (TCDD), simply called dioxin that was released by the ICMESA reactor[7] is an extremely dangerous molecule due to its very high toxicity, persistence, and stability. Nevertheless, dioxin was little known at the time of the accident.

In 1976, knowledge of the extremely harmful effects of dioxin on human health was mostly based on suppositions resulting from toxicological evidence. There had been few epidemiological studies and they had been limited to following up on cohorts of industrial workers. Dioxin environmental contamination affecting an entire population was without precedent. Scientists were unable to anticipate the damage to be expected (to the environment, animals, or human beings of varying sex and age) and neither were they able to provide decontamination methods. Besides, there were no technical instruments to measure human blood dioxin levels (Mocarelli 2001). The result was a "radical uncertainty" (Callon, Lascoumes, and Barthe 2001) surrounding the consequences of the contamination to be expected for human health and the environment, and their extent in both space and time. There was just one certainty: the extreme toxicity of dioxin proven in laboratory tests. This led to fears of catastrophic scenarios.

These catastrophic scenarios, however, did not materialize immediately after the accident. The toxic cloud passed largely unnoticed, with the Seveso and Meda people considering it just a "usual" nuisance, one in a long series. A "week of silence" (Fratter 2006) passed, but in the meantime various alarming phenomena were noticed in the area near ICMESA: a sudden fall of leaves; the deaths of small animals (birds and cats); and a "mysterious" skin disease (chloracne) affecting children. Anxiety grew among the population. On 19 July, Roche experts informed the Italian public authorities that the accident at the ICMESA plant had caused widespread dioxin contamination and highly recommended the evacuation of part of Seveso's and Meda's populations as a precautionary measure.

On 24 July the evacuation began: 736 inhabitants of Seveso and Meda were forced to leave their houses with all their personal belongings inside. Two hundred people never returned to their houses, which were demolished during the decontamination. "Risk zones"[8] were created, officially on the basis of the estimated trajectory of the toxic cloud and of random dioxin concentration tests on the ground. In fact, the criteria adopted to delimit risk zones also included practical feasibility and the reduction of the negative social side effects that were to be expected in the case of massive displacements.

The design of the risk zones implied a delimitation of the area officially considered "at risk." Faced with widespread contamination probably affecting a large and difficult-to-define area, the public authorities tried to reduce the "risk" area to the minimum. This reduction of the crisis area had the effect of producing an overlap between the district of Seveso—and its population—and the area at risk. Of the municipalities affected, it was Seveso that became the only one constantly associated with the crisis, particularly in the media. The association of the name Seveso with dioxin was considered a form of injustice by its citizens. It appeared to them that the regional and national authorities had decided to sacrifice them in order to reduce the extent of the crisis.

This clear-cut definition of the area at risk was just one of the measures adopted to reduce the uncertainty that the public authorities were confronted with. In fact, the authorities further decided to reduce uncertainty by denying it, by acting "as if" there were none. Another measure was the creation of Technical-Scientific Committees of experts in charge of deciding on the steps to be taken to manage the dioxin health risk, decontamination, and the socioeconomic implications of the crisis. The definition of the problems at stake was delegated entirely to the experts. These committees were in fact taking decisions of a political nature and were therefore not just advisory committees (Centemeri 2006: 87–96).

With the public authorities embracing a "paternalistic stance" (Conti 1977), the citizens—and their political representatives at the municipal level—were not allowed to participate in decision making. Nevertheless, decisions were made that greatly affected them, as persons and as a community. In particular, given the suspected teratogenic effects of dioxin, pregnant women from the contaminated area (within the third month of pregnancy) were "left free" to ask for a medical abortion. Abortion was still illegal in Italy, and in fact the Italian social movements' fight for its depenalization was at its peak.[9] About thirty women from the contaminated area—although the precise number is not known—decided to interrupt their pregnancies (Ferrara 1977).

From Scientific Controversy to Cultural Conflict: Rival Local Interpretations of the Dioxin Crisis

Given the radical uncertainty surrounding the effects of dioxin, it was clear to the citizens that public decisions could not rely on any kind of scientific "truth." In fact, the scientific controversies over dioxin risk were widely discussed in the media.

The scientific uncertainty surrounding the effects of dioxin implied that the decisions taken in response to the crisis were not just technical, but political. Nevertheless, the public authorities insisted on denying uncertainty. No public debate involving the communities affected took place to define the problems to address in response to the contamination or how to address them. Nevertheless, conflicting definitions of the problems involved in responding to the crisis emerged. This happened through the mobilization of the people affected and of national social movements, resulting in contentious public controversies.

In particular, one controversy was centered on the question of whether potential malformation of embryos caused by dioxin should be prevented through abortion. In fact, abortion rapidly became the central issue in the national public debate concerning the dioxin effects in Seveso. In this debate, the Catholic Church, whose influence was very strong locally, opposed left-wing movements. Other controversies concerning the uncertainty of long-term dioxin health effects slipped into the background. The centrality gained by the abortion controversy largely explains the shift of the dioxin risk from being a public health problem to a moral-cultural problem.

Another controversy contributing to this same shift was related to the issue of what should be considered "safe." The public authorities defined safety by starting from the detached standpoint of experts and laboratory science. In this view, safety is the condition of not being exposed to risk and so displacement from the contaminated area was considered the solution guaranteeing the highest level of safety. Local committees of Seveso citizens supported a different definition of safety. They argued for the relevance of a specific risk: that of the Seveso community disappearing as a result of the way the public authorities were responding to the contamination. This response sought to preserve not only individual safety but also the "attachment to the territory" that was shared by the population affected in terms of being a community. This *attaccamento al territorio* (attachment to the territory) refers to the feelings of familiarity with people and spaces held both individually and collectively by the inhabitants. This familiarity is acquired over time, through the everyday experience of living together in a specific place, and is transmitted from one generation to the next.

Attachment to the territory also refers to the fact that place is considered a constituent of the collective and individual identity, at the same time bearing the traces of a specific way of organizing individual and collective life. It thus refers to both active participation by the territory in shaping social life and at the same time the shaping of the territory by the activities of the community inhabiting it (Berque 2000; Breviglieri 2002; Thévenot 2006).

Arguing for the relevance of attachment to the territory as a public good to be preserved while responding to the dioxin crisis, the local Seveso committees found themselves opposing not only the public authorities but also the national social movements mobilized in Seveso to support the victims.

Social movements already active in the Italian political scene on the issue of environmental health, together with left-wing political parties, mobilized in Seveso. They organized a Scientific Technical Popular Committee (STPC) to help those they considered "victims" obtain justice for their suffering. One of the most important actors in this movement was Medicina Democratica (MD).[10] For MD, the Seveso disaster required a large coalition of citizens and workers to impose the issue of environmental health on the political agenda. The concept of environmental health involved health damage caused by industrial production both inside and outside plants. Underlying this, there was a discourse of social criticism of capitalist exploitation (Boltanski and Chiapello 1999). Capitalist exploitation entailed "hidden costs"—"hidden" because of the control exerted over scientific knowledge production by hegemonic forces. MD's struggle was oriented toward democratizing the production of knowledge to make those responsible for the negative consequences of industrial society socially accountable.

The call for widespread mobilization of the people affected by the contamination and their participation in the production of knowledge about dioxin damage found little response from Seveso's population, thus reducing the critical force of MD's public arguments.

This lack of support from the affected people can be explained if we consider that MD interpreted the dioxin contamination in Seveso and its effects in terms of criticizing capitalism. The Seveso disaster was considered a typical "capitalist crime" (Maccacaro 1976). What was happening in Seveso was a clear example of capitalist injustice, which needed to be denounced. The Seveso people were being asked to join the preexistent cause of the workers and their class struggle. There was no place for more local or even personal definitions of the issues at stake in the disaster situation. In this respect, the activists were as incapable as the public authorities of understanding what mattered to the Seveso people in responding to the dioxin crisis.

For a large majority of these people, the priority in responding to the crisis was to maintain their previous way of life, to preserve the specificity of the relationship between their community and their territory—but neither the public authorities nor the left-wing activists were able to take this dimension of attachment to the territory into account.

The scientific uncertainty about dioxin risk implied that no clear evidence was available to support the public authorities' and social movements' interpretation of the dioxin risk. Appealing to this uncertainty, a grassroots mobilization of strong Catholic background took shape and urged the public authorities to consider not only the seriousness of the health risk but also the fear that Seveso as a community might disappear. The collective damage caused by dioxin was thus interpreted as damage to the community. However, the public authorities opened no arenas for public discussion of these issues, and neither did they propose any mediation. This caused the grassroots movement to radicalize its protest. This radical turn became visible in the central role assumed within it by the militants of the Catholic movement Comunione e Liberazione (CL). For CL, the disaster was not a "crime," but a "test" of the community's ability to stick together, and to its values, in responding to the crisis. CL asked the public authorities to recognize a right of the local community to self-organization in its response.[11]

This interpretation of the dioxin damage as a cultural threat to the community and its values contributed to obscure the controversial implications of the contamination: those jeopardizing community cohesion and in particular the long-term health effects. Moreover, the way the Swiss multinational corporation Roche managed compensation to the victims in the immediate aftermath of the disaster also contributed to downgrading the public health consequences to the level of personal problems. The compensation issue was dealt with through instruments of private settlement in the form of individual contracts agreed on between victims having suffered material losses and the multinational corporation. No public discussion took place on how to compensate for the negative consequences of the disaster that were to be expected in the future.[12]

Long-Term Dioxin Health Effects in Seveso: A Scientific Problem, an Invisible Public Health Issue

The interpretation of dioxin as a threat to the community instead of a public health problem led to a situation in which scientists alone were left in charge of exploring and assessing the health consequences of the contamination.

The design of the research on dioxin health effects was heavily influenced by laboratory science and by the controversies surrounding the carcinogenic effects of dioxin. There was no involvement of the population affected in terms of participation in the production of knowledge. Furthermore, the people affected never asked to be directly involved in the design of this scientific research.

As Wynne (1996: 52) remarks, an absence of criticism of expert knowledge does not automatically equal trust. The relationship between lay people and experts is in fact ambivalent: dependency and lack of agency might both explain lack of voice. In the case of Seveso, the dioxin damage was interpreted by the population affected as a cultural threat affecting a community. This is an important dimension that should be taken into account to understand why dioxin never became a public health issue in the area affected. At the same time, the case of Seveso tells us much about the specific kind of knowledge that is assumed to be relevant in the debate over toxicants and carcinogenesis, and is consequently funded and supported by research institutions and public agencies that are looking for solid evidence to guide regulation.

Today, research on dioxin effects has partially assessed the damage from the contamination in Seveso, revealing that it is not limited to cancers but also includes transgenerational effects, in particular, thyroid dysfunction linked to maternal exposure (Baccarelli et al. 2008). Nevertheless, the scientific controversies remain acute because the Seveso data are insufficient to establish clear-cut cause-effect relations.

For science, dioxin is still an incomplete jigsaw puzzle because of the complexity of the mechanisms of its interaction with the human body. As Douglas (2004) notes, dioxin challenges current models for assessing the carcinogenicity of toxicants and shows how regulating toxicants, relying on carcinogenic effects, cannot be just a matter of uncontroversial scientific evidence. This uncontroversial scientific evidence is a chimera and conceals decisions made by scientists (in terms of research priorities, or of data excluded as irrelevant) in order to reduce uncertainty (Latour 1987). Uncertainties about toxicant carcinogenicity abound, and they are related to the complexity of the interactions involved when investigating carcinogenesis. Nevertheless, the dominant paradigm, which explains carcinogenesis on the basis of one single factor accounting for its insurgence, is still the reference point for defining what knowledge should count for regulation.

Despite being widely mentioned in the literature concerned with dioxin carcinogenicity, the Seveso population shows no interest at all in knowing more about how this knowledge is produced and what it means in terms of the consequences for public health. It sees scientists as "people who made their careers exploiting our misfortune and using us as guinea-pigs."[13] To quote Massimo Donati, a family physician in Seveso who personally spent ten years trying to organize dioxin victims to start a legal action against Roche:

> You cannot speak about all the scientific results concerning dioxin effects here in Seveso. It is a taboo: public administrators and citizens don't want to speak about it. I'm in contact with Seveso people on an everyday basis, because of my activity as a physician, and I can tell you that people are divided in two categories. You have people, the large majority, who don't care about dioxin, because they are fine or because they were not exposed. Then you

have people who were exposed to the dioxin contamination and who are now sick: immediately they ask if dioxin could be a possible cause of their disease. It would be necessary to organize an epidemiological study in parallel to those already in place with the data collected by family physicians, with a geographical representation of the distribution of pathologies. This is necessary to see if there are localized concentrations of pathologies.[14]

What Donati complains about is the fact that the scientific output on the dioxin effects in Seveso focuses exclusively on issues and questions defined within the generalized detached frame of understanding how dioxin (in general) interacts with the human body (in general). This is the kind of research promoted and funded by public and private actors, "because then you can publish your article in a scientific journal. But I need knowledge on dioxin effects that allows me to act for public health in this area, and this kind of knowledge is lacking."[15]

There is no link between the scientific work on dioxin effects based on the Seveso case and the territory of Seveso. First of all, no epidemiological research has been done starting from health concerns defined as such within the area, such as for example the perceived presence of possible anomalies in the concentrations of pathologies reported by isolated actors (physicians, ordinary citizens). Second, there is no link between the epidemiological studies on dioxin in Seveso and the implementation of preventive action in terms of public health in the area affected by the disaster. Donati's idea of starting an epidemiological study using the geographical area as the central reference point (rather than the individuals exposed according to risk zones) has not found support, either from the regional authorities or from the population.

In fact, the epidemiological studies on the effects of the dioxin contamination in Seveso are mainly focused on using the Seveso case to explore the biochemical mechanisms through which dioxin can affect human health. It is not by chance that these studies have focused progressively on the populations of the three risk zones. This population is in fact of particular scientific interest because it was exposed to high concentrations of pure dioxin with no other relevant forms of exposure to toxic sources. Data concerning the ICMESA workers and the workers employed in the decontamination activity were only collected until 1985. In this case, other kinds of exposure might severely interfere with the dioxin exposure, making this cohort scientifically less interesting. This fact reveals the specific logic that underlies the epidemiological research on dioxin exposure in Seveso: to identify the specific way in which dioxin interacts with the human body by trying to "purify" this effect from possible interferences related to situations of multiexposure. We can define this logic as a laboratory logic. It is detached from the territory and it is focused on the interaction of the toxicant with a partially decontextualized human being. It is considered fundamental to defining the forms of regulation that should be applied across different contexts.

The aim here is not to say that a "health disaster" went unnoticed in Seveso because of the incapacity of the laboratory logic dominant in epidemiological studies to detect it, but rather to highlight how the knowledge produced on the dioxin effects in Seveso is based on a very specific model of the production of epidemiological data. The issue to be discussed is the consequences of this lack of pluralism in the epidemiological investigation into dioxin effects in Seveso.

In fact, a problem of environmental health such as the dioxin contamination caused by ICMESA can be explored by starting with at least four different and complementary epidemiological approaches: a molecular approach, an individual approach, an approach in terms of population, and an approach in terms of ecosystem (Pekkanen and Pearce 2001). In the case of Seveso, the epidemiological studies were first driven by an individual approach, studying the individuals living in the risk zones. They then evolved toward investigations exploring the molecular mechanisms involved in dioxin toxicity. As previously stressed, there has been no epidemiological study defined in terms of studying the present state of the area affected by the disaster, looking at the pathologies observed, rather than exclusively following the individuals living in the risk zones defined in 1976.[16] The problem involved in relying exclusively on epidemiological studies interested in exploring environmental risks at the individual and molecular levels is that of disconnecting epidemiological studies from a public health goal, from the production of knowledge useful for prevention on the territory in a locality (Pekkanen and Pearce 2001). The rules prevailing in the scientific community thus create a space for debate that is autonomous in a way, that is, it is guided by hypotheses, methods, and investigation procedures defined as such within a specific paradigm of knowledge in which the individual (with her genes, her behaviors) is considered the reference point.[17]

The Seveso case shows how the prevailing model in the debate concerning the effects of toxicants on human health and regulating them is that of looking for a direct cause-effect relationship in terms of carcinogenicity, assuming the individual as the reference. In doing this, by limiting itself to the pursuit of knowledge relevant to global regulation and legal norms for compensation (both dominated by the logic of univocal cause-effect), epidemiology seems to abdicate the role of also providing knowledge relevant to acting to guarantee public health in the areas at risk.[18]

Final Remarks

The dioxin contamination caused by the ICMESA accident has never emerged in the area affected as an issue of environmental health. This can explain the specific direction taken by the epidemiological research on long-term dioxin effects in Seveso, which was mainly guided by a laboratory logic (focused on carcinogenesis) but not connected with a prevention logic. Although this laboratory logic may be crucial for setting regulation standards, as the Seveso case shows, ongoing controversies can lead research to be monopolized by the internal logic of these issues and to reduce to marginality, to the point of complete obscurity, any perspective oriented toward the production of knowledge of use in implementing actions beneficial to local environmental and public health.

The absence of involvement by the citizens affected in the production of knowledge about dioxin effects is crucial for explaining how research on the dioxin effects in the Seveso area developed and progressively detached itself from the territory.

We have tried to explain this lack of citizen involvement by linking it to the interpretation given to dioxin risk by the grassroots groups mobilized following the ICMESA disaster. In particular, dioxin was seen as a threat to the very existence of the community. The fact that the public authorities opted for an authoritarian approach to managing the crisis increased this fear. At the same time, the movement for environmental health active in Italy in the 1970s took a highly ideological approach to the disaster situation. This created difficulty in integrating into its agenda the point of view of the victims and their fears concerning the disappearance of Seveso as a community. The centrality acquired by the issue of therapeutic abortions highlights the conflicting values that became an obstacle to the dialogue between the population affected and the activists. Moreover, we should not overlook the contribution by Roche to the individualization of the dioxin damage in reducing its attention to the dimension of material losses.

No "uneasy alchemy" (Allen 2003) among citizens, activists, public authorities, and scientists took place in the aftermath of the disaster, thus causing the issue of dioxin as a problem of public health to become progressively invisible in the public space.[19] Dialogue among these groups seems in fact to be a necessary condition for the production of knowledge about environmental health problems that can help the design and implementation of prevention policies at a local level.

The dioxin disaster was a moment of high visibility of the hidden costs of industrial production in terms of environmental and human health. Nevertheless, in the area affected, environmental health never became an issue. The Seveso disaster turned out to be, paradoxically, an event that contributed to the invisibility of the issue of environmental health in the heavily industrialized *Brianza Milanese* area. This should make us aware of the difficulties that are always present in the construction of environmental health issues as public problems.

A lesson we can learn from the Seveso case is the central role of public participation in decisions concerning how to respond to environmental risks arising from toxicants. When public authorities rely exclusively on experts to define what a risk is, and what the priorities should be in responding to it, they fail in their role to aid the collective construction of the specific risk as a public problem. They fail to take into account the existence of the different concerns the risk raises at the local level and the different kinds of knowledge that should be considered legitimate in shaping the orientation of research. An absence of participation results in turning the risk into an external object that communities are not able to appropriate and turn into an actual concern. This risk "externalization," in turn, heavily affects the kind of knowledge produced about the risk itself, promoting a vicious circle of separation between knowledge for global regulation and that relevant to local situations.

Joint involvement of citizens, activists, scientists, and public authorities is necessary in order to promote the production of knowledge about environmental risks related to toxicants that is not exclusively guided by laboratory logics but that seriously takes into account the local dimension of environmental health. In this process, the role of public authorities of guaranteeing the conditions for participation is crucial. Power inequalities have a key effect on the process of making things visible. These inequalities have to be addressed in order to create the conditions for collectively dealing with the harmful consequences of industrial activities.

Notes

1. The compound 2,3,7,8-tetrachlorodibenzo-p-dioxin (TCDD) is produced as an unwanted by-product in various chemical reactions and combustion processes, including the manufacture of chlorinated phenols and derivatives.
2. Milena Sant, speaking about her experience in Seveso at the public event organized by the feminist group Maistat@zitt@, "Topo Seveso. Produzioni di morte, nocività e difesa ipocrita della vita." 14 April 2007, Milan.
3. The analysis I develop in this contribution is based on my Ph.D. research on the collective responses to the Seveso disaster (Centemeri 2006). The research was designed to investigate the legacy of the ICMESA accident in the community affected through historical analysis of the 1976 event and an ethnographic study concerning the ongoing construction of a collective memory of the disaster, namely, the project "Bridge of Memory" run by a group of local activists. The data discussed here were collected through the analysis of documents, interviews with people affected by the disaster, local activists, representatives of public institutions, scientists, and participative observation of events related to the legacy of the disaster.
4. Stavros Dimas, Member of the European Commission, Responsible for Environment, "Seveso: The Lessons from the Last 30 Years," European Parliament, Brussels, 11 October 2006, SPEECH/06/588.

5. A question that remains open is the doubt concerning the true destination of the TCP produced by ICMESA in the early 1970s. According to journalist Daniele Biacchessi (1997), it was transported to the United States and used in the production of chemical weapons for the Vietnam War.
6. The air emission originated from a TBC (1,2,3,4-tetrachlorobenzene) alkaline hydrolysis reaction vessel of sodium 2,4,5-trichlorophenate, an intermediate compound in the preparation of trichlorophenol. The direct cause of the emission was excessive pressure induced by an exothermic reaction in the TCP vessel, which occurred a few hours after suspending operations and caused the disk of a safety valve to break down: the disk broke when the pressure reached 4 atmospheres at 250°C, and TCDD—together with the above-mentioned products, and with ethylenic glycol and soda—burst out of the roof and spread directly in the air due to the lack of an expansion chamber (Ramondetta and Repossi 1998).
7. The mixture inside the vessel at the moment operations were suspended was probably composed of about 2,030 kg of sodium 2,4,5-trichlorophenate (or other TCB hydrolysis products), 540 kg of sodium chloride, and over 2,000 kg of organic products. In recovering the vessel, 2,171 kg of material, mainly sodium chloride (1,560 kg) were found. It can therefore be concluded that the air emission, composed of a mixture of several different pollutants including dioxin, was about 3,000 kg. As for the dioxin content in the toxic cloud, technical literature reports different evaluations, ranging from 300 g to 130 kg (Ramondetta and Repossi 1998).
8. In zone A (108 hectares, 736 inhabitants), the authorities decided on the evacuation of the whole population; in Zone B (269 hectares, 4,600 inhabitants) there was no evacuation, but the inhabitants were forced to follow strict rules of conduct; in the Prevention Zone (1,430 hectares, 31,800 inhabitants) there was no evacuation but inhabitants were forced to follow some precautionary rules of conduct, less constraining than those in zone B.
9. Voluntary pregnancy terminations were finally permitted in Italy by law 194 in 1978.
10. Medicina Democratica (Democratic Medicine) was an Italian social movement born in the 1970s on the initiative of industrial workers, scientists, and intellectuals. MD argued for the importance of developing participative forms of knowledge production on health problems related to industrial activities.
11. Comunione e Liberazione is a Catholic movement born in Italy in the 1950s and particularly active in Lombardy. Its main trait is the charismatic dimension that goes with the promotion of what are called *opere*, that is, the supply of social services through associative organizations. The relationship between CL and the state has always involved a measure of conflict. In the opinion of CL, the state cannot and ought not take part in social organization: "The welfare State must limit its intrusion into people lives" (Abruzzese 1991: 171). On the "fundamentalism" of CL, see Zadra (1994).
12. The compensation issue remained open in Seveso until 2007, when the two proceedings instituted against Roche on the initiative of two groups of dioxin victims were declared invalid as a result of the statute of limitations. The two groups of victims never succeeded in gaining local support for their initiative (Centemeri 2006: 135–58). It is important to note that Roche has never admitted its responsibility for the disaster in any court of law.
13. Interview with L.S., resident of Seveso (April 2004).

14. Interview with Massimo Donati (June 2004).
15. Ibid.
16. As Barbara Allen remarks in her study on the mobilization of citizens for environmental health in the Louisiana "chemical corridor": "by placing the specific resident or community at the centre of an investigation, science is constructed around what is happening to the people, rather than people being constructed to fit mathematical scientific models" (Allen 2003: 148).
17. For a critique of this approach in the field of job-related cancers, see Thébaud-Mony (2007).
18. On this point see also the chapters of Barbara L. Allen and of Paul Jobin and Yu-Hwei Tseng in this volume.
19. Another case of this virtuous alchemy is discussed by Paul Jobin (2006) in his study of the Minamata disease.

Bibliography

Abbruzzese, Salvatore. 1991. *Comunione e Liberazione*. Roma-Bari: Laterza.
Allen, Barbara L. 2003. *Uneasy Alchemy: Citizens and Experts in Louisiana's Chemical Corridor Dispute*. Cambridge: MIT Press.
Baan, Robert, Yann Grosse, Kurt Straif, Béatrice Secretan, Fatiha El Ghissassi, Véronique Bouvard et al. 2009. "A Review of Human Carcinogens—Part F: Chemical Agents and Related Occupations." *Lancet Oncology* 10: 1143–44.
Baccarelli, Andrea, Sara M. Giacomini, Carlo Corbetta, Maria Teresa Landi, Matteo Bonzini, Dario Consonni, Paolo Grillo, Donald G. Patterson Jr., Angela C. Pesatori, and Pier Alberto Bertazzi. 2008. "Neonatal Thyroid Function in Seveso 25 Years after Maternal Exposure to Dioxin." *PLoS Medicine* 5(7): 1133–42.
Bagnasco, Arnaldo. 1977. *Tre Italie: la problematica territoriale dello sviluppo italiano*. Bologna: Il Mulino.
Berque, Augustin. 2000. *Écoumène. Introduction à l'étude des milieux humains*. Paris: Belin.
Boltanski, Luc, and Eve Chiapello. 1999. *Le nouvel esprit du capitalisme*. Paris: Gallimard.
Biacchessi, Daniele. 1997. *La fabbrica dei profumi. La verità su Seveso, l'ICMESA, la diossina*. Milano: Baldini Castoldi Dalai.
Breviglieri, Marc. 2002. "L'horizon du ne plus habiter et l'absence du maintien de soi en public." In *L'héritage du pragmatisme. Conflits d'urbanité et épreuves de civisme*, ed. Daniel Cefaï and Isaac Joseph, 319–36. La Tour d'Aigues: Ed. de l'Aube.
Callon, Michel, Pierre Lascoumes, and Yannick Barthe. 2001. *Agir dans un monde incertain: essai sur la démocratie technique*. Paris: Seuil.
Centemeri, Laura. 2006. *Ritorno a Seveso. Il danno ambientale, il suo riconoscimento, la sua riparazione*. Milano: Bruno Mondadori Editore.
Cole, Philip, Dimitrios Trichopoulos, Harris Pastides, Thomas Starr, and Jack S. Mandel. 2003. "Dioxin and Cancer: A Critical Review." *Regulatory Toxicology and Pharmacology* 38(3): 378–88.
Consonni, Dario, Angela C. Pesatori, Carlo Zocchetti, Raffaella Sindaco, Luca D'Oro Cavalieri, Maurizia Rubagotti, and Pier Alberto Bertazzi. 2008. "Mortality in a Population Exposed to Dioxin after the Seveso, Italy, Accident in 1976: 25 Years of Follow-Up." *American Journal of Epidemiology* 167(7): 847–58.

Conti, Laura. 1977. *Visto da Seveso: l'evento straordinario e l'ordinaria amministrazione*. Milano: Feltrinelli.

De Marchi, Bruna. 1997. "Seveso: From Pollution to Regulation." *International Journal of Environment and Pollution* 7(4): 526–37.

De Marchi, Bruna, Silvio Funtowicz, and Angela Guimaraes Pereira. 2001. "From the Right to be Informed to the Right to Participate: Responding to the Evolution of European Legislation with ICT." *International Journal of Environment and Pollution* 15(1): 1–21.

De Marchi, Bruna, Silvio Funtowicz, and Jerry R. Ravetz. 1996. "Seveso: a Paradoxical Classic Disaster." In *The Long Road to Recovery: Community Responses to Industrial Disaster*, ed. James K. Mitchell, chap 4. New York: United Nations University. http://www.unu.edu/unupress/unupbooks/uu21le/uu21le09.htm (last accessed 12 November 2012).

Douglas, Heather. 2004. "Prediction, Explanation, and Dioxin Biochemistry: Science in Public Policy." *Foundations of Chemistry* 6(1): 49–63.

Eskenazi, Brenda, Paolo Mocarelli, Marcella Warner, Steven Samuels, Larry Needham, Donald Patterson, Paolo Brambilla, Pier Mario Gerthoux, Wayman Turner, Stefania Casalini, Mariangela Cazzaniga, and Wan-Ying Chee. 2001. "Seveso Women's Health Study: Does Zone of Residence Predict Individual TCDD Exposure?" *Chemosphere* 43(4–7): 937–42.

Ferrara, Marcella. 1977. *Le donne di Seveso*. Roma: Editori Riuniti.

Fratter, Massimiliano. 2006. *Memorie da sotto il bosco*. Milano: Auditorium.

Jasanoff, Sheila, ed. 1994. *Learning from Disaster: Risk Management After Bhopal*. Philadelphia: University of Pennsylvania Press.

Jobin, Paul. 2006. *Maladies industrielles et renouveau syndical au Japon*. Paris: Editions de l'EHESS.

Kohler, Pierre. 2002. *L'imposture verte*. Paris: Albin Michel.

Latour, Bruno. 1987. *Science in Action: How to Follow Scientists and Engineers through Society*. Cambridge, M.A.: Harvard University Press.

Maccacaro, Giulio Alfredo. 1976. "Seveso, un crimine di pace." *Sapere* 79(796): 4–9.

Mills, Charles Wright. 1959. *The Sociological Imagination*. New York: Oxford University Press.

Mocarelli, Paolo. 2001. "Seveso: a Teaching Story." *Chemosphere* 43(4–7): 391–402.

Pekkanen, Juha, and Neil Pearce. 2001. "Environmental Epidemiology: Challenges and Opportunities." *Environmental Health Perspectives* 109(1): 1–5.

Ramondetta, Miriam, and Alessandra Repossi, eds. 1998. *Seveso vent'anni dopo. Dall'incidente al Bosco delle Querce*. Milano: Fondazione Lombardia per l'Ambiente.

Steenland, Kyle, Pier Alberto Bertazzi, Andrea Baccarelli, and Manolis Kogevinas. 2004. "Dioxin Revisited: Developments Since the 1997 IARC Classification of Dioxin as a Human Carcinogen." *Environmental Health Perspectives* 112(13): 1265–68.

Thébaud-Mony, Annie. 2007. *Travailler peut nuire gravement à votre santé. Sous-traitance des risques, mise en danger d'autrui, atteintes à la dignité, violences physiques et morales, cancers professionnels*. Paris: La Découverte.

Thévenot, Laurent. 2006. *L'action au pluriel. Sociologie des régimes d'engagement*. Paris: La Découverte.

van Eijndhoven, Josee. 1994. "Disaster Prevention in Europe." In *Learning from Disaster: Risk Management After Bhopal*, ed. Sheila Jasanoff, 113–32. Philadelphia: University of Pennsylvania Press.

Werner, Marcella, Paolo Mocarelli, Steven Samuels, Larry Needham, Paolo Brambilla, and Brenda Eskenazi. 2011. "Dioxin Exposure and Cancer Risk in the Seveso Women's Health Study." *Environmental Health Perspectives* 119(12): 1700–05.

Wynne, Brian. 1996. "May the Sheep Safely Graze? A Reflexive View of the Expert-Lay Knowledge Divide." In *Risk, Environment and Modernity: Towards a New Ecology,* ed. Scott Lash, Bronislaw Szerszynski, and Brian Wynne, 44–83. London: Sage.

Zadra, Dario. 1994. "Comunione e Liberazione: A Fundamentalist Idea of Power." In *Accounting For Fundamentalism: The Dynamic Character of Movements,* ed. Martin E. Marty and R. Scott Appleby, 124–48. Chicago: The University of Chicago Press.

 CHAPTER 7

From Suspicious Illness to Policy Change in Petrochemical Regions
Popular Epidemiology, Science, and the Law in the United States and Italy

Barbara L. Allen

Louisiana's chemical corridor (United States) and northern Italy's Porto Marghera chemical region are both sites of long-term, highly visible citizen struggles. In both locations the debates about, and the shaping of, environmental health knowledge related to toxicants was key to the emergence of the controversy as well as its outcome. This chapter examines these dynamics, particularly those of citizen-expert alliances, to develop an understanding of the construction of policy-relevant or "actionable" science.

Specifically, my research focuses on the intersection of citizen activism, environmental health science, the public use of science, industrial regulation, and policy change related to toxicants in petrochemical regions primarily through the lens of legal controversies. One reason for using court cases is that it allows a view inside the construction and use of science within environmental controversies. I am interested in how social structures and networks, legal systems, professional and nonprofessional cultures, NGOs, and labor unions impact the making of policy-relevant environmental knowledge directed toward determining regulatory action and effecting positive environmental change.

This comparative work on shaping environmental health regulation counters the often totalizing globalization discourse regarding technology. However, comparative research does not mean to suggest the application of successful strategies of citizens, experts, and governmental institutions cross-nationally, ignoring the social and cultural situatedness of local dynamics. Knowledge making leading to policy change should be contextualized, understanding the interaction of knowledge and power from the ground up, with clear acknowledgement of both similarities and differences within dynamic social and cultural milieus. Any cross-national extrapolation or even nation-state comparisons are performed with caution such that differences are not lost in the move to level arguments and approaches.

The Italian and U.S. case studies reinforce this claim: while the technologies, hazards, and pollution are the same, the social and cultural landscape of the struggles and their outcomes are quite different. The physical similarities of these two places are striking. They both: (1) are industrial regions that have between 85 and 130 chemical plants and petro-processors, (2) produce a large quantity of particularly noxious chloro-chemicals, (3) are located adjacent to important historic and cultural tourism locations and on important waterways used for fishing as well as other commercial and recreational activities, and (4) were home to plants operated by one or more of the same multinational chemical corporations. There are also some notable differences that emerged in the case narratives. The national and regional context of the struggles over toxics conditioned both the process and result. A discussion of these variances follows the presentation of the two case studies.

Theoretically, my project is situated within the interdisciplinary field of Science and Technology Studies (STS), particularly the Public Use of Science (PUS) (Irwin and Wynne 1996). One of the trajectories of this analytic lens is to understand how public and civic cultures shape the formation of scientific knowledge. I also draw on approaches from the New Political Sociology of Science (NPSS) as a way to examine, comparatively, case studies as well as to understand the political and institutional dynamic that has shaped environmental health knowledge (Frickel and Moore 2005). Comparative studies can illuminate how social arrangements and institutional networks can influence the production of policy-relevant knowledge and better reveal the effects of power relationships inherent in knowledge production systems. And finally, I engage with Sheila Jasanoff's work on U.S.-Europe science policy; specifically, her notion of civic epistemologies, or "how democratic polities acquire communal knowledge for purposes of collective action" (Jasanoff 2005: 9). This research expands Jasanoff's research by beginning at the local level and, using ethnographic methods, attempts to fully understand the local formation of scientific and technical knowledge that eventually counts as policy-relevant at the national level. Furthermore localism and the particularities of place often supersede national concerns in the coproduction of politics and knowledge formation (Fischer 2000; Agnew 2002; Jasanoff and Martello 2004; Jasanoff 2005; Hess 2009). This "epistemology from the bottom" has great implications for just distributions of power as its effects counter-percolate at times reinforcing and/or calling into question the ability of regulators and legal systems to meet the needs of citizens (Jasanoff 2005: 14). I track the flow of this construction through meso- and macro-level institutions to more fully understand the process from emergence through codification. By tracking science in this manner, the power implications of knowledge are revealed through a lens wide enough to glimpse possibilities for social justice and participatory democracy.

The Grand Bois Case (Louisiana, United States)
Beginning of the Controversy

Grand Bois is a small town of about 300 residents, 50 percent of whom are Native American and most of the remaining of whom have Cajun French ancestry. The town is in the petroleum and chemical parishes of south Louisiana, in what is sometimes referred to as "Cancer Alley."[1] In the 1980s the oil companies, having a very powerful industry lobby, were successful in having the waste from oil and natural gas exploration, production, and refining exempted from environmental regulation by the U.S. Environmental Protection Agency (EPA). Oilfield waste was automatically classified as nonhazardous waste according to EPA regulations even though this type of waste can contain large concentrations of toxins such as heavy metals, benzene, and hydrogen sulfide. Regulation was, and still is, left entirely up to the individual states.

In 1982 the Campbell Wells site in Grand Bois was granted a state permit to accept oil and gas processing waste. The facility was a kind of landfill consisting of a series of cells or large rectangular pits dug out of the ground into which liquids and semisolid sludge were dumped. The cells were located adjacent to the St. Louis canal portion of the Intracoastal Waterway in a region home to some of the most productive fishing areas in the United States. The dumpsite was separated from the town by a chain link fence, the closest cell being 300 feet (100 meters) away from the nearest residence.

The local citizens were told that the waste was primarily salt water, and though there were occasional complaints of bad odors and headaches, the townspeople lived with the facility until things got very bad in March 1994. That month, the facility accepted eighty-one tanker-truck loads of production-pit sludge from Exxon's natural gas plant operation in Alabama. Alabama had classified the sludge as hazardous waste due to its high content of hydrogen sulfide, benzene, arsenic, and other heavy metals and demanded that Exxon properly dispose of the waste. Exxon's choices appeared to be to process the 5,600 barrels of hazardous waste in Alabama at a cost of over $100 a barrel or to ship the waste to Louisiana at a cost of $8 a barrel, thus saving $500,000 (Dunne 1998).

The tanker trucks rolled into the community during a ten-day period in the spring of 1994 bearing "hazardous material" labels. The townspeople said that the odor was overwhelming throughout the entire community and complained to Louisiana's Department of Environmental Quality (DEQ) that they smelled strong "chemical" odors such as the pungent odor of rotten eggs, a sign of deadly hydrogen sulfide gas. While the DEQ did show up and take air samples during the Exxon waste disposal, the inspectors were specifically instructed not to test for hydrogen sulfide (Roberts and Toffolon-Weiss 2001). The DEQ also did not sample the hazardous sludge that was later found to contain benzene and heavy metals. Almost immediately following this large disposal of Exxon waste, the town began complaining not only of the awful, lingering stench but, eventually, of headaches, sinus problems, fatigue, dizziness, and an array of other nervous system disorders. Within a month after the last tanker of oilfield waste was dumped, all three hundred residents of the town joined in filing a lawsuit against Exxon and Campbell Wells.[2] The citizens needed help and they were eventually directed to two local scientists.

Two Local Scientists, Two Approaches to Science

Wilma Subra is a biochemist and toxicologist who, in 1999, won a prestigious MacArthur award for her work on behalf of environmental and environmental justice groups. Subra began her career in the late 1960s at the Gulf South Research Institute in south Louisiana and eventually became head of the environmental science and analytical chemistry sections of the institute. When the citizen disputes arose around the effects of hazardous waste at Love Canal, New York, Subra was the scientist in charge of the on-site lab studies. Once the lab results were recorded, none of the residents were ever allowed to know their own toxicity or the toxicity of their family members. This bothered Subra and she quit the firm after her Love Canal experience.

Patricia Williams was, at the time, the director of the Louisiana State University (LSU) medical school Occupational Toxicology Outreach program, as well as the laboratory director of their in-house medical surveillance lab that focuses on the effects of chemical exposure. She is also a biochemist with postdoc work in epidemiology. Williams had worked within the institutional framework of the LSU medical school for almost thirty-five years. Her studies follow published research protocol and typically took a long time, even years, to complete. In the past, she had conducted tests on exposed populations and had testified in court as an expert witness on behalf of a number of Louisiana communities facing toxic threats.

While both scientists work on local pollution issues with the community's best interests at heart, their differing approaches to doing science have sometimes clashed. What counts as science in both women's eyes is dramatically different and would have unfortunate consequences to the community at risk, Grand Bois.

Wilma Subra's Heterogeneous Public Science Network

The community immediately asked for Subra's help, and she began collecting all the information she could on the waste dump and holding meetings with the community to help them understand the issues and the regulatory rules. Her method of working as a scientist also involved connecting with a wide network of state and federal agencies to do the sampling and studies, because often the communities affected have no funding for this type of background work. But even with technical assistance from state and federal agencies, the citizens still needed a scientist such as Subra working on their side to help them properly utilize the outside expertise. She explains: "They [the outside agency–funded scientists] go into a community and they want the community to engage them. The community doesn't know the questions and what to ask them to do. They [outside agencies] do a piece of something but they can't be advocates."[3] Subra's goal, following the community's wishes, is typically to get the site cleaned up, shut down, or to stop the siting of a new hazardous facility. She does not strive for definitive scientific proof of causation for the residents' illnesses as this kind of science is lengthy, expensive, and tends to put industry on the offensive. She explains that "you need the health part in the equation to drive it [the community push to clean up the site], but it's not the main driver."[4]

Media publicity is an important tool, according to Subra, as media exposure is typically good for the victims' causes. She submitted the Grand Bois controversy story to the popular television show *60 Minutes*, and they agreed to produce a special hour-long program on the community hosted by the late Ed Bradley. "It was their story, so I made a point of never appearing on camera ... just doing all the legwork,"[5] Subra said. So in Subra's "public participation" approach to science, the more actors in the network the better; the more media exposure and the more agencies doing sampling, the more likely it is that the state will properly regulate industry or that industry will improve its practices.

Subra, because of her position outside of institutional science and her pro-community work on behalf of exposed populations, has a conflicted status within her scientific community. While her training in science has an elevated status that provides her authoritative cultural capital, her involvement in controversial issues makes her position within the ranks of scientists insecure (Downey 1988). But Subra is unmoved by the issue of status among her peers. She has a vision of environmental science as a public-based heterogeneous enterprise consisting of many actors, both inside and outside of what is traditionally termed "science." She sees her work as both developing a "relationship between local people as agents and [using] scientific understanding as a force for change" (Irwin et al. 1996: 59). Her version of public science includes actively constructing networks and alliances that intentionally blur the boundaries between local and cosmopolitan knowledge as well as between science and politics (Irwin 1995; Irwin and Wynne 1996; Fischer 2000).

Patricia Williams's Institutional "Science for the People"

In 1997, a few years after Subra had begun working in Grand Bois, Williams was contacted by Dr. Mike Robichaux, a local physician and state senator for the region. In his professional practice he had seen many health problems in the citizens from Grand Bois firsthand. He was particularly concerned about the various nervous system disorders that seemed exceedingly high for such a small community. He did propose several bills on the floor of the senate, first to close the oilfield waste disposal loophole statewide, and later to provide an exemption for Grand Bois: both were firmly defeated because of concerns that such a bill would harm the oil industry in Louisiana. Dr. Robichaux did secure state funds for a formal study of the community and Williams was chosen to do the study.

The first stage of her project was a community self-study asking the community members questions regarding their health, occupation, lifestyle, and behavior. Williams also did a self-study of a control community for comparison purposes. The self-study results showed that Grand Bois residents had statistically significant higher rates of gastrointestinal disorders, neurological symptoms, muscle and joint pain, as well as increased fatigue (Williams 1997). Lead and other heavy metal exposure can produce all of these symptoms. This prompted Williams to gather the past medical records for children participating in the study. These records were from the state Office of Public Health (OPH), the Women Infants and Children (WIC) program, and the local hospital. Out of about 24 children, 4 had elevated blood lead levels and another 6 had blood indicators that warranted further testing and/or monitoring.

With questions regarding heavy metal exposure in at least 37 percent of the town's children, Dr. Robichaux went to the state senate and asked for additional funding for laboratory testing. The study called for pregnant women and children in the Grand Bois area to be tested for exposure to twenty-six different chemicals once a month for twelve months. The first blood samples were to be drawn on 7 April 1998; the case was set to go to trial three months later on 13 July.

Science Under Litigation Pressure

Williams and her team of physicians and experts continued to track the problem and do studies in the community with the full cooperation of the residents. When asked, however, Williams declined to testify in the Grand Bois lawsuit, saying it would have been a violation of her Institutional Review Board (IRB) confidentiality agreement to do so. Litigants were individually given copies of their children's blood work and their own medical test results that they could use as they saw fit. She told the litigants of both sides that they would have to wait until the complete scientific study was finished before receiving a copy of her final report. Williams's attorney, defending her position to have her raw data remain confidential, stated that, "Dr Williams does not want her objective science to be tainted by the appearance of bias" (McMillan 1998).

By April 1998, the second phase of Williams's study was in progress—she needed one more year to complete her study that followed careful scientific protocol and procedures for proving causation and effects of toxic exposure. Having a complete study that met currently acceptable scientific standards would provide evidence of exposure, not only for the Grand Bois community, but also for other communities with similar hazardous waste exposures. The lawsuit, however, was moving much faster than the scientific study.

There was already some tension between the two scientists, as according to Subra, Williams discouraged residents from participating in some of the state-sponsored public health research other than her own. Williams did not trust the Louisiana OPH due to many past interactions with the state agency and said that her survey team would not work with representatives of the OPH (Daugherty 1997). The fear was that they would design a study to show that nothing was wrong in an attempt to negate Williams's careful medical surveillance study.[6]

All parties tried to force Williams to produce the raw data of her study-in-progress. She refused and fought back. At this point the media attention was so great, the Louisiana state legislature passed a law mandating state-funded researchers (i.e., LSU employees) provide their in-progress experimental data when asked. Unfortunately, in the end, the residents of Grand Bois had to settle for far less than they had hoped for. They received only a small cash payment instead of the medical monitoring, treatment for life, and the closure of the waste facility, which is what they had asked for. Williams' partially finished study was incomplete and was used by Exxon to show that there was not clear evidence of exposure and damage caused by the chemicals Exxon dumped at the site. Incomplete was equated with inconclusive in the context of the lawsuit. Eventually, Williams' final study, which came out too late to help the citizens of Grand Bois, did show exposure and health issues related to the oilfield site. The study was used to, at least partially, close the state loophole on oilfield waste in places other than Grand Bois where such facilities were still operating.

The Case of Porto Marghera (Venice, Italy)

Marghera, a small town on the Venice Lagoon, housed one of the largest chemical complexes in Italy, and in the 1990s provided about half of the polyvinyl chloride (PVC) produced in the country (Bortolozzo 1994a: 42–45). The regulatory plan that industry submitted to the government, the Piano Regalatore Generale di Venezia 1962–1990, described the kinds and amounts of hazardous pollution that would be produced and emitted in the air, water, and soil—and won government approval (Bettin 1998). The concentration limits were set by industry, unrelated to research on human health or environmental protection. No provision was made for progressive concentration or toxic buildup over time in either the human body or the lagoon.

In the 1970s chemical workers, and subsequently residents, expressed suspicions about the dangers posed by the petrochemical plants. In 1971 the Institute of Medicine in Padova conducted a pediatric study in Marghera, finding 90 percent of children between the ages of six and eleven had respiratory ailments and other illnesses (Bettin 1998). This occurred at a time when the local environmental movement dedicated to cleaning up the Venice Lagoon, the Fronte per la difesa di Venezia e della sua Laguna, was also at its height. Citizen protest attributed blame for the polluted state of the local waterways to Porto Marghera and to chemical industry practices (Mencini 2005). The movement gained greater visibility by adopting Greenpeace-style tactics such as blocking oil tankers and waste disposal ships with small boats. This citizen-led movement was successful in stopping the building of the final, and largest, phase of the Porto Marghera chemical complex.

The Production and Publication of Health Science

The years from the mid 1960s to the early 1970s was a fertile period for Italian environmental health research on the effects of PVC and VCM (vinyl chloride monomer) on human and animal health. Initially, the research was carried out by company doctors employed by industry. Their findings were so alarming that in 1966 a private meeting of U.S. and European chemical companies was held to discuss health problems related to the production and use of PVC and VCM. Such problems included skin thickening, bone disease, and some liver pathologies. Soon after this meeting the major cancer research institute in Italy, Regina Elena Institute, began studies of exposure to VCM, subjecting rats to VCM gas at the approved limits of 500 ppm. Its research was presented at an international cancer congress in Houston in 1970 and published the following year in *Cancer Research;* it revealed cancer in rats with exposures as low as 50 ppm.[7]

Next, the European vinyl chloride producers commissioned their own research by Cesare Maltoni of the Bologna Center for the Prevention and Detection of Tumors and Oncological Research. In 1972 Maltoni also found cancer from VCM occurring in rats at very low exposures and in various sites on their bodies, including the liver. The European and American PVC manufacturers, concerned about these findings, made a pact of secrecy during the early 1970s regarding all research about the toxicity and carcinogenicity of the PVC/VCM chemical processes (Markowitz and Rosner 2002: 183). By early 1973, Maltoni was sure that there was a relationship between angiosarcoma of the liver and low-level VCM exposure. In the spring of 1974, Maltoni's work on the carcinogenicity of PVC and VCM appeared in print. He claimed that epidemiological proof only after deaths have occurred was insufficient to address the issue of safe levels of human exposure to substances. He further asserted that animal studies such as his were a preferable mechanism to predict effect and set safe exposure levels before damage was done (Markowitz and Rosner 2002: 201).

Industry disagreed, claiming that human disease, not animal studies, must be the proof of harm. However, on discovering the deadly effects of their chemicals, the industry *had* made some adjustments to production. The growing environmental awareness and consumer vigilance threatened to make VCM's use as a propellant in aerosol sprays (hairspray, room deodorants, insecticides, paint) a public liability nightmare. As a consequence they voluntarily discontinued the use of VCM in spray products and in the packaging of alcohol without notifying any public authorities of their concerns (Markowitz and Rosner 2002: 185).

At the beginning of 1974, VCM was suspected in the deaths of four workers at a plant in the United States—they died of angiosarcoma of the liver, a rare cancer linked to heavy metal poisoning and identical to the cancer reported in Maltoni's rat studies. The National Institute for Occupational Safety and Health (NIOSH) was quickly informed of the potential link between PVC/VCM exposure and cancer and they sided with Maltoni's conclusions. In publicizing their concerns, they went further and suggested that epidemiology was not an adequate measure of the induction of cancer, only of death, and a poor measure at that, because of the potential for underestimation of effect due to the post hoc nature of exposure and disease (Markowitz and Rosner 2002: 191–92, 202–3).

One consequence of Maltoni's published research was a radical reduction in the industry standard for worker exposure to VCM over an eight-hour period from 500 ppm to 50 ppm in 1974, then to 1 ppm by 1976 (Bortolozzo 1994b: 46–56). The response of Montedison—the largest PVC/VCM producer in Marghera—to the sharp lowering of exposure limits was creative. According to one technical expert, Montedison changed the monitoring system in 1975 and dramatic drops in ambient VCM concentrations in the workplace occurred overnight (Rabitti 1998). The new system averaged concentrations across different spaces and did not give point source readings at the locations of maximum exposure. The new monitors had a maximum reading of only 25 ppm and were incapable of measuring spikes in chemical releases. Most importantly, they also neglected to monitor the exposure of individual workers' bodies on either a daily or cumulative basis.

The late 1970s and 1980s was a period of production efficiency and workforce decline in Marghera's chemical industry due largely to plant automation. Communities were left with fewer jobs and accumulating pollution. The scarcity of work and fear of plant closures reduced the inclination of many workers, unions, and even townspeople to criticize industry (Bettin 1998). Extreme exposures were also concentrated in certain parts of the plant, PVC/VCM being produced in perhaps fifteen of the eighty-five facilities. The chemical worker's union, Federazione Unitaria Lavoratori Chimici (FULC), investigated the claims once it became clear that PVC/VCM was harmful. But it appeared satisfied with industry's claim that levels were immediately reduced to meet new standards.

From Worker to Activist to Expert

One person who remained unsatisfied with Montedison's response to the known hazards of vinyl chloride production was Gabriele Bortolozzo, a laborer at Montedison since 1956. He had worked in the PVC/VCM unit for some years with five other men cleaning the vats where the chemical was mixed at exposures reaching 32,000 ppm or more: all were dead, four of them from a rare liver cancer. In 1979 Bortolozzo discussed his complaints with others, keeping a log of deaths and illnesses at the plant. Perceived by management as a "problem," he was moved and isolated from other workers.

Bortolozzo continued his campaign even after retirement in 1990, obtaining information from public health officials as well as the Montedison company (Rabitti 1998). He made the study, collection, and assimilation of information regarding pollution and health problems in Porto Marghera his retirement passion. In the spring of 1994, Bortolozzo wrote nine short articles for a thematic issue of *Medicina Democratica,* a "red-green" professional journal dedicated to topics of health and environmental safety as well as the rights of workers.[8] Wide-ranging in their scope and meticulously detailed with charts and statistics, Bortolozzo's contributions explained the vinyl chloride production process and the history of VCM production in Italy, as well as mortality rates for Marghera's VCM workers and the wider environmental impacts of pollution. In the years following his tireless inquiries and the eventual publication of his concerns in *Medicina Democratica,* Bortolozzo was acknowledged as an expert and authority on PVC/CVM exposure among workers in Porto Marghera.[9] In August 1994, armed with his volume of *Medicina Democratica,* he finally succeeded in meeting Felice Casson, the attorney general of Venice, who would eventually become the government prosecutor in the Montedison case. Casson quickly assembled his own group of experts to investigate health and pollution claims. In September 1995, Bortolozzo, an avid naturalist and bicyclist, was struck and killed by a truck while biking near Marghera. It was deemed an accident by local officials.

The momentum in favor of a prosecution was sustained after his death. In 1997 the Venice attorney general's office moved to preliminary trial, indicting twenty-eight executives and managers of petrochemical companies on charges ranging from manslaughter to environmental pollution. The next year, the case went to trial having about 550 plaintiffs including: state government, local government, several environmental organizations, trade unions, 103 sick workers, and the families of 156 dead workers. Before the court reached a decision, the chemical companies settled with the state, agreeing to pay to clean up the Venice Lagoon. They also settled with many of the worker plaintiffs, effectively clearing the courtroom.[10]

The trial ended in 2001; the chemical plant managers and executives were found "not guilty." The court's siding with the defendants in the most heinous allegation, manslaughter, hinged on "who knew what when" with respect to the effect of certain levels of PVC/VCM on human health.

Casson filed an appeal in court. The facts, including those relating to worker protection regulations, were re-presented and reanalyzed. In 2004 a verdict was reached—five chief executive officers and former managers were convicted on many counts including 156 counts of manslaughter.

In the Italian legal system, defendants may avoid punishment if too much time has passed or if they are elderly, as judges have considerable discretion in sentencing. The five convicted executives never served a day in prison. However, the judgment did allow for additional civil suits for damages, both by individuals and by the government for environmental harm. It also served to verify and legally reinforce the health effects of vinyl chloride production (nationally and internationally) as well as serve as a deterrent to future negligent behavior.

The catalyst for the lawsuit brought by the attorney general of Venice might have been attributed to the tenacity one worker, Gabriele Bortolozzo. However, the history of the activism, legal action, and results in Porto Marghera are not that straightforward. Bortolozzo did not act alone, either in the acquisition of information or as the solo plaintiff in the lawsuit. Initially, his allies were the medical institutions that offered their research to enable him to write a series of articles he published in a union-affiliated medical journal. Additionally, he joined in the efforts of national environmental groups that were concerned about pollution in the Venice Lagoon, both a World Heritage Site and major international tourist destination. After enlisting the Venice attorney general in his cause, other international environmental groups, such as Greenpeace, joined the legal actions, as did unions, governmental agencies, and many individuals. The lawsuit gained wide attention and became a public media circus with all three national newspapers and many smaller local and regional papers regularly reporting on its progress. Both scientists and journalists wrote books in the time leading up to the various stages of the lawsuit. It was no longer the cause of Bortolozzo alone, but had grown to a large heterogeneous "movement" with an accompanying media entourage. Many actors, such as scientists and citizens, workers and residents, environmentalists and union officials, all held a stake in finding the chemical companies and their officials guilty. The boundaries had been blurred between local and cosmopolitan knowledge and well as between science and politics. The overall impact of this unwieldy network and their alliances was positive for both environmental knowledge and policy action regarding PVC/CVM production on the Venice Lagoon and beyond.

Comparing and Contrasting Citizen-Expert Alliances and Environmental Knowledge Formation in the U.S. and Italian Cases

These two cases reveal the problems and possibilities of making of policy-relevant environmental health science about toxic chemical exposure. In particular, they make apparent the social construction of credible knowledge and the role that citizens have in this process. Civic epistemology, or "culturally specific, historically and politically grounded, public knowledge-ways" evidence the diverse dynamics of how science is "articulated, defended and represented" in Italy and the United States (Jasanoff 2005: 249). Citizen relationships with state institutions and civil society organizations, including the forms of their social networks, are contributors to regional and national differences. From these examples, faced with a similar industry producing similar hazards, the process, choices, and outcomes of the people were divergent.

First, the time frame of the two cases was different. The trajectory of the Italian controversy took place over several decades, growing out of the labor activism of the late 1960s and early 1970s and culminating in the legal action that began in 1997 and ended in 2004. The Louisiana (U.S.) dispute trajectory was significantly more condensed; while low-level citizen dissatisfaction began in 1982, the heated controversy began in 1994, followed by a more rapid lawsuit that was finally settled in 1998. But even larger was the difference between the legal cultures of the two countries and regions.

The state judges in Louisiana (as in many states) are elected, thus, lobbying and campaign funding could potentially bias judicial decision making. Furthermore, the political-legal climate was particularly inflamed when the Grand Bois case went to trial. Only a year before, a huge controversy in the same region over the siting of a new PVC chemical plant had pitted university law clinic students, representing poor citizens, against a legion of corporate attorneys. The state supreme court, pressured by industry lobbyists, ruled on

July 1998—just as the Grad Bois trial was beginning—that student law clinics would be severely restricted in the support they could offer citizens (Allen 2003: 101–5). The climate for environmental justice cases in the state had become decidedly anti-plaintiff.

The judiciary in Italy, also far from being apolitical, has more autonomy than that in any other democratic nation (Koff and Koff 2000: 165). As a backlash against the conservative judges inherited from the fascist era in postwar Italy, the 1960s and 1970s brought a new wave of radicalized professionals (Koff and Koff 2000: 177). Judicial activism for political change emerged at the same time as other left social labor movements such as *Potere Operaio* and the growth of the unions, as well as the Communist Party.[11] The power of the judiciary was openly demonstrated in the early 1990s when a number of Italian magistrates tried to end rampant corruption among political leaders, leading to the fall of the government (Koff and Koff 2000: 175–78; Ginsborg 2003b: 267–69). In the mid 1990s, a new, more conservative government emerged from the conflicts; however, the power of a strong judiciary branch had been solidified. Another element of difference from the United States is that the public prosecutor in Italy is similar to a judge and is a member of the judiciary. She is both the accuser and investigator in a legal case and this has implications for science in the courts.

Another legal difference between the two cases was the ability to link civil and criminal cases in the Italian court. Besides the scientific evidence introduced by both defendants and plaintiffs, the judges in the case also hired technical experts to provide research, analysis, and opinions. The chemical firms were found guilty, not only of worker safety violations that led to many deaths, but also of damage to the air, land, and water of the Venice Lagoon ecosystem. Notably, corporate executives were found guilty and convicted for their negligent behavior. Tying the criminal and civil cases together was a productive way to bracket corporations as an amalgam of responsible persons *and* company policies/practices.

In the United States, Exxon and other oil companies managed to use their lobbying clout to omit their waste from governmental scrutiny. Thus, it was difficult to accuse them of any wrongdoing as they were within their legal rights. But even given the same case in a state with stricter regulations, like Alabama, there would be no possibility of trying both civil and criminal cases together. A civil case could have been pursued, but the possibility of actually holding a company executive personally and criminally liable would be highly unlikely. While the Italian executives never served prison time, the court sent a strong message that, I believe, has served as a deterrent. The fact that a significant number of the chemical plants in the Porto Marghera have downsized or closed in the following years might indicate how serious companies and their CEOs took the threat of punishment by the Italian state. While the U.S.

practice of fining companies and awarding damages is an appropriate action, this can often be offset by the profits made during years of illegal and/or irresponsible corporate practices (Allen 2003). Placing the legal responsibility on persons, the actual decision makers in corporations, will increase the effort that companies and their employees make to obey the law and avoid damaging human health.

The corporate constitution of the chemical industry was also very different at the time of the case studies. In the United States, Exxon was (and still is) a wholly private corporation that is regulated by both state and federal agencies. The relationship between the regulations, regulators, and chemical companies in Louisiana is fraught with conflicts of interest among the parties and suffers from lax enforcement, low penalties, and a culture of corporate hiring of former regulators and vice versa (Allen 2003). In Italy, during the last half of the twentieth century, the government-owned sector of business was the largest of any noncommunist state (Bull and Newell 2005: 172). Ente Nazionale Idrocarburi (ENI), or National Hydrocarbons Corporation, established in 1953, was the umbrella state holding company under which ENIChem (formerly Montedison) operated. The 1990s and the emergence of the Second Republic (after the fall of the government under the weight of excessive corruption) also came with the abolition of the Ministry of State Holdings (Bull and Newell 2005: 181). During the early 1990s, the government privatized ENI and the other large state-owned corporate conglomerates. This opened a new "negotiating space" between government and industry that provided opportunity for directing blame and responsibility in a more transparent way.

But the most important difference was that in the Italian case there was an acceptance of a more participatory citizen science. The boundary between what was considered science and nonscience for both court and regulatory purposes was more permeable. Medical workers from the university were willing to share information on citizen exposure with the town as well as research on exposure limits for workers. Former corporate scientists were willing to form nonprofits (i.e., Istituto Ramazzini) to further exploration of exposure science and publish results openly. Traditional scientific results could travel across a variety of venues to hybrid publications such as the union-sponsored health journal *Medicina Democratica*. Citizens performing popular epidemiology studies could be published alongside traditional medical research and thus be recognized as expert knowledge in both court and policy arenas.

In the U.S. case, the boundaries between traditional science and public/participatory science were firmly drawn. This was due, in part, to one scientist's desire to legitimate her research both within her profession and within the court. Sharing or hybridization of any kind would mark her work as less scientific and open it to denial of admissibility in the trial (or future exposure lawsuits).[12] Additionally, strict medical privacy laws in the United States further hindered the collection of public health data.[13] Exxon was able to win by strategically using the court's heavily weighted reliance on traditional science to argue: (1) the universal problem of scientific uncertainty, and (2) the particular problem of incomplete research.

Conclusion

Today there are striking differences in the development of the two regions. In Porto Marghera, many of the plants have been shut down. While this, along with stricter regulations and fines, has led to a cleaner environment, it has also led to fewer high-paying industrial jobs in the chemical industry. The state did have the foresight to found a new "green chemistry" research institute (INCA) in Marghera, but it has had little impact on actually changing chemical production processes, as companies have chosen to move rather than retool at some expense.[14] The environmental social movement that pressured to plants to take responsibility for worker and lagoon health, did not also use its momentum to pressure/enable industry (via government incentives and regulation) to take a new "green" path. Now, the momentum is lost, and the collective memory of the greater Marghera area is of jobs lost, fueling an animosity among many residents toward environmentalists.

In "Cancer Alley," the region along the Mississippi River that includes Grand Bois, industry, particularly noxious chloro-chemical production, has grown. Citizen groups, particularly poor and minority residents, organized to enable industry to get permits and further expand. The reason for this was the promise of jobs. Some of these groups have actively asked that citizen-oriented scientists, such as Wilma Subra and others, not attend their meetings or give them information and advice.[15] They consider Subra's work to be political and even incendiary and prefer to consider the economic health (i.e., employment) of their community first and foremost. The new, post-Katrina, more conservative governor and regulators in the state are pleased with this new alliance: the poor/minority citizens, industry, and the state have all come together to further petrochemical expansion in the region.

Cross-national comparative case studies about regulating toxicants and understanding their effects are instructive. They are less about developing "best practices" that can be uncritically extrapolated to other locations and cultures and more about, in a Tocquevillian fashion, understanding one's own. As Sheila Jasanoff eloquently states: "The aim of comparison is to reveal, with critical detachment but epistemic charity, what gives significance to another culture's distinctions and differences, not forgetting in the process to reflect on the commitments encoded in one's own. It is not the divine prerogative of producing universally valid principles of knowledge or governance that comparison should strive for. It is to make visible the normative implications of different forms of contemporary scientific and political life, and show what is at stake" (2005: 291).

Notes

1. This region is commonly defined as the communities that lie along the east-west stretch of the Mississippi River between New Orleans and Baton Rouge and are surrounded by over 130 chemical- and petroleum-processing plants.
2. The fact that every citizen of Grand Bois joined in the lawsuit is unusual and goes somewhat against the research of Couch and Kroll-Smith (1994) showing that exposure controversies tend to divide communities. In this case the community was a multigenerational, long stable community united through both intermarriage and religion. They had also witnessed a similar exposure fight in an adjacent parish (county) where one of the problems was a rise in childhood leukemia cases. And lastly, the fight was to close the site and locate it elsewhere, so the dispute was not only an "exposure" dispute. These factors likely led to community unity in fighting the waste site in their town.
3. From the online interview entitled "Wilma Subra." http://www.commonweal.org/wilmasubra.html (last accessed 3 November 2010).
4. Interview by author (16 May 2003).
5. Ibid.
6. Louisiana state agencies had been complicit a number of times in designing studies that showed nothing was wrong. For example, in the 1980s the state Department of Health and Hospitals commissioned a study that was designed in such a way that it disproved popular epidemiological evidence of a high miscarriage rate around chemical plants. Similarly, the Louisiana Tumor Registry has produced data that could be interpreted to mean that it is actually healthier for some populations to live near chemical plants (Allen 2003).
7. For a complete history of what the corporations knew regarding the health effects of PVC/VCM and their pact of secrecy, see Markowitz and Rosner (2002: 168–94).
8. This journal is also discussed in some detail in Stefania Barca's chapter in this volume.
9. For an excellent discussion of the "expertification" of the layperson in an activist movement see Epstein (1996).
10. From an interview by the author with Felice Casson, now a senator in the Italian legislature, conducted 13 November 2006 in Venice, Italy.
11. Potere Operaio (Workers' Power) was a radical social movement based in Porto Marghera, consisting of both workers and academics, such as Antonio Negri, one of the leaders. *Porto Marghera—The Last Firebrands* is an excellent documentary film of the movement available on DVD distributed by Wildcat (Germany).
12. In the 1990s there were several U.S. Supreme Court decisions limiting the admissibility of scientific evidence in the courts, the first being *Daubert v. Merrell Dow Pharmaceuticals, Inc.* See Solomon and Hackett (1996) for an STS analysis of *Daubert* and Carl Cranor (2006) for a legal-philosophical analysis of relevant cases and their implications for toxic torts.

13. For more on the problem of the right to privacy hindering the collection of public health data see Allen (2008).
14. From an interview by the author (1 February 2008) with Pietro Tundo, professor of organic chemistry, University of Venice, and researcher at INCA.
15. From an interview by author with Wilma Subra (25 January 2006).

Bibliography

Agnew, John A. 2002. *Place and Politics in Modern Italy.* Chicago, I.L.: University of Chicago Press.

Allen, Barbara L. 2003. *Uneasy Alchemy: Citizens and Experts in Louisiana's Chemical Corridor Disputes.* Cambridge, M.A.: MIT Press.

———. 2008. "Environmental Health and Missing Data." *Environmental History* 13(4): 659–66.

Benatelli, Nicoletta, Gianni Favarato and Elisio Trevisan, eds. 2002. *Processo a Marghera. L'inchesta sul Petrolchimico. Il CVM e le morti degli operai. Storia di una tragedia umana e ambientale.* Venice, Italy: Nuova Dimensione.

Bettin, Gianfranco. 1998. *Petrolkimiko: Le voci e le storie di un crimine di pace.* Milan, Italy: Baldini & Castoldi.

Bortolozzo, Gabriele. 1994a. "La produzione di CVM e PVC al Petrolchimico di Porto Marghera." *Medicina Democratica* 92/93(January–April): 42–45.

———. 1994b. "La cancerogenesi da CVM." *Medicina Democratica* 92/93(January–April): 46–56.

Bull, Martin J., and James L. Newell. 2005. *Italian Politics: Adjustment Under Duress.* Cambridge, U.K.: Polity Press.

Couch, Stephen R., and Steve Kroll-Smith. 1994. "Environmental Controversies, Interactional Resources, and Rural Communities." *Rural Sociology* 59(1): 25–44.

Cranor, Carl F. 2006. *Toxic Torts: Science, Law, and the Possibility of Justice.* New York: Cambridge University Press.

Daugherty, Christi. 1997. "Legal Poison? Grand Bois Residents Want the State to Shut Down the Oil Pit Next Door." *New Orleans Gambit Weekly,* 25 November.

Downey, Gary Lee. 1988. "Structure and Practice in the Cultural Identities of Scientists: Negotiating Nuclear Waste in New Mexico." *Anthropological Quarterly* 61(1): 26–83.

Dunne, Mike. 1998. "Exxon Engineer Testifies in Lawsuit." *Baton Rouge Advocate,* 22 July.

Epstein, Steven. 1996. *Impure Science.* Berkeley, C.A.: University of California Press.

Ferstel, Vicki. 1998. "Tests Reveal Abnormal Levels of Lead in Grand Bois Residents." *Baton Rouge Advocate,* 22 December.

Fischer, Frank. 2000. *Citizens, Experts, and the Environment.* Durham, N.C.: Duke University Press.

Frickel, Scott, and Kelly Moore, eds. 2005. *The New Political Sociology of Science: Institutions, Networks, and Power.* Madison, W.I.: University of Wisconsin Press.

Ginsborg, Paul. 2003a. *A History of Contemporary Italy: Society and Politics, 1943-1988.* New York: Palgrave MacMillan Press.

———. 2003b. *Italy and Its Discontents: Family, Civil Society, and State: 1980-2001.* New York: Palgrave MacMillan Press.

Harding, Sandra G. 1991. *Whose Knowledge? Whose Science?* Ithaca, N.Y.: Cornell University Press.
———. 2008. *Sciences From Below: Feminisms, Postcolonialities, and Modernities.* Durham, N.C.: Duke University Press.
Hess, David J. 2009. *Localist Movements in a Global Economy: Sustainability, Justice, and Urban Development in the United States.* Cambridge, M.A.: MIT Press.
Irwin, Alan. 1995. *Citizen Science.* New York: Routledge.
Irwin, Alan, Alison Dale, and Denis Smith. 1996. "Science and Hell's Kitchen: The Local Understanding of Hazard Issues." In *Misunderstanding Science?*, eds. Alan Irwin and Brian Wynne, 47–64. New York: Cambridge University Press.
Irwin, Alan, and Brian Wynne, eds. 1996. *Misunderstanding Science?* New York: Cambridge University Press.
Jasanoff, Sheila. 2005. *Designs of Nature: Science and Democracy in Europe and the United States.* Princeton, N.J.: Princeton University Press.
Jasanoff, Sheila, and Marybeth Long Martello, eds. 2004. *Earthly Politics: Local and Global in Environmental Governance.* Cambrigde, M.A.: MIT Press.
Koff, Sondra Z., and Stephen P. Koff. 2000. *Italy: From the First to the Second Republic.* London: Routledge.
McMillan, John. 1997a. "Records: Exxon Told to Clean Up Act, then Used La. as a Dump Site for Waste." *Baton Rouge Advocate,* 3 November.
———. 1997b. "Toxicologist Grilled about Study Results." *Baton Rouge Advocate,* 5 November.
———. 1998. "EPA Heeds Request on Grand Bois." *Baton Rouge Advocate,* 29 April.
Markowitz, Gerald, and David Rosner. 2002. *Deceit and Denial: The Deadly Politics of Industrial Pollution.* Berkeley, C.A.: University of California Press.
Mencini, Giannandrea. 2005. *Il Fronte per la difesa di Venezia e della Laguna: e le denunce di Indro Montanelli.* Venice, Italy: Supernova.
Rabitti, Paolo. 1998. *Cronache dalla Chimica: Marghera e le altre.* Naples, Italy: Cuen.
Roberts, J. Timmos, and Melissa M. Toffolon-Weiss. 2001. *Chronicles for the Environmental Justice Frontline.* New York: Cambridge University Press.
Robichaux, Michael R. 1999. "The Story of Grand Bois." http://senate.legis.state.la.us/Senators/Archives/1999/Robichaux/topics/grandbois.htm (last accessed 3 November 2010).
Solomon, Shana M., and Edward J. Hackett. 1996. "Setting Boundaries between Science and Law: Lessons from *Daubert v. Merrell Dow Pharmaceuticals, Inc.*" *Science, Technology & Human Values* 21(2): 131–56.
Williams, Patricia. 1997. *Grand Bois: Community Health Assessment, Volume 1.* Baton Rouge, L.A.: Occupational Toxicology Outreach Office.

 CHAPTER 8

Guinea Pigs Go to Court
Epidemiology and Class Actions in Taiwan

Paul Jobin and Yu-Hwei Tseng

This chapter describes the first two major cases of industrial diseases brought to justice in Taiwan, with the support of an original citizen mobilization and a network of lawyers. The first case was brought in the north of the island near Taipei. The 450 plaintiffs had been exposed to a wide range of organic solvents like trichloroethylene and other toxins while they were working for the Radio Corporation of America (RCA), a U.S. manufacturer of television sets. More than a thousand people identified with this case have developed various sorts of cancer. The second case was brought near Tainan, in the south of the island, where tremendous concentrations of dioxin were left by a former chemical plant. In both cases, the plaintiffs complained that they were used as guinea pigs for the sake of science. The court hearings and the interviews of the various actors involved suggest that the scientific uncertainty inevitably generates various forms of compromises, between "perhaps" and "probable," epidemiology and toxicology, humans and animals, and thus, generates all sorts of possibilities for a legal decision or a policy.

In the United States, many critics of *epidemiology as usual* in the case of industrial diseases have emerged, both from within the discipline and outside of it. Epidemiologist Carl Shy (1997) reproduced the proceedings of an imaginary court of law, where epidemiology is charged with "failure to serve as the basic science of public health"; no decision is rendered, but this somewhat humorous article points out the discontent with the field's limited perspective. Outside the discipline, as early as the mid 1980s and with much more consistency, the sociologist Phil Brown (1987) brought the *public* back into public health, through his conceptualization of "popular epidemiology," in which non-experts initiate the search for scientific evidence of the causes and effects of toxic issues. Brown would later identify "critical epidemiologists" as scientists not only eager to cooperate with the public, but also ready to challenge the methodological blind angles of their discipline (Brown 1997). Finally, he synthesized the various forms and levels of cooperation between laypeople and scientists (Brown 2007: 14–39). Incidentally, Brown's seminal fieldwork on popular epidemiology was on the case of Woburn, Massachusetts[1] and also dealt with trichloroethylene, the major contaminant at issue in the RCA case in Taiwan. But in contrast with Woburn, where the victims were residents' children affected by leukemia, the RCA case mainly concerns former workers of the plant—predominantly female—affected by all sorts of cancers and mutagenic or reproductive disorders. Moreover, the two cases that we present here highlight the tensions between classical epidemiologists, who nevertheless share some of the insights of popular epidemiology, and their more hardcore peers. Many of them were trained in the world's top universities, like the Harvard School of Public Health, which eased their path to success in their respective fields and helped them to accumulate impressive lists of publications in the best scientific journals. However classical these epidemiologists may be, they have played a decisive role in constructing the scientific truth used to determine the impact of the chemical toxins used or generated by the industrial process on residents and former workers of the plant. But the translation of this epidemiological truth into a judicial decision—eagerly awaited by the victims of the contamination—is yet to come and does not depend only on the will of these epidemiologists.

The Toxicant between "Possible" and "Probable"

RCA was founded in 1919 in Camden, New Jersey, with the backing of General Electric (GE). The history of RCA has been described as a "70-year quest for cheap labor" (Cowie 1999). In 1939, after the company succeeded in developing the United States' first all-electronic television system, it opened a new plant in Bloomington, Indiana, a more rural area with less unionized labor, and then in Memphis in 1965, where African-Americans made up the core of the labor pool. As U.S. environmental regulation was becoming more stringent, RCA was among the first big American corporations to move abroad, initially to Ciudad Juarez, Mexico, in 1964. In 1970, RCA founded two factories in Taiwan for the production of television sets, both to the south of Taipei, one in Taoyuan and the other one in Chupei. A third factory would later be built in Ilan county, northeast of Taipei. But Taoyuan would remain the biggest plant and the future center of mobilization concerning the hazards issue. The company easily recruited thousands of workers, mostly young women who had just finished junior high school. The prestige of the company and the fascination with its output (brand-new models of TV sets) made it very attractive for the rural populace. RCA was a nice "family," providing workers with social activities, dormitories, etc. The pliable workforces in Taiwan and Mexico would eventually be used as a sort of blackmail to prevent American workers from striking. In 1986, RCA was acquired by General Electric, which, two years later, sold its consumer-electronics branch to the French corporation Thomson. In 1992, however, after examination of the groundwater and soil, Thomson sold the Taiwanese plants to local companies, then moved its own production to China and Singapore.

Fifteen Years of Investigation to Measure RCA's Hazards Legacy

Before it became an emblematic case of occupational hazards, the RCA issue started as a matter of ex post facto environmental concern around the vicinity of the former factory. In June 1994, a legislator and former director of the Environmental Protection Administration (EPA) pushed forward a survey that concluded with the presence of extremely high concentrations of several organic solvents like trichloroethylene (TCE) in the soil and groundwater. After another survey, in June 1998, the EPA announced that the RCA site in Taoyuan was a "permanently contaminated area" (Wu 2009: 206–7). Meanwhile, it happened that many former workers were suffering from various sorts of cancer. In 1998, they launched the RCA Self-Help Association (RCA-SHA), which soon received the support of the Taiwan Association of Victims of Occupational Injuries (TAVOI) founded in 1992 by intellectuals and labor activists with Christian or leftist sensibilities. Together, TAVOI and the SHA started to lobby the government to get compensation. In April 2001, the two associations conducted an investigation with the help of the government's Council of Labor Affairs (CLA). They found 1,395 former workers with cancer (226 had died already) and 100 with various tumors. In 2002, members of TAVOI and RCA-SHA went to the United States for a two-week campaign that sought the support of the U.S. Labor Department, members of Congress, the GE Labor Union, etc., for their cause (Ku 2006).

Under pressure from the media coverage and a critical report of the Control Yuan[2], the government launched an inter-ministry task force to set up epidemiological and risk assessment surveys among former workers and neighbors of the site, and to identify the contamination source with hydrological checks. One study was conducted by the Institute of Occupational Safety and Health (IOSH), which is affiliated with the CLA. The other survey was sponsored by the government's Environmental Protection Agency (EPA) and conducted by the College of Public Health of National Taiwan University (NTU), under the leadership of Professor Wang Jung-Der, a prominent figure in Taiwan in occupational and environmental medicine.

The IOSH team produced three reports in Chinese between 1999 and 2001,[3] then three articles in English in international scientific journals—the last one in 2005—while the team from NTU wrote two reports in Chinese in 1999 and 2000, then submitted six articles to international scientific journals—the last in 2009. The final results of the IOSH team (Chang et al. 2005) concluded that there was no significantly elevated cancer incidence nor any "standardized incidence ratio" (SIR) for any type of cancer in exposed workers, arguing that the numerous short durations of employment might bias the cancer risk toward false positives. The authors presume that the cancers could only appear after a long period of exposure, neglecting the possible increase in toxicity from the combination of the various carcinogens involved and their massive use. On the basis of IOSH's reports, and at the time that TAVOI was campaigning in the United States in May 2002, GE made a statement to the press that the company could not be held liable, since the Taiwanese government itself had confirmed that the cancers were not related to RCA (Ku 2006).

The results of the NTU team draw a much different picture. The collaboration with toxicologists for experimentation on mice showed that the mixture of organic solvents (including trichloroethylene) present in the underground water near the factory was a potential carcinogen to both male and female mice (Wang et al. 2002). Other NTU articles were epidemiological surveys. The first results could only suggest evidence for liver cancer among male residents (Lee et al. 2002; Lee et al. 2003). In their last series of articles (Sung et al. 2007; Sung et al. 2008; Sung et al. 2009), it is apparent that the authors had investigated all possible means to find evidence, but that they were limited by the methodological constraints of classical epidemiology. The first one concerned the consequences for the workers themselves. It was based on a cohort of 63,982 female workers covering the period 1973–1997 (Sung et al. 2007). Despite a total of 1,572 cancer cases for the period 1979–2001, and despite an extensive review suggesting an association of TCE exposure with kidney cancer, liver cancer, and non-Hodgkin's lymphoma, as well as with cervical cancer, Hodgkin's disease, and multiple myeloma, no increase of SIR could be found. The authors could only conclude that workers first employed prior to 1974, with exposure to trichloroethylene and/or a mixture of solvents, "may have an excess risk of breast cancer" (Sung et al. 2007).

To account for such limitations, the authors stressed that the analysis was "limited by the lack of detailed exposure information"—a reference to the fact that RCA, GE, and Thomson not only refused to disclose job histories and other archives, but eventually tried to hide or destroy all potential evidence. As the authors point out, "the factory had been inspected eight times by the Taiwanese government's inspection agency, with multiple violations of the regulations having been recorded" (Sung et al. 2007). The last two articles focused on the possible consequences for the workers' offspring. It seems as if the authors were finally forced to conclude much less than what they intuitively felt was there, as in both papers they emphasize the lack of data and the multiple violations of solvent regulations by the company. At least they could report an increased incidence of leukemia in the children of female workers (Sung et al.

2008) and a relative increase in infant mortality due to congenital malformations, especially for cardiac defects, for the children of male workers (Sung et al. 2009).

Besides these surveys, two literature reviews have been carried out, one for a public report in Chinese conducted by Wang Jung-Der (Taiwan Bureau of health Promotion, 2003), the other one in English by the Chinese-American epidemiologist Otto Wong (2004). The former found short-term high exposure in female workers during the early 1970s. Though the latter pretends to be an exhaustive analysis including most of the articles related to the RCA issue in Taiwan available at that time, yet Wong's conclusion seems to take into account only those supporting an absence of risk. Historians Gerald Markowitz and David Rosner have included the author, Otto Wong, among the "damn liars" more inclined to serve the interests of industry than those of public health, as he was instrumental for both the vinyl chloride industry and the chemical polluters of "Cancer Alley" in Southern Louisiana (Markowitz and Rosner 2002). Both Wang Jung-Der and Otto Wong have accepted to stand at the bar as expert—the first one at the demand of the plaintiffs' lawyers, the second at the demand of the defendants; their court hearing, which is expected in the last period of the trial (2013–14), might play a major role in the decision.

Challenges to the Pax Epidemiologica

For the last decade, as the former workers of RCA faced various sorts of cancer but received no compensation, this issue has generated growing criticism of the conservative conclusions of the Taiwanese *pax epidemiologica*—to borrow from what Christopher Sellers (1997) defined as the *pax toxicologica* in America in the 1930s. Indeed, in this case, the Taiwanese bureaucrats have only considered scientific data, overemphasizing epidemiology in particular and completely disregarding the testimonies of the workers.

Inspired both by the gender studies and the popular epidemiology of Phil Brown, Lin Yi-Ping (2006) stressed that the surveys of both teams (IOSH and NTU-CPH) were also distorted by a male-dominated methodology that ignored or minimized the specificities of the majority of former workers, i.e., women. As she was a doctoral candidate at NTU at that time, she joined the team of Wang Jung-Der for their next survey (Sung et al. 2007) and was instrumental in correcting those weaknesses. Nevertheless, the authors could show causal links only for breast cancer. Also inspired by Phil Brown and by other alternatives to dominant epidemiology, Wu Yi-Ling (2009) performed a critique of the IOSH surveys of RCA from a sociological perspective. She found a number of methodological weaknesses and stressed that the routine comparison of data through a "one cause, one effect" approach leads to conclusions of false negatives and is characteristic of what she terms a "politics of scientifically inconclusive results." Such an approach may thus serve to undermine preventive measures, as well as the subsequent payment of compensation. But all the scholars should not be lumped together.

The researchers of the government-controlled IOSH might be inclined to minimize the problems that the government has to deal with, especially as the polluter had already left the country. NTU scholars, on the other hand, as long as they publish in renowned international journals, are assured of getting research credits, even from state-related institutions like the National Science Council, EPA, or Labor Department, no matter what their findings. And as new clusters or significant issues may benefit from "publication bias" or rather good quotation scores, their authors are more easily incited to feel sympathetic to the victims. Because they lacked company data, NTU researchers tried animal experimentation to find and demonstrate a causal link, but RCA's former workers viewed their research as just another useless attempt to accumulate data; their perception of the process summed up by the lament: "How many of us shall die until we shall be recognized as statistically significant?"[4] The slogan "We're not guinea pigs!" that appeared on a placard during a protest action around the same period expressed a similar misunderstanding of what the scientists from NTU were trying to do, conflating them with the IOSH team, which repeatedly denied any possible cause-effect relationship. While the IOSH team did not pursue its investigation after its last publication in 2005, the NTU team did continue a systematic quest for more evidence. Despite their limitations, the latter's epidemiological and toxicological results are now considered to be valuable arguments by the plaintiffs' lawyers, whose challenge will consist in translating these "*inconclusive* results" into a judicial *decision*.

Five Years after It Started…the Lawsuit Is Just Beginning

On January 2001, the inter-ministry investigation task force was dissolved. Accompanied by TAVOI, the RCA-SHA launched a protest action before the government; they also petitioned the Legislative Yuan (Parliament) and the Ministry of Foreign Affairs and met with some lawyers who formed a voluntary group to help in the legal battle to come. One year later, the lawyers used secret documents that they had obtained from the CLA to urge the court to seize the assets left by RCA in Taiwan. It would later appear in the financial documents of the company that, in 1998, RCA had already moved abroad a bank deposit of 2.8 billion New Taiwan Dollars.[5] After hesitating to start a lawsuit in the United States, some 200 members of RCA-SHA decided finally to launch a lawsuit in the Taipei District Court in April 2004. This suit was rejected for procedural reasons. The association appealed the decision to the High Court of Taiwan, which also rejected the case—on the same grounds—in August 2005. The association then brought the case to the Supreme Court,

Figure 8.1. TAVOI's General Secretary Hwang Hsiao-ling addressing the media in front of the Taipei District Court on the day of the first court hearing, 11 November 2009. On her right is Lin Yong-song, lead counsel for the plaintiffs.

which, in December 2005, declared the original judgment unsuitable and ordered the High Court to reexamine the case. In March 2006, the High Court rejected the previous decision of the Taipei District Court, which was ordered to reexamine the case.[6] So the plaintiffs had to start all over again! In the meantime, forty-seven of them had died. Despite their long latency, occupational cancers often kill before justice can be meted out and compensation paid. For the current lawsuit, the first court hearing occurred in March 2009, and the last hearing is planned for August 2014.

During the year 2007, the association received the support of the Legal Aid Foundation (*Fafu* in Chinese), an organization that was launched in 2004, thanks to a mobilization of lawyers and the democratization of the country. *Fafu* established a support group of around fifty lawyers with a core group of ten devoted to the RCA issue. After investigating other evidence, these "cause lawyers" (Sarat and Scheingold 2006) also considered suing General Electric and Thomson for compensation in the amount of 2.4 billion NT$ (New Taiwan Dollars),[7] for a total number of 438 plaintiffs registered under three distinct groups. The ten lawyers clearly established that in 1987, one year after the sale of the plant to GE, RCA and GE had jointly conducted an environmental survey but failed to disclose the results. In 1994, following the sale of the plant to Thomson and then to local owners, Taiwan's EPA insisted that RCA, Thomson, and GE act jointly to clean up the pollution, but the companies demanded that the Taiwanese government agree not to pursue them for liabilities. The government consented in order to speed the clean-up process. Under the strong leadership of the lawyer Lin Yong-Song, the Taipei branch of *Fafu* has held regular brainstorming discussion meetings with the RCA-SHA and TAVOI, inviting experts to join when necessary. Due to the highly technical aspects of the debate on causality, health experts play a crucial role in these meetings, but the friendly atmosphere allows the most engaged plaintiffs (a core group of six women and one man, all of them former workers at RCA) to participate actively in the discussion.

Animal and Human Experimentation

A document submitted to the court by the defendants (GE, Thomson) in March 2009 argued that "plaintiffs must present expert testimony demonstrating that exposure to (a particular chemical) more than doubled the risk of their alleged injuries."[8] They further asserted that "a possible cause only becomes 'probable' when ... it becomes more likely than not that the injury was the result of the action." Their document was based mainly on verdicts in the American courts, except for its quotation of the three IOSH reports to reject causality for the various cancers in the specific case of Taiwan RCA former workers. The document also stipulated that "epidemiology is the best evidence of causation in the mass torts context," as if toxicology and animal experimentation were not appropriate sources of evidence. Against this simplistic reasoning and caricature of epidemiology, the plaintiffs' lawyers deployed in their response the complexity of carcinogenesis, stressing the absence in the literature of any threshold of exposure and the possible combined effects of the cocktails of toxicants the RCA workers had been exposed to.[9] As a means to valorize toxicology and animal experimentation as legitimate complements to human epidemiology, they pointed out that the system of classification and labeling of chemicals established by the United Nations Economic Commission for Europe (UN-ECE) assumes that animal experimentation is sufficient to determine human carcinogenicity unless proven otherwise. Therefore, products that have been proven to be toxic or carcinogenic through animal experimentation do not necessarily need to be "tested" by epidemiology in order to prove their toxicity/carcinogenicity for humans.

The lawyers then showed that the surveys conducted by the NTU team of Professor Wang Jung-Der, both in their epidemiological and toxicological dimensions, provide a sufficient body of evidence, congruent with the standards of such organizations as the International Agency for Research on Cancer (IARC), the National Toxicology Program (NTP) of the U.S. Department of Health, and the U.S. National Institute for Occupational Safety and Health (NIOSH).

Concretely speaking, because most of the workers had to clean PC boards with organic solvents, they have been exposed in massive quantities to trichloroethylene, tetrachloroethylene, and chloroform, which are all recognized as occupational carcinogens by NIOSH and are classified as "reasonable" carcinogens by the NTP and as "probable" or "possible" by the IARC. Besides, they were exposed to naphtha, which contains benzene, a certified carcinogen for the IARC and the NTP. Furthermore, the water that they were given for drinking or washing themselves contained not only the solvents already mentioned, but also vinyl chloride, a certified carcinogen according to the IARC, as well as 1,2-dichloroethane and methylene chloride, classified as "reasonably" carcinogenic by the NTP and "possibly" carcinogenic by the IARC. Besides these recognized carcinogens, workers have also been exposed to other strong toxins like xylene, toluene, isopropyl alcohol, acetone, and ethyl acetate. Such complex combinations of toxins should therefore invalidate any attempt at a "one cause, one effect" approach.

Therefore, the surveys conducted by the NTU team tend to show that solvents and chemicals used at RCA are carcinogenic both through animal experimentation and human epidemiology. These surveys may yield a "not statistically significant" result, but this does not mean that the relationship between exposure and disease is insignificant or nonexistent. A lack of data presenting complete job histories of the plaintiffs (to prove the exact location of their exposure) might be a greater obstacle to proving their case. Moreover, the judges have considerable latitude for interpreting the *probabilities* that founded the current classification of the toxicants involved.

Between "Possible" and "Probable"

Concerning the list set by NIOSH, there is no graduation of probability; all the toxicants listed are *occupational* carcinogens. But the meta-categories set by IARC and NTC aim at a different purpose: to provide an accurate synthesis of the available international literature. As the defense may argue, the major toxicants at issue in the RCA case—trichloroethylene (TCE) and tetrachloroethylene—are *only* Class 2 in the NTP ratings ("reasonably anticipated to be a human carcinogen"), and Class 2A in IARC's ("*probably* carcinogenic to humans"). Class 2A is based on animal experimentation, but with limited or insufficient human epidemiological evidence for the toxicant's inclusion in Class 1, substances that are definitely proved to be "carcinogenic to humans." The category 2B further designates substances that are "*possibly* carcinogenic to humans." In October 2012, IARC shifted trichloroethylene from category

Table 8.1. Carcinogens Implicated in the RCA Case

Categories of carcinogens	NIOSH	NTP	IARC
1: Certified as "occupational carcinogens" (NIOSH), or "known to be human carcinogens" (NTP) or "carcinogenic to humans" (IARC)	Trichloroethylene, tetrachloro-ethylene, chloroform	Benzene, vinyl chloride	Benzene, vinyl chloride, trichloroethylene (since October 2012),
2: "reasonably anticipated to be a human carcinogen" (NTP)		Trichloroethylene, tetrachloroethylene, chloroform, 1,2-dichloroethane, methylene chloride	
2A: "probably carcinogenic to humans" (IARC)			Trichloroethylene (until September 2012), tetrachloroethylene
2B: "possibly carcinogenic to humans" (IARC)			Chloroform, 1,2-dichloroethane, methylene chloride
3: "not classifiable as to its carcinogenicity to humans" (IARC)			Xylene, toluene, isopropyl alcohol
4: "probably not carcinogenic to humans" (IARC)			?

2A to category 1, which could give another advantage for the plaintiffs of the RCA case in Taiwan.

It appears as if the line of demarcation between Classes 1 and 2A is a distinction between humans and animals, and Class 2A a sort of waiting room for Class 1. But until someone has the opportunity to create an ethnography of these organizations and their decision processes, no one really knows how the toxicants are shifted from one category to the other. So, the difference between 2A and 1 might be as thin as the one between 2A and 2B, or between "probably" and "possibly." As the criteria for such decisions are unclear, the decision-making process results perhaps less from the smart probability reasoning inherited from Pascal and Bayes than from all sorts of *compromises* between one hypothesis and another, or perhaps between science and economic priorities. Like Shapiro (1991), who has described the "probable cause" standard as both a cornerstone and a "talismanic formula" of the American judicial system, IARC's categories from 2A downward (to the Class 3 "not classifiable" and 4 "probably not") also appear to be talismanic formulas to soothe our anxious *ignorance*[10] as incomplete moderns, as posited by Latour (1993).

Similarly, such uncertainty leaves plenty of room for the judges to make up their own minds. As identified by Jasanoff (1995: 114–37) in her analytical framework of toxic torts, they may lean toward the arguments of the "radical reformists," who favor hard epidemiological data, or they may be more sensitive to those of the "incrementalists," who draw from a larger repertory of evidence, from clinical data to limited—though not insignificant—statistical significance in toxicology or epidemiology. Moreover, the concrete testimony of the plaintiffs and/or their physicians may gain more attention from the judge than the strictly abstract figures favored by the "radical reformists"; that would be considered as the "human factor" in the decision. In other words, so as to decide between these different regimes of truth, the judges also have to make some sort of intellectual compromises. In the next part, we will explore further aspects of these compromises.

The Toxicant as a "Resource"

The Anshun area is located in the rural suburbs of Tainan city, in the south of Taiwan. It is a beautiful, quiet area, between seashores and hectares of former salt ponds converted into oyster and fish farms. Just a few hundred meters from the "dioxin hot spot" is one of Taiwan's oldest and most magnificent Mazu temples, visited by pilgrims from all over the island. The industrial hazards that struck this lovely place can be seen as a legacy of its colonial and postcolonial modern past. In 1938, when Taiwan was still part of the Japanese empire and the Japanese army was expanding its control over China, the chemical company Kanegafuchi Sōda, a subsidiary of the firm Kanebō, received land—confiscated from local salt farmers—to open a plant in Anshun. After inauguration by Shinto priests and military officers in 1942, the factory began the production of caustic soda, hydrochloric acid, liquid chlorine, and toxic gas to be used in the war effort. Caustic soda was made through electrolysis of the chloralkali process using large quantities of mercury (Chang et al. 2008). This marked the first phase of occupational and environmental hazards in this area. After Japan's defeat in 1945, another era of colonialism started for the Taiwanese people when the island was taken over by the troops of General Chiang Kai-Chek. In Anshun, the company was renamed Taiwan Alkali Industry, and despite its partial destruction, the Anshun factory relaunched production of its three core products. In 1965, the factory began producing pentachlorophenol (PCP), which has been used extensively as an herbicide and wood preservative. By the 1970s, Taiwan Alkali had become the largest PCP maker in East Asia. However, its production was halted in 1978, and four years later, the entire factory was closed down. PCP has been documented per se as a hazardous occupational and environmental toxin, and it would later appear that the production of PCP might also incidentally generate dioxin. The hazards left are therefore a complex cocktail of mercury, PCP, and dioxin.

To add institutional complexity, as is often present with industrial pollution, control of Taiwan Alkali passed from hand to hand following the war. In 1966, by order of the Ministry of the Economy, the company was placed under the umbrella of China Petroleum, a public company. In 1983, after the closure of the plant, it became part of its subsidiary, the China Petroleum Development Company (CPDC), which was reprivatized in 1994. This ping pong game between the public and private sectors has created many pitfalls for the victims of this industrial pollution.

Some twenty years after the plant's closing, a confluence of scientific concern and grassroots mobilization transformed the dormant cocktail of hazards into a local and national issue. A doctoral thesis submitted at the National Tsing Hua University, followed by a journal article in 1997, examined the Anshun case (Soong et al. 1997). The survey established an exhaustive list of the various sorts of dioxins found around the plant, with one sample showing a concentration one hundred times higher than the sediments from the Er-Jen River, a known dioxin-polluted river in the south of Tainan. However, it was not until 2002–2003 that it would become a wider matter of concern. Then, in 1993, 1995, and 2004, the main author of this survey, Professor Soong Der-kau, conducted or participated in a series of systematic surveys commissioned by the EPA. However, despite thousands of pages of accumulated results, Professor Soong is not the most visible scientist in this affair.

The Complementary Narratives of Two Local "Kings"

Two key players really emerged from the activity surrounding this issue, each representing a different group in Tainan City. One is Lee Ching-Chang, professor of environmental sciences at National Cheng-Kung University (NCKU); the other is Hwang Hwan-Jang, who also teaches environmental sciences, but in more humble institutions: the Chung Hwa College of Medical Technology and the Tainan Community College. Cheng-Kung University, which ranks as the second best university in the country, is located on a wide and beautiful campus in the center of the city, with thousands of elite students and researchers and a lot of money in research funding. In contrast, Tainan Community College occupies a much smaller building and provides night classes for all sorts of citizens, yet, under Hwang's leadership, it has helped to launch and sustain the grassroots mobilization in Anshun.

When high levels of dioxin were discovered around 2002, the local population was reluctant to accept the facts, so Hwang and his comrades had to convince both the residents and the local media of the potentially dramatic impact on the environment and on their health. As Hwang discovered that there had also been significant emissions of mercury into soil and fishponds, he contacted Professor Harada Masazumi, a world-renowned specialist on Minamata disease. As he frequently did all over the globe, Harada came to Anshun to take some measurements and try to gauge its potential similarities of the situation there with the contamination of the food chain that occurred in Minamata. This sudden visit helped Hwang and his colleagues attract more media attention to the issue. Hwang and friends also went to Japan to attend a conference and learn more about Minamata's long and tragic story.[11] As word spread, with the help of Harada, other Japanese environmental specialists would come to Anshun. So far, however, the main focus of the Anshun issue remained principally dioxin. Borrowing from Latour (2005), we could say that Hwang and his young colleagues at the community college *translated* the scientific discoveries of Lee et al. into words that spur action at the grassroots level. And sometimes they must bear the various "translation costs" of this role, like incurring the anger of the public or state. Without Hwang's forceful explanations, the ordinary people of Anshun would not be able to understand the complex scientific conclusions of Lee's research (which, moreover, is mostly in English). Hwang also displays a talent for attracting and communicating with the media. So Hwang plays the role of local intermediary, while Lee stands as a sort of "imperial scholar" or "scientific autocrat," without, however, being an "at-your-service expert" (*yuyong zhuanjia*) of either industry or the state. Yet Hwang is doing more than mere *translation*. Although they are not published in English or in international journals, but in Chinese, in activist publications, Hwang's narratives on the Anshun issue are more than a simple vulgarization of Lee's surveys; they provide different insights. Through mappings and interviews of the local people, along with comparisons of various international standards on the control or treatment of dioxin, biochemical hypotheses, etc., Hwang's narratives develop a comprehensive understanding of the complex trajectories of the toxins and their impact on fish, oysters, vegetation, and people. Unlike the expert-activists described by Barbara Allen (2003) in the case of Louisiana's "Cancer Alley," Hwang has not yet trained any local activists in Anshun to develop a popular epidemiology, strictly speaking. But he has played a valuable role as a whistle-blower, attracting attention at the local, national, and even international levels, something that international journal articles alone will not produce.

Turning to the international literature on the Anshun issue, the name of Lee Ching-Chang is indeed unavoidable. With his research team from NCKU's Department of Environmental and Occupational Health (DEOH), he has designed and directed most of the surveys on the matter. Along with reports in Chinese to the Tainan City office, one of the main sponsors of those surveys, he has authored five important articles in well-known international reviews (Chen et al. 2006; Lee et al. 2006a; Lee et al. 2006b; Lee et al. 2006c; Chang et al. 2008). From time to time, he has collaborated with researchers from the National Health Research Institutes (NHRI), which is attached to the Ministry of Health. Other teams competed for research funding from Tainan City or the EPA, but Lee and his colleagues succeeded in getting most of it, thus securing his access to the cohort population. Despite such a hegemonic position on the Anshun issue, and compared to the NHRI's rather inconclusive first report, Lee's reports more firmly establish the causal links between the former plant and a large set of diseases among the population of Anshun. They also make concrete recommendations both for medical follow-up and for quick treatment of the dioxin in the soil. After his education in public health and environmental engineering at National Taiwan University (1978–1992), Lee worked at the EPA (1986–1988). This prior connection may explain why, in 1999, the EPA asked him to conduct a survey to determine serum levels of polychlorinated dibenzo-p-dioxins and dibenzofurans (PCDD/Fs) in the general population living around nineteen incinerators. They discovered incidentally that the population of two villages, Hsien-Gong and Lu-Erh, in the immediate vicinity of the former Taiwan Alkali plant, had much higher levels. Further investigation on larger human cohorts along with analysis of fish and

Figure 8.2. Placard forbidding the use of the fish farms, Anshun, August 2008.

Figure 8.3. Containers of soil contaminated by the dioxin, stored in a former factory of Taiwan Alkali Industry.

soil sediments would confirm that the Anshun area was a "hot spot" for dioxin, "the first one reported in Taiwan" (Lee et al. 2006a; 2006b; 2006c).

The Compromise: Necessary for Some, Impossible for Others

Lee's articles were in English, however, and the people of Anshun were not really informed of what was at stake. Hwang was therefore urging the Tainan City office to disclose the epidemiological surveys in Chinese. In 2008, Lin Ji-Jin, a resident of Hsian-gong who had initiated the Self-Help Association [of the victims] of Dioxin from Taiwan Alkali Anshun, sued the City Office to get those reports disclosed. He was helped in this by Wang Yu-Cheng, an assistant professor of environmental law at Cheng-Kung University. Within a few months, the city office chose to make those reports public to prevent further protest from the people of Anshun, who were being unusually restive. Lee Ching-Chang, the main author of those surveys, expressed his discontent to Wang concerning this judicial offensive: Why make such a fuss?[12]

Meanwhile, in 2005, as dioxin also became a controversial issue concerning milk and duck eggs in Taiwan (Chou 2008), Chang Kuo-Lung, director of the EPA, pushed the government to launch a program to provide medical and economic assistance to the victims and to pave the way for the cleanup/removal of the dioxin.[13] The Tainan city office was also under pressure from growing discontent among the population of Anshun; no doubt anxious that it would lose support in the next county elections, it then established a healthcare unit for the residents. A total sum of 1.3 billion New Taiwan Dollars (around 28 million Euros) was allocated for five years (2005–2010), the major part of it for the "relief" (not *compensation*) of the population. Residents of the three villages (Hsien-gong, Lu'er, and Sicao) that were most exposed could apply for a low monthly allowance if they had a blood rate of 64 picograms (pg, 10^{-12} g) of dioxin, and the equivalent of a monthly minimum wage, if they had developed serious diseases.[14] The criteria of 64 pg was presented as based on the result of the epidemiological surveys, but some residents in Anshun were not convinced by this arbitrary decision. When Lin Ji-Jin, the representative of the Anshun Self-Help Association, requested that this scale be lowered to 32 pg, which is the safety criterion recommended by the WHO, the Tainan city mayor proposed a compromise of 48 pg, but it was rejected by the expert committee. As the mayor confessed frankly: "Some of them wanted 32, others wanted 64. Well, then, I proposed a compromise. Of course, a compromise is not science, but everyone has good reasons, haven't they?"[15] The selected value of 64 pg was proposed by Lee: "The mayor finally admitted that it was not a political matter, but a problem that was strictly scientific."[16]

Bringing the "Hot Spot" to Court

Even more problematic was the fear that the "relief plan" would end in June 2010, no matter how stricken the adult population was by dramatic levels of diabetes and various cancers, and what the consequences were for their children. By 2007, Hwang Hwan-Jang therefore convinced the local chapter of the Legal Aid Foundation (*Fafu*) to make a public call to the residents to initiate a lawsuit. It was difficult for the three young female lawyers, headed by Lin Hsuan-Chi, to convince the rather elderly population of Anshun, but by July 2008 they had established a group of 85 plaintiffs who matched the financial criteria to receive legal aid. In addition, a group of ten attorneys would progressively set up another group of 115 out-of-pocket plaintiffs. The plaintiffs accuse the China Petroleum Development Company of tortuous conduct, while the Ministry of Economic Affairs, the Tainan City office and its Bureau of Environmental Protection are being sued on the basis of negligent violation of official duty. In a secondary claim, the Ministry of Economic Affairs is also being targeted as a joint tortfeasor. It requests compensation for medical care and moral suffering, and consolation payments for the relatives of those who have already died.

The litigation focuses on three issues: *liability* (who is responsible and who is not); *causation* (whether dioxin has caused physical damage); and *validity* (if the claim is made within two years after damage is known by the plaintiff, or ten years after the pollution is known to have happened). According to *Fafu*, the serum level of mercury is not particularly high on average among the residents, and is hard to prove in court; the dioxin level is comparatively higher, so it is easier to establish the exposure-disease causation. As compared to organic solvents, as in the RCA case, general causation has been more strongly established for dioxin by the international literature, notably for diabetes and cancer. However, the plaintiffs' attorneys must not only prove that CPDC's former PCP plant is the source of the dioxin in their bodies, but that the dioxin does increase their morbidity.

The first court hearing, at which both parties presented their positions, was held in February 2009. The judge tried to convince the defendants to settle with the plaintiffs by offering a settlement, arguing that the state had already given NT $1.4 billion and that the plaintiffs were claiming only a few million (100–200 million); moreover, it was clear that there was pollution, so it would be better to avoid spending so many social resources, along with the fees of six attorneys. But the CPDC said that they had nothing to do with the health of the plaintiffs, because among the 17 dioxins involved, the company only generated OCDD and not the TCDD that was found in the victims. They even

Figure 8.4. Pilgrims at Mazu Temple, Anshun, October 2012. The health checks take place in the temple.

argued that OCDD was not as fatal as TCDD. According to Lee Ching-Chang, there was no basis for such arguments and he stated his readiness to declare this in court.[17] In August and October 2010, on the invitation of the plaintiffs' attorneys, Soong Der-Kau, the author of the first report on the issue, attended two court hearings as an expert witness. Despite all the evidence that he was able to provide, he hesitated to declare unequivocally that the PCP must have emanated from the plant. The people of Anshun, whom this trial affects the most, have an ambivalent relationship with these scientists who will in large part determine their fate. Although some have had negative reactions to the repeated drawing of blood[18] (see Figure 8.2), many of them have high expectations of the resultant science, and particularly of its emblematic figure, Lee, as expressed by Mr. Su, one of the plaintiffs: "We have little chance of winning this suit against the state, and *the only resource we have is the toxin in our bodies*. ... The government will just delay and delay until all the plaintiffs die! Just within one year, ten people have died already. ... Lee Ching-Chang is the one who can determine our life and death, but he doesn't... I don't say this to attack him, but I mean he's the one who can make the State give us compensation or not."[19]

Conclusion: To Be or Not To Be a Guinea Pig

Our purpose in this article was to clarify the role played by epidemiology and toxicology in the specific case of industrial hazards. As shown by Desrosières (1998), the birth of probabilities in the seventeenth century had a lot in common with gambling (among other things); it later played a decisive role in the development of modern statistics and the Public Health Movement of nineteenth-century England. This can be considered as the positive side of what the author called the "politics of large numbers." But there is also a very dark side if we look at the sort of "sinister lotto" that many industries have been playing (Thébaud-Mony 2007). As is palpable in the argumentation of RCA lawyers, many industries have bet that not all workers exposed to toxins would be hurt, and for those who will get hurt, the long latency will help to dilute the evidence. If some epidemiologists provide evidence and are ready to testify for the victims, like Wang Jung-Der and Lee, others (the researchers of NIOSH or NHRI) minimize the cost of compensation for the sake of the state's finances, and still others (like Otto Wong) directly serve the interests of the polluting company. Epidemiology, therefore, presents a "plurality" of faces. But all those epidemiologists agree on the principle of a truth that would be indivisible, single, and unique. The idea of compromise as inherent to the very practice of science would be considered blasphemy by all of them. The political executives would be less resistant to it, even though they would not confess it as frankly as the mayor of Tainan did. And this is where the polluters can take the upper hand.

We saw that, in their quest for compensation, the people of Anshun perceive the toxin in their body to be their sole "resource." As a human cohort, they also feel that they are treated as a resource for the sake of scientific knowledge, which does not lead to fair compensation and medical care for them or to a safe solution for the future of their land. In the case of RCA, just after the disclosure of toxicological results attained through animal experimentation, the former workers protested: "We are not guinea pigs!" Conversely, in Anshun, although there was no animal experimentation, people protest: "We *are* guinea pigs!" Of course, in both cases, the meaning is the same: they feel that they are treated *as if they were* guinea pigs. And in both cases, it has motivated them to go to court. As Hwang Hwan-Jang, furious about the blood serum test organized by Tainan City, remarked: "Nature also serves as a guinea pig!" This reminds us that "ecology is not about a naturalization of politics as if one wanted to 'treat humans like plants and animals'; it's about the immense complexity involved for any entity—human or non-human—to have a voice" (Latour and Weibel 2005: 458). Yet, we think it important to highlight that in this hybrid parliament of *Res Publica* (Latour and Weibel 2005), some humans may be forced to reduce their right to speak through the *thing* (*Res*) that invaded their body, and which is measured in *invisible* quantities as small as picograms. Not only do toxins become their sole resource with which to negotiate and build their future, but their final recourse is to go to court to *publicly* voice the intimate details of their bodies' suffering. Because of the long periods of waiting between court hearings, both the cases of RCA and Anshun are still far from a conclusion. At Taipei District Court, the former RCA workers who appear in court to testify are compelled to give details about their whole life, their family, and of course about their physical problems—from relatively minor diseases to extremely delicate issues related to gender identity, such as breast or uterine cancer, as well as to ill or stillborn children. Conversely, at the Tainan District Court, the plaintiffs have not yet even been given a chance to express their suffering, all the hearings being devoted to debates between medical experts. The court therefore allows only very limited self-expression, so that, even if the plaintiffs should win the case with severe sanctions for the polluters, they might still be left with feelings of great frustration.[20] There is much that is wrong with this state of affairs, between human experimentation by epidemiology and frustrating condemnation at court. In the process, however, all the actors can contribute to rebuilding that thing called "public health." In the two cases that we presented here, many former workers of RCA, and to a lesser extent the residents of Anshun, are clearly engaged in this process. Despite all their frustrations, the efforts of these victims, and their lawyers, to make use of the various scientific studies for *their* cause may result in legal breakthroughs, which might change the way these "guinea pigs" are treated in the future.

Acknowledgments

We are grateful to the lawyers of *Fafu* in Taipei and Tainan, TAVOI, RCA-SHA, Anshun-SHA, the interviewees in Tainan and Taipei, to Rebecca Fite, Soraya Boudia and Nathalie Jas for their editing of the drafts, and to the following for their precious comments on different drafts: Yves Cohen, Nancy Langston, Ho Ming-sho, Marylène Lieber, Ellen Hertz, Sheila Jasanoff, Brian Wynne, Lin Yi-Ping, and Christelle Gramaglia.

Notes

This chapter is based on research which led to two other publications: Jobin, Paul. 2010. "Les cobayes portent plainte. Usages de l'épidémiologie dans deux affaires de maladies industrielles à Taiwan." *Politix* 23(91): 53–75; Jobin, Paul and Yu-Hwei Tseng. 2011. "Bailaoshu shang fayuan: cong liangli gongye wuran susong tanqi", *Taiwanese Journal for Studies of Science, Technology and Medicine* 12: 159–203.

1. This story was also made famous by a fascinating best seller by Jonathan Harr (*A Civil Action*, New York: Vintage, 1995), then a movie starring John Travolta.
2. The Control Yuan is an investigatory agency that monitors the other branches of the Republic of China government.
3. For the complete references in Chinese, see our article in the special feature on the RCA issue in the *Taiwanese Journal for Studies of Science, Technology and Medicine*, January 2011.
4. Ku Yuling (former general secretary of TAVOI), "The RCA Case and Other Occupational Hazards in Taiwan," oral presentation at the Centre de Recherches sur les Enjeux Contemporains en Santé Publique (CRESP), University of Paris, 13 September 2003. See also TAVOI, "Questions and Answers about RCA" (in Chinese), four-page leaflet printed by TAVOI, 2001.
5. By today's rates, approximately U.S. $84 million.
6. Our interview with Lin Yong-Song in Taipei, 23 November 2009.
7. Around U.S. $72 million.
8. *Taiwan RCA former workers v. RCA/GE/Thomson,* Preparatory document of the defense No. 9, 26 March 2009, 22 pages.
9. *Taiwan RCA former workers v. RCA/GE/Thomson,* Preparatory document of the plaintiffs No. 22, 26 March 2009, 35 pages.
10. Here we take *ignorance* at large. Felt et al. (2007: 36) has identified more precisely several levels: *risk,* when we know the probabilities of possible harmful events and their associated kinds of damage; *uncertainty,* when we know the types and scales of possible harms, but not their probabilities; *ambiguity,* when the meaning of the different issues are themselves unclear; full *ignorance,* where we don't know what we don't know (the "unknown unknowns"); and *indeterminacy,* when what we know is conditioned by our preference.

11. Interview with Harada in Kumamoto, 7 July 2008, and with Hwang in Tainan, 28 July 2008. Concerning Harada and Minamata, see Jobin (2005; 2006).
12. Our interview with Wang Yu-Cheng, Tainan, 30 October 2009.
13. A physicist of international reputation, Chang had been also a pioneer of the antinuclear movement in Taiwan and was concerned about all sorts of industrial hazards (our interview in Taipei, February 2002).
14. All the residents could apply for a monthly allowance of NT $1.814 (approximately U.S. $60); NT $3,000 (U.S. $100) if they have a blood rate of 64 picograms, and the equivalent of a monthly minimum wage (NT $15.840, U.S. $500) if they have serious diseases.
15. Our interview with the mayor at Tainan City office, 13 April 2010.
16. Our interview with Lee at Cheng-Kung University, 26 November 2009.
17. Ibid.
18. Our fieldwork in Anshun, 26 November 2008, and our observation of the medical check offered by the Tainan City office on 27 December 2008..
19. Our interview in Anshun, 30 October 2009.
20. See Jobin (2006: chaps. 4–5).

Bibliography

Allen, Barbara L. 2003. *Uneasy Alchemy*. Cambridge: MIT Press.
Brown, Phil. 1987. "Popular Epidemiology: Community Response to Toxic Waste-Induced Disease in Woburn, Massachusetts and Other Sites." *Science, Technology, and Human Values* 12(3–4): 76–85.
———. 1997. "Popular Epidemiology Revisited." *Current Sociology* 45(3): 137–56.
———. 2007. *Toxic Exposures: Contested Illnesses and the Environmental Health Movement*. New York: Columbia University Press.
Chang, Jung-Wei, Ming-Chyi Pai, Hsiu-Ling Chen, How-Ran Guo, Huey-Jen Su, and Ching-Chang Lee. 2008. "Cognitive Function and Blood Methyl Mercury in Adults Living near a Deserted Chloralkali Factory." *Environmental Research* 108: 334–39.
Chang, Yung-Ming, Chi-Fu Tai, Sweo-Chung Yang, Chiou-Jong Chen, Tung-Sheng Shih, Ruey-Shiong Lin, and Saou-Hsing Liou. 2003a. "A Cohort Mortality Study of Workers Exposed to Chlorinated Organic Solvents in Taiwan." *Annals of Epidemiology* 13: 652–60.
Chang, Yung-Ming, Chi-Fu Tai, Ruey-Shiong Lin, Sweo-Chung Yang, Chiou-Jong Chen, Tung-Sheng Shih, and Saou-Hsing Liou. 2003b. "A Proportionate Cancer Morbidity Ratio Study of Workers Exposed to Chlorinated Organic Solvents in Taiwan." *Industrial Health* 41: 77–87.
Chang, Yung-Ming, Chi-Fu Tai, Sweo-Chung Yang, Ruey-Shiong Lin, Fung-Chang Sung, Tung-Sheng Shih, and Saou-Hsing Liou. 2005. "Cancer Incidence among Workers Potentially Exposed to Chlorinated Solvents in an Electronics Factory." *Journal of Occupational Health* 47: 171–80.
Chen, Jein-Wen, Shu-Li Wang, Hui-Yen Yu, Po-Chi Liao, and Ching-Chang Lee. 2006. "Body Burden of Dioxins and Dioxin-like Polychlorinated Biphenyls in Pregnant Women Residing in a Contaminated Area." *Chemosphere* 65: 1667–77.

Chou, Kuei-Tien. 2008. "Glocalized Dioxin; Regulatory Science and Public Trust in a Double Risk Society." *Soziale Welt* 59: 181–97.
Cowie, Jefferson. 1999. *Capital Moves, RCA's Seventy-Year Quest for Cheap Labor.* Ithaca, N.Y.: Cornell University Press.
Desrosières, Alain. 1998. *The Politics of Large Numbers.* Cambridge: Harvard University Press.
Felt, Ulrike, Brian Wynne, Michel Callon, Maria E. Gonçalves, Sheila Jasanoff, Maria Jepsen, Pierre-Benoît Joly, Zdenek Konopasek, Stefan May, Claudia Neubauer, Arie Rip, Karen Siune, Andy Stirling, and Mariachiara Tallacchini. 2007. *Taking European Knowledge Society Seriously.* Brussels: European Commission.
Harr, Jonathan. 1995. *A Civil Action.* New York: Vintage.
Jasanoff, Sheila. 1995. *Science at the Bar.* Cambridge: Harvard University Press.
Jobin, Paul. 2005. "The Tragedy of Minamata." In *Making Things Public,* ed. Bruno Latour and Peter Weibel. 988–93. Cambridge: MIT Press.
———. 2006. *Maladies industrielles au Japon.* Paris: EHESS.
Ku, Yuling. 2006. "Former RCA Workers Contaminated by Pollution." In *Challenging the Chip,* ed. Ted Smith, David A. Sonnenfeld, David Naguib Pellow, and Leslie A. Byster, 181–90. Philadelphia: Temple University Press.
Latour, Bruno. 1993. *We Have Never Been Modern.* Cambridge: Harvard University Press.
———. 2005. *Reassembling the Social: An Introduction to Actor-Network-Theory,* Oxford: Oxford University Press.
Latour, Bruno, and Peter Weibel. 2005. *Making Things Public.* Cambridge: MIT Press.
Lee, Ching-Chang, Wu-Ting Lin, Pao-Chi Liao, Huey-Jen Su, and Hsiu-Ling Chen. 2006a. "High Average Daily Intake of PCDD/Fs and Serum Levels in Residents Living near a Deserted Factory Producing Pentachlorophenol in Taiwan: Influence of Contaminated Fish Consumption." *Environmental Pollution* 141: 381–86.
Lee, Ching-Chang, Yei-Jen Yao, Hsiu-Ling Chen, Yue-Liang Guo, and Huey-Jen Su. 2006b. "Fatty Liver and Hepatic Function for Residents with Markedly High Serum PCDD/Fs Levels in Taiwan." *Journal of Toxicology and Environmental Health, Part A* 69: 367–80.
Lee, Ching-Chang, Yue-Liang Guo, Chun-Hsiung Kuei, Ho-Yuan Chang, Jing-Fang Hsu, Shan-Tair Wang, Pao-Chi Liao. 2006c. "Human PCDD/PCDF Levels near a Pentachlorophenol Contamination Site in Tainan, Taiwan." *Chemosphere* 65: 436–48.
Lee, Jyuhn-Hsiarn Lukas, Chang-Chuan Chan, Chih-Wen Chung, Yee-Chung Ma, Gan-Shuh Wang, and Jung-Der Wang. 2002. "Health Risk Assessment on Residents Exposed to Chlorinated Hydrocarbons Contaminated in Groundwater of a Hazardous Site." *Journal of Toxicology and Environmental Health, Part A* 65: 219–35.
Lee, Jyuhn-Hsiarn Lukas, Chih-Wen Chung, Yee-Chung Ma, Gan-Shuh Wang, Pau-Chung Chen, Yaw-Huei Hwang, and Jung-Der Wang. 2003. "Increased Mortality Odds Ratio of Male Liver Cancer in a Community Contaminated by Chlorinated Hydrocarbons in Groundwater." *Occupational and Environmental Medicine* 60: 364–69.
Lin, Yi-Ping. 2006. "Women and Water: Looking at the Health Research on RCA from a Gender Analysis." *Journal of Women's and Gender Studies* 21: 185–212. (In Chinese)
Markowitz, Gerald and David Rosner. 2002. *Deceit and Denial: The Deadly Politics of Industrial Pollution.* Berkeley: University of California Press.
Sarat, Austin, and Stuart Scheingold, eds. 2006. *Cause Lawyers and Social Movements.* Stanford: Stanford Law and Politics.

Shapiro, Barbara. 1991. *"Beyond Reasonable Doubt" and "Probable Cause."* Berkeley: University of California Press.
Shy, Carl. 1997. "The Failure of Academic Epidemiology: Witness for the Prosecution." *American Journal of Epidemiology* 145(6): 479–84.
Sellers, Christopher. 1997. *Hazards of the Job.* Chapel Hill: University of North Carolina Press.
Soong, Der-kau, Chien-chou Hou, and Yong-Chien Ling. 1997. "Dioxins in Soil and Fish Samples from a Waste Pentachlorophenol Manufacturing Plant." *Journal of Chinese Chemistry Society Taipei* 44: 545–52.
Sung, Tzu-I, Pau-Chung Chen, Lukas Jyuhn-Hsiarn Lee, Yi-Ping Lin, Gong-Yih Hsieh, and Jung-Der Wang. 2007. "Increased Standardized Incidence Ratio of Breast Cancer in Female Electronics Workers." *BMC Public Health* 7: 102.
Sung, Tzu-I, Jung-Der Wang, and Pau-Chung Chen. 2008. "Increased Risk of Cancer in the Offspring of Female Electronics Workers." *Reproductive Toxicology* 25(1): 115–19.
Sung, Tzu-I, Jung-Der Wang, and Pau-Chung Chen. 2009. "Increased Risks of Infant Mortality and of Deaths Due to Congenital Malformation in the Offspring of Male Electronics Workers." *Birth Defects Research (A): Clinical and Molecular Teratology* 85(2): 119–24.
Taiwan Bureau of Health Promotion. 2003. *Literature Evaluation and Policy Analysis of Health Care for RCA Employees.* Taipei: Department of Health, Executive Yuan, Republic of China. (In Chinese)
Thébaud-Mony, Annie. 2007. *Travailler peut nuire gravement à votre santé.* Paris: La Découverte.
Wang, Fun-In, Min-Liang Kuo, Chia-Tung Shun, Yee-Chung Ma, Jung-Der Wang, and Tzuu-Huei Ueng. 2002. "Chronic Toxicity of a Mixture of Chlorinated Alkanes and Alkenes in ICR Mice." *Journal of Toxicology and Environmental Health, Part A* 65: 279–91.
Wong, Otto. 2004. "Carcinogenicity of Trichloroethylene: An Epidemiologic Assessment." *Clinics in Occupational and Environmental Medicine* 4: 557–89.
Wu, Yi-Ling. 2009. "The Political Economy of Occupational Disease in Taiwan." Ph.D. diss., University of Sussex.

 PART III

Putting Knowledge, Ignorance, and Regulation into Perspective

 CHAPTER 9

Reckless Laws, Contaminated People
*Science Reveals Legal Shortcomings
in Public Health Protections*

Carl F. Cranor

Based on the analysis of the U.S. law, this chapter argues that a much more systemic approach with appropriate premarket testing is needed to reduce exposures to toxicants. I describe the contamination of citizens, sketching some findings from developmental toxicology. I review failures of reckless postmarket laws and diagnose some of these failures. Learning from the ethics of medical experimentation and premarket laws, I suggest more prudent legal structures.

The chapter first highlights that the adverse health effects of toxicants are much wider than cancer. People are at risk for reproductive effects, immune system dysfunction, and neurological problems, among others. Prenatal and early childhood exposures to toxicants can make such contributions. A variety of substances are known developmental toxicants: lead, mercury, diethylstilbestrol (DES), thalidomide, pesticides, anti-convulsive drugs, sedatives, arsenic, tobacco smoke, alcohol, and radiation. Two hundred known human neurotoxicants may also be developmental toxicants with appropriate exposures. Experimental studies point to a wider range of toxicants, including brominated fire retardants, BPA, phthalates, other pesticides, and cosmetic ingredients.

This chapter then discusses American public health law. Although far too many industrial chemicals, some toxic, are released each year that may affect the function and development of our reproductive organs, hormones, immune system, or brain, no public health law requires routine product testing of the vast majority of chemical compounds before they enter the market. If products are found to be risky or harmful, they must be forcibly reduced or removed—but only after risks or injuries have are apparent. Postmarket laws permit chemical inventions into commerce without any required routine toxicity testing. These govern about 80 to 90 percent of all chemicals. Once they are in commerce, individual self-help to try to avoid them has only modest protective effects. The chapter finally argues that just as medical testing cannot be conducted with preliminary testing for the safety of the experiment and just as pesticides and pharmaceuticals cannot be sold without premarket testing, industrial chemicals and other products should be subject to similar safety measures.

Scientific developments are revealing how outmoded the United States' and probably other countries' laws are for addressing and controlling carcinogenic, developmental, and reproductive threats to people and threats to the environment. Moreover, scientists have not merely identified various adverse effects in human or animals, but also have found more fundamental explanations for them. Scientists, governments, and national populations need to understand recent scientific developments and respond to these by modifying legal protections accordingly. In the end this will necessitate transformation of laws, some scientific practices, and perhaps the funding of relevant research. Unless the laws reflect the science concerning the developmental origins of disease, it will be one more example of how powerless science is in the face of political inertia and vested interests.

Scientific Developments

Bullets, knives, and blunt objects are often philosophers' examples of risk- or harm-bearing entities. We have worried about differences between homicide and attempted homicide, often using guns as the cause of death. We and legal theorists have sometimes been concerned about which of two people should be held accountable if, acting independently, one stabs the victim and the other hits him with a rock. Should the person who is the actual cause of death be the one held for murder or should both be held? If only one is charged with murder, this does not mean the other escapes liability because he still assaulted the victim. While such sources of harm and moral and legal concerns remain with us, recent developments in science have revealed much more subtle sources of harm, namely, molecules contributing to disease, dysfunction, and death.

Of course, scientists and public health agencies have long realized that molecules can harm adults and children alike. People have long been aware of the toxic effects of lead, mercury, and some poisons. More recently, vinyl chloride monomers caused a rare form of liver cancer in a number of employees who worked in a Goodyear polyvinyl chloride manufacturing facilities in the mid 1970s (Heath et al. 1975) and benzene exposures in the workplace have long been recognized as a source of various forms of leukemia and other blood disorders.[1] Molecules in air pollution have caused morbidities and premature death.

Molecularly caused harms and diseases, thus, are not new. However, scientific research is revealing that children exposed in utero or shortly after birth to various industrial chemicals, pesticides, pharmaceuticals, and cosmetics ingredients can suffer diseases or dysfunction, and sometimes death as a consequence. Diagnosing their adverse effects is much more subtle and difficult than identifying the causal agents of harm from macro objects such as bullets, knives, or blunt objects. In what follows I argue that recent scientific developments reveal (a) hitherto unexpected contamination by industrial chemicals, (b) a new source of some diseases, and (c) a plausible (epigenetic) mechanism for some of them, and that these in turn point to failures of existing legal systems to protect us from these newly revealed sources of disease and dysfunction. The scientific developments also show the need for greater urgency to protect the public health and provide a powerful argument to improve protections. How can we utilize the law and science to reduce the risks to children from toxic molecules?

Contamination

Our bodies are permeated by pesticides and industrial chemicals, many of which are toxic. Use some cosmetics and absorb some phthalates through your skin. Protect yourself with sun block and experience a similar result. Apply certain lipsticks, add to the lead in your body that may already be present because of the former presence of leaded gasoline in the environment.

Phthalates may contribute to premature breast development, sex organ problems in males, and some reproductive and developmental risks (Swan et al. 2005; Bothwell 2008; Rawlins 2009). Lead is a potent neurotoxicant for adults and children alike, adversely affecting learning, IQ, and behavioral controls (Wigle and Lanphear 2005). It also contributes to cardiovascular disease. All this occurs at surprisingly low concentrations (Navas-Acien et al. 2007).

Tap water and vegetables contain small amounts of a former rocket fuel, fireworks, or munitions component, perchlorate. This contaminant can be especially problematic for pregnant women, children developing in utero, or even newborns. Perchlorate can interfere with thyroid hormones needed for brain development. If pregnant women have too little circulating thyroid hormone, their children's brains will not develop properly. Similarly, if young children have too little thyroid hormone, this interferes with brain development (Woodruff et al. 2008).

Much furniture, drapes, and electronic equipment, such as television sets and computers, contains some brominated fire retardants, polybrominated diphenyl ethers (PBDEs). They are not chemically bound to the fabrics or plastics, but merely mixed in, so over time they can disperse into your home, house dust, and ultimately into your body. In the United States, concentrations of PBDEs in citizens' bodies are rapidly increasing even though some steps have been taken to reduce the production and use of some of these chemical products. As products with PBDEs deteriorate or are discarded in waste dumps, the PBDEs will disperse throughout the environment in much the same manner that other persistent chemical products have, traveling around the world, entering the ocean, and many ultimately highly contaminating Arctic ecosystems and animals. Indeed, PBDEs have been found to contaminate Tasmanian devils, hundreds of miles from any industrialized society (Denholm 2008; Hanford 2009).

Of course, we do not need to be concerned only about the most recent chemicals in domestic and international markets; there are considerable legacies from industrial chemicals already in the environment and our bodies. Eat beef in the form of steak or hamburger and you will likely ingest some molecular remnants from polychlorinated biphenyls. These come from industrial insulating and lubricating fluids. These have been long banned in the United States, Europe, and many industrialized countries, yet they remain present in the environment and in our bodies. PCBs have also contaminated many fish and marine mammals as well (Langston 2010).

PCBs and PBDEs reach ecosystems and populations at the ends of the earth by at least two different routes. They can enter the bodies of fish and sea mammals, which in turn are part of Inuit diets in Alaska, Canada, and the northern European and Asian countries. PCBs and other chlorinated compounds can also be volatilized and then transported long distances through the atmosphere. Typically, animals and people of the northern latitudes receive greater doses of such chemicals than people in the lower forty-eight states of the United States. People can also accumulate greater concentrations of PCBs than those in the animals they eat (Cone 2005).

People and animals are not exposed merely to the substances described above. Hundreds of pesticides, industrial chemicals, cosmetics, and numerous other products contaminate each of us. They enter our bodies and then reach our tissues, organs, and blood. The Centers for Disease Control and Prevention (CDC) has developed reliable techniques for detecting some industrial chemicals by measuring the amounts in our blood or urine; this is called biomonitoring. In its latest report the CDC reliably identified 246 substances in the bodies of U.S. citizens. However, as it develops reliable protocols for detecting such substances, this number will only increase.[2] The CDC has chosen to investigate these particular substances because they constitute substantial exposures or are known or suspected toxic hazards, or both. Most of the compounds have intrinsic toxic properties or a "built-in ability to cause an adverse effect." These are known as toxic "hazards" (Faustman and Omenn, 2001; Heinzow 2009).

Each person is contaminated to a greater or lesser degree. Much more worrisome is that industrial chemicals can contaminate the very tissues that go into creating a child. For example, these substances taint parents' bodies before they ever decide to have a child. Women's eggs and men's sperm, the very genetic sources of children, along with many other tissues in their bodies have intimate contact with industrial chemicals.

In addition, once a woman is pregnant, most industrial chemicals, pesticides, and pharmaceuticals can cross the placenta and enter the womb, depending upon such properties as size, electric charge, fat solubility and so on. As one of the leading experts puts the point, "It is clearly evident that there really is no placental barrier *per se:* The vast majority of chemicals given the pregnant animal (or woman) reach the fetus in significant concentrations soon after administration" (Schardein 2000). Once a child is born and begins nursing, most substances can similarly enter the breast milk, be conveyed to the child, and transfer some of a mother's body burden of industrial chemicals to the child (Heinzow 2009).

These concerns are not merely theoretical. Children are born already carrying a body burden of pesticides, cosmetic ingredients, and industrial chemicals, as a recent study of ten newborns revealed: it found 232 industrial chemicals in their umbilical cords (Environmental Working Group 2009; Fimrite 2009).

Adverse Health Effects

Developing children in particular are much more vulnerable to adverse health effects than adults because they are in one of the most vulnerable life stages. Whatever organ system one considers—the brain, the immune system, the reproductive system, or the lungs—each is much more vulnerable to toxic perturbations that can result in adverse consequences than the same system in adults.

Moreover, developing children are typically subject to *greater exposures* than adults. According to the first conference on the developmental origins of disease, "the mother's chemical body burden will be *shared with her foetus or neonate,* and the child may, in some instances, be exposed to *larger doses relative to the body weight*" (Grandjean et al. 2008). For example, methylmercury concentrations in the fetal brain can be as much as five times higher than concentrations in the mother's blood (Honda et al. 2006; Grandjean et al. 2008). Breast-fed infants may have greater concentrations of lipophilic (fat soluble) toxicants, since breast milk contains considerable fat. Researchers have estimated that a nursing child's daily dose of PCBs in the breast milk "may be 100-fold higher" than the concentration of the PCBs in the mother's blood, "resulting in much greater toxic concentrations in the child than in the mother" (Grandjean et al. 2008). Not all toxicants will show similar increases in breast milk, but some clearly do.

In addition, during development children have lesser defenses than adults do. A child's immune system is not developed in utero or at birth. A mother's immune system offers some degree of protection in utero, but her immune system offers less protection for each of them considered separately than it would for the mother alone.[3] The blood-brain barrier, which ordinarily protects the brain and other neurological tissue from some toxicants, does not develop until about six months after birth. Once developed it imparts protection against some chemicals entering the brain and other neurological tissues. Similarly, many enzymes that can detoxify toxic substances are often poorly developed in young children, resulting in greater toxic insults to children than adults from industrial contamination. At the same time and in contrast, some enzymes that *increase* the toxicity of other chemicals may not have matured, so on this dimension, children can have greater protection than adults.

These are a few of the general or typical biological tendencies of developing children that increase their vulnerability to toxic insults. However, when genetic variability and diversity are considered, the range of adverse effects can increase.

For example, polycyclic aromatic hydrocarbons (PAHs), which are formed during incomplete combustion of organic compounds from such things as sidestream and secondhand tobacco smoke as well as from combustion of coal, gas, and oil, can cross the placenta and create adducts on DNA (Perera et al. 1999). When such substances bind to the DNA, they usually alter its function and cause mutations or incorrect repair, leading to cancers or other diseases. Urban areas are higher in PAHs than regions with few combustion by-products. Subpopulations of fetuses with more PAH-DNA adducts show increased sensitivity to genetic damage compared to the mother and compared to others (Miller et al. 2002; Perera et al. 1999). This can lead to smaller head circumference, associated with other adverse effects, as well as genetic damage in the newborn (Perera et al. 1999).

Similarly, vulnerability to organophosphate pesticides can "vary by age and genotype." Children as well as adults with a variant of a particular gene have lower levels of an enzyme that assists in metabolizing organophosphate pesticides. Having less of this enzyme puts them "at higher risk of health effects from organophosphate exposure" (Eskenazi et al. 2008). Potential effects include neurotoxic effects as well as some cardiovascular endpoints (Ecobichon 2001).

As a consequence, even if an average or typical child might not be as susceptible to a particular contaminant compared to an adult, human genetic variability can increase or decrease the extent of sensitivity. This fact of biology increases the range of susceptibility of developing children to adverse effects compared with adults.

Furthermore, because young children have more years of future life ahead of them than adults, if children are contaminated in utero or shortly after birth, and disease processes are triggered early in their lives, this provides more time for diseases or dysfunction to develop during a lifetime. A disease process might require one, two, or three critical steps to occur before the disease is fully initiated. However, if one or two steps occur in utero, as they likely did with DES, or in early childhood, as occurs with lead, then fewer steps would need to occur later in life for full-fledged disease or dysfunction to appear (Heindel 2008). For instance, "Cancer is a multistage process and the occurrence of the first stages in childhood increases the chance that the entire process will be completed, and a cancer produced, within an individual's lifetime" (Miller et al. 2002). The earlier all the biological processes needed for the development of cancer are completed, the earlier the disease will appear in a person's life.

Generic vulnerability, greater exposures, and (generally) lesser biological defenses than those of adults have resulted in risks of diseases for developing children. Moreover, these usually occur at concentrations of toxicants much lower than those that cause adverse effects in adults. In addition, the timing of exposure can be quite important. Consider a few examples.

Human exposure to methylmercury from eating fish contaminated by this substance in Minamata Bay in Japan resulted in some adverse effects to adults, but in catastrophic effects to children who were contaminated in utero (Honda et al. 2006). Adults developed some neurological problems, ranging from numbness and loss of feeling to being permanently disabled (McCurry 2006). And, some died (Honda et al. 2006). Children contaminated in utero by MeHg were at greater risk. They contracted cerebral palsy at ten times the rate of unexposed children and a number died (Weiss 1994). In part this occurred because they had much greater exposures to methylmercury in the brain, which has a selective affinity for it, and, of course, they were in general much more susceptible to adverse effects than adults (Honda et al. 2006).

In utero exposure to DES caused vaginal cancer in comparatively young women (about twenty years of age) and also increased breast cancer in DES daughters as they reached middle age (Kortenkamp 2008). DES mothers also appear to have experienced a modestly elevated rate of breast cancer because of DES exposures decades earlier (Titus-Ernstoff et al. 2001). While thalidomide caused some peripheral neuropathy in women who took it, by and large it may have had some benefits for them. However, developing children exposed in utero suffered terrible physical abnormalities and birth defects along with neurological problems (Landrigan et al. 2004). Some anti-convulsive drugs can reduce convulsions in women prone to them (for example, because of epilepsy), but can cause birth defects in children exposed to them in utero. (Landrigan et al. 2004).

Sometimes adults take Coumadin, an anti-coagulant blood thinner, in order to help prevent heart attacks and strokes when they have artificial heart valves. Yet, children exposed in utero to the drug can be born with underdeveloped cartilage and sometimes weakened optical nerves (Landrigan et al. 2004). Children have higher rates of leukemia and thyroid cancer from radiation exposure than adults at similar exposures. Teenage women exposed to radiation tend to have higher rates of breast cancer than older women similarly exposed (Miller et al. 2002). In addition, women younger than fourteen who were exposed to greater concentrations of DDT when it was in widespread use in the United States contracted breast cancer at a fivefold higher rate than older women with similar exposures (Cohn et al. 2007).

Other substances recently becoming the objects of study also raise health concerns. Polychlorinated biphenyls, long banned in many countries, have been known for contributing to neurological problems in children, e.g., resulting in lesser Intelligence Quotients (IQs) and hearing problems. Other children whose mothers ate fish likely contaminated with PCBs exhibited learning and memory deficiencies (Kuratsune et al. 1972).

PCBs have been widely studied in various animal species—mice, rats, and monkeys. Animals exposed in utero and neonatally have exhibited impaired learning, decreased cognitive function, sensory deficiencies, and even behavior similar to Attention Deficit Hyperactivity Disorder (Kodavanti 2005). Many of these adverse effects are also seen in humans. Such results have been known and of concern for some time.

However, a much more recently introduced class of chemicals, polybrominated diphenyl ethers, used as flame retardants, have a number of chemical and biologically active properties that are quite similar to PCBs and other compounds known to cause human developmental effects. There are good reasons to believe that they will pose similar problems in humans once they are well studied. However, we should not wait that long to eliminate them from commerce.

The chemical structure of PBDEs and their toxic effects resemble PCBs in experimental animal studies. PCBs have been documented as causing adverse effects in humans. What has been missing is that there has been too little time to fully study the effects of PBDEs in humans, but the body of scientific evidence to date suggests there will very similar toxic effects from PBDEs in humans (Kodavanti 2005). As this chapter goes to press, researchers at the University of California, Berkeley, have found that "children exposed to PBDEs tend to have poorer attention, motor skills and IQ scores" (Lee, 2012). As just noted, this is what one would have expected based on what was known about the similarities between PCBs and PBDEs (Cranor, 2011).

Nonetheless, researchers are beginning to find that PBDEs cause developmental and reproductive effects in people. Kim Harley and Kathleen M. McCarty (2009) from Brenda Eskenazi's lab have found that women with higher concentrations of PBDEs in the blood take longer to successfully get pregnant than women with lesser concentrations of PBDEs. A small Dutch study reported both adverse and positive neurological effects for children exposed in utero. PBDEs can cross the placenta and expose children in utero. Higher exposures to the flame retardants in utero were associated with decreased fine motor skills and decline in attention, both echoing effects in animal models. However, the children also showed improved coordination, visual perception, and behavior at age six. Because the Dutch study is small, it will need to be replicated in larger studies for scientists to be more fully persuaded of the results.

While the effects of PBDEs in humans are not unexpected based on experimental studies, PBDEs are of great concern because their high concentrations in U.S. citizens. For example, young women (from age 25–29) as a cohort and in prime childbearing age have higher concentrations of PBDEs in their breast milk than do older women (Harley 2010). Thus, their children will receive a substantial dose of PBDEs in utero and after birth.

To this point I have reviewed adverse effects in animals and humans considering one substance at a time. However, realistically, we are all routinely contaminated by multiple industrial chemicals, many of them toxic. Some substances add to the toxic effects of other compounds, typically seen with estrogen-mimicking compounds, dioxin-like compounds, and androgen antagonists. Some substances add their toxic effects together because toxicants and naturally occurring biochemicals in the body attach to the same cellular receptor. This is true of natural estrogens and xenoestrogens that can cause harm in the body. Substances similar to dioxin, dioxin-like compounds, can attach to the same cellular receptors and join together to cause harm (Simon et al. 2007). For estrogens, it appears that a woman exposed to more estrogen over her lifetime is at greater risk for breast cancer (Kortenkamp 2008). Thus, to the extent that each of us is contaminated by substances that attach to cellular receptors and increase the toxicity of other biochemicals that attach to the same receptors, these increase our risks to any diseases they cause.

There is a much more general issue concerning additive effects that pose concerns. Woodruff and her coworkers (2008) have identified several compounds that can disturb different pathways that cause adverse effects. For example, pregnant women need sufficient levels of thyroid hormones to facilitate proper neurological development, including brain development, of their children. If thyroid hormones are too low, a child can suffer from poor brain development. Women might have insufficient thyroid hormones because of their circumstances, e.g., too little iodine in their diets. However, even if they did not, Woodruff and her coworkers (2008) have shown that one class of substances, e.g., dioxins, dibenzofurans, and dioxin-like PCBs, adversely affect one group of liver enzymes, while another class of compounds, e.g., non-dioxin-like PCBs affect other liver enzymes that also cause adverse effects to circulating thyroid hormones. These two classes of chemicals cause similar adverse effects to thyroid hormones by means of different biological mechanisms. In addition, it now appears that the brominated fire retardants add to such adverse outcomes. Exposure to perchlorate, a discarded rocket fuel and fireworks component, can also contribute to these adverse effects (Woodruff et al. 2008). These researchers also indicate that different classes of substances produce "a dose-additive effect on [thyroid hormones] at *environmentally-relevant* doses ... demonstrating exposures to chemicals acting on different [biological] pathways can have cumulative effects" (Woodruff et al. 2008). They conclude, "Its is appropriate to presume cumulative effects unless there is evidence to the contrary, and it is important for risk assessments to consider real-life exposure mixtures."

Experimental animal studies have shown that animals exposed to mixtures of phthalates exhibit penile digenesis syndrome (Hass et al. 2007). At a minimum they might exhibit penises with the urethra mislocated to the bottom or top of the penis or elsewhere than on the penis. More worrisome, they found animals with split penises and other malformations. At the same time human studies have also revealed that mothers who are more highly exposed to phthalates gave birth to baby boys that exhibited the early stages of penile digenesis syndrome at a higher rate than that seen in mothers with lower exposures (Swan et al. 2005). These baby boys mainly showed mislocated urethras and a shorter ano-genital distance than less exposed boys. These data suggest that the results from animal studies are concurring with human data on the effects of phthalates on male reproductive systems. Citizens of the United States, and likely other countries, are highly exposed to phthalates.

As a final point, researchers have begun to understand a mechanism by which diseases are triggered and during development. A person's genetic sequence little changes, but how that sequence is expressed and when it is expressed turns out to be quite important. This generic area of study is called epigenetics (Jirtle and Skinner 2007). It is in its infancy but appears to hold great promise for understanding such disease processes (Cranor 2011).

Human data have shown that DES, thalidomide, various pesticides, sedatives, arsenic, anti-convulsive drugs, tobacco smoke, alcohol, and radiation are developmental toxicants. More recent research strongly suggests that phthalates, BPA, and PBDEs are also human developmental toxicants. Many of these substances have long been recognized as causing adverse effects in experimental studies; they are beginning to be documented in humans. There are about two hundred human neurotoxicants that likely would adversely affect developing children if there were appropriate exposures (Grandjean and Landrigan 2006). For some, there is no lowest level that is safe: many carcinogens, radiation, tobacco smoke, lead, and water disinfectant by-products (Wigle and Lanphear 2005).

Reforming the Law

We need not accept the existing world, which burdens our children and us with toxicants, some of them resulting in diseases, dysfunctions, or deformities that are likely not yet fully understood. The developmental or fetal origins of disease reveal how early toxicological "hits" can result in adverse effects and how reckless existing laws and policies are toward our children. The scientific developments sketched above strongly suggest the need for a more prudent approach to industrial chemicals, which would result in a paradigm shift in the law. This shift is required because many existing environmental health laws are postmarket laws, permitting human-created chemicals into commerce without legally required testing. Moreover, many of them drew some of their inspiration from older judge-made law that is now quite incongruous with the circumstances in which we find ourselves.

Legal theorists suggest that current environmental health laws in the United States drew inspiration from nuisance laws of the early Anglo-American tradition. Yet brief reflection on these legal paradigms shows their inadequacy for addressing toxic threats to our health.

In an earlier historical period when communities were exposed to environmental pollutants, these might have been smoke, dust, possibly arsenic. Nuisance laws that existed at that time would have provided a typical means of protection. According to one of the leading texts on the tort or personal injury law, a nuisance is a "hurt, annoyance or inconvenience," and in the law this refers to "interference with the use or enjoyment of land, and thus was the parent of the law of private nuisance as it stands today" (Keeton et al. 1984). Typical nuisances might be smoke, dust, loud noises, vibration, blasting, pollution of a river, flooding, unpleasant odors, excessive light. Even activities that spoiled the quiet or peacefulness of one's life, such as houses of prostitution, vicious dogs, or even funeral homes, could be considered nuisances (Keeton et al. 1984). Nuisances had to be "substantial and unreasonable ... [as well as] offensive or inconvenient to the normal person" (Keeton et al. 1984). If one brought a legal action in nuisance he or she had to show that the offending activity interfered with the enjoyment and use of property. The typical remedy would be for it to cease (Keeton et al. 1984).

Environmental law scholars trace aspects of several current environmental protection laws to the earlier law of nuisance. However, insofar as environmental health laws model their legislation on nuisance, much of that older law is inapt and deeply misleading.

First, molecules differ extensively from most nuisances. The well-recognized unpleasant odors, excessive light, blasting and vibrations, and smoke or dust that are typical nuisances are readily perceptible by normal human senses (Keeton et al. 1984; Cranor 2011). In this they differ substantially from silent, odorless, undetectable toxic invaders. Nuisances typically reveal and announce, often quite dramatically, their existence; molecules, toxic or not, typically do not. Stinks, vibrations, impaired visibility, and loud noises typically do not accompany or announce the presence of toxicants.

Nuisances also tend to *immediately* disturb one's use or enjoyment of property or one's life. Molecules, because they are undetectable, typically do not. Moreover, substantial latency periods may well delay the discovery of toxicants or molecular-caused diseases. Typical nuisances are usually known immediately (Cranor 2011).

Identification of a cause-effect relationship between an event and its consequences is facilitated by the proximity with which effects follow their causes (Cranor 2011). Long latencies between an initial event and undesirable causal consequences make it difficult to trace the causal paths. Furthermore, molecular hits that result in the fetal origins of disease might hide for years until there is a second or third contributing event. For example, experimental studies have shown that exposures to pesticides in utero are likely to be the first hit needed to cause Parkinson's disease in middle age, but that must be followed by a second hit prior to the manifestation of the disorder (Heindel 2008). In contrast, nuisances, because of their public nature, tend to be apparent the entire time they are present.

Second, legal procedures that seem quite appropriate for nuisances are inapt for toxic substances. Because nuisances publicize their presence, or even noticeably intrude on people's senses, postmarket legal action to reduce or remove them makes sense. Injured parties can quickly detect and marshal evidence of their presence. They can then initiate legal action supported by evidence that is normally readily available to the public .

In order to detect the toxic effects of molecules, subtle scientific studies, many of which can take considerable time, will be needed. The absence of quick production of evidence of risks or harms can easily frustrate protective legal action.

When the U.S. Congress initially passed legislation addressing toxicants and their cleanup, it might have been enamored of postmarket laws, thought that even chemicals should be considered "innocent until proven guilty," or perhaps did not consider the issue much at all. In most of the legislation aimed at removing toxicants from the water, the air, drinking water, or in laws seeking to address chemicals in a more generic manner, postmarket laws were the legislation of choice.

Congress in passing this legislation created blueprints for public health and environmental agencies to follow. These were both procedures to be followed in addressing quite specific problems and legal standards that had to be satisfied when a public health or environmental regulation was issued to correct an existing problem. However, common to all these laws is the view that industrial compounds may enter commerce without any required toxicity testing of their properties and with almost no agency review of any data submitted to the agency. The exceptions to this are laws concerning pharmaceuticals, pesticides, and to some extent new food additives.

Risk assessment was central to using postmarket laws to prevent public health and environmental harms. If there were procedures for quickly identifying risks before they materialized into harms to people, the thought was that postmarket laws utilizing risk assessments would successfully serve preventive aims for protecting the public health. Typically, public health agencies attempting to identify risks sought to utilize studies that did not rely upon human data, but used animal and other nonhuman studies.

Unfortunately, this idea did not pan out well. Early risk assessments were quickly and efficiently conducted, resulting in protective health and environmental rules. However, it did not take long for public health advocates to realize that risk assessments could be exceedingly slow. For example, under the Clean Water Act amendments in 1972 the EPA was charged with issuing requirements to issue ambient water quality standards to reduce toxicants in the water and to make rivers and harbors "fishable and swimmable." However, by 1975 there had been so little progress that environmental organizations sued the EPA for failure "to regulate toxic pollutants" (U.S. Congress OTA 1987). This resulted in a court-sanctioned consent degree in which the EPA, in an effort to expedite the cleanup, had to utilize technology controls "to place specific 'numerical limits on the quantities of 65 toxic pollutants in 21 industrial categories'" (Gaba 1984). Frequently after that environmental organizations had to return to the court to keep the EPA working in a timely manner on this task, while often the agency sought more time to complete its job.[4]

This particular experience with risk assessments for setting ambient exposure concentrations of toxicants was fairly typical. Moreover, over time companies learned various strategies for slowing data requirements and assessments to a snail's pace or slower. They might, for example, try to insist on more and better evidence before protective health standards are issued. Some have urged high standards of proof of risks or harms before health protective rules were issued. Some submit data that are ultimately irrelevant, but the submission means public health officials must take it seriously before they can reject it. If public health agencies yield to some of these arguments, the tactics can greatly delay improved health protection (Bohme et al. 2005; Cranor 2011). In addition, too often Congress has reduced funding for agencies, further handicapping their efforts, and Congress, under a rationale of concerns about costs to companies or American consumers, even created legislative barriers that might be invoked, further slowing health protections. A major consequence is that there is little toxicity data on the vast majority of industrial chemicals and commerce (Guth et al. 2007; Claxton et al. 2010).

This suggests that postmarket laws can be gamed by companies seeking to keep their products in commerce longer or delay having to clean up wastes or pollutants for a longer period of time. For example, the U.S. EPA began examining the toxicity risks of dioxin in 1984, but as of late 2012, twenty-eight years later, that risk assessment had not been completed and some argued that "more study is needed" (Cranor 2011). In addition, these laws have led to a great deal of ignorance about the toxicity of products in the market. Except for pharmaceuticals, pesticides, and new food additives, there is little toxicity data about the vast majority of substances in commerce.

Finally, scientific research on the developmental origins of disease shows that postmarket laws are much too late to prevent risks and some harms to developing children. Since substances subject to postmarket laws (about 80–90 percent of the chemical products created for commerce) can enter commerce without any legally required testing and no meaningful agency review of their toxicity, adults, children, and developing children are exposed to the compounds and any toxic properties well before any public health agency can address them.

Toward an Improved Legal Strategy

The ethics of medical experimentation and premarket testing and approval laws, typically used for pharmaceuticals and pesticides in the United States, all provide aspects of a model for an alternative to postmarket laws.

When humans volunteer to participate in medical experiments there are number of conditions that must be satisfied before they may voluntarily participate. They must volunteer, must be informed about the research project and any risks to volunteers, and must be legally capable of consenting to participate.[5] They must in fact give their informed consent.[6] A partial philosophic justification of this provision seems to be a matter of recognizing their authority over what happens to them and provides a means by which they can protect themselves from risks and harm. Participants should be at liberty to end the experiment at any time.[7] This also seems to be recognition of their continued authority over what happens to them and protections against any risks they perceive.

When participants, such as children, cannot give informed consent, a legally authorized representative should be required to decide whether or not to grant the consent.[8] However, and quite importantly, children "should not be included in research unless the research is necessary to promote the health of the population represented and [it cannot] be performed on legally competent persons."[9] Thus, experiments on children come close to being forbidden because of their age and inability to grant consent. In order to ensure that the above conditions are satisfied, the experiment must be overseen and approved by an ethical review committee.[10]

In order for the experiment to proceed and to have information that volunteers can use to consent to participate, there must be proper preparations and assurances of their safety. This includes prior research on the safety and risks of the experiment.[11] Such research should be conducted to protect the experimental subjects against even *remote* possibilities of injury, disability, or death.[12] Moreover, the ethics of medical experiments specify that the well-being of the human subject should take precedence over the interests of science and society.[13] In short, concern for the participant who volunteers to participate in the experimental procedure is central.

As a final consideration, in order to ensure that the experimental procedure itself is sound there must be scientific oversight and management. This covers not only the science but constitutes an independent review of the experiment to ensure compliance with safety, aims, and informed participation.

The above brief foray into the ethics of medical experiments puts into sharp relief how reckless our laws are with regard to exposure from and contamination by industrial chemicals. Release of industrial chemicals without any required data about their toxicity stands in jarring contrast to the requirements of medical experiments. We now know that individuals will be exposed to industrial chemicals in the environment or in products just as we know they are exposed to medical experiments. However, with the release of untested industrial chemicals there are no prior preparations or reasonable assurances of safety. There are no careful assessments of safe exposures. There is no special concern for children. Concern for contaminated persons is not central. And there is no independent scientific or ethical oversight concerning the resulting contaminations of people that are almost certain to occur. Moreover, the United States, and likely other countries, require testing of pharmaceuticals and pesticides for possible risks before human exposures are permitted. In the instance of pharmaceuticals, these tests must occur even before there are *voluntary* experimental exposures to determine beneficial and adverse effects on humans before they are considered for approval for sale to the general public. Although both pharmaceuticals and pesticides likely need testing improvements for developmental effects—because drugs and pesticides cause developmental effects—they are vastly superior to postmarket laws.

Legislatures should authorize laws allowing products into market and citizen contamination only if there is *routine prior testing* and reasonable *assurances of safety* for children and others in sensitive life stages. Moreover, testing that would likely need to be conducted on experimental animals should aim to mimic as nearly as possible real world conditions. People are already contaminated—animals used in experiments should copy human contamination, especially for substances similar to the new compound. Substances add to the effects of other substances to which people are exposed. They might act through similar cellular receptors, e.g., estrogenic receptors or dioxin-like receptors. They might act via similar mechanisms as some pesticides all inhibit cholinesterase, which stops nerve cells from firing, thus contributing to continual nerve reactions (Ecobichon 2001). Some substances affect different mechanistic pathways, but result in the same adverse outcomes, as several substances adversely affect the thyroid cycle. People show a wide range of biological variability; this should be assumed in permitting substances into commerce. These various exposures and susceptibilities should be mimicked in premarket testing (Cranor 2011).

There are some limitations to a testing and agency review procedure that might replace existing postmarket laws. As the U.S. Academy of Sciences has pointed out "[I]t is neither practical nor desirable to attempt to test every chemical (or mixture) against every end point during a wide range of life stages. The committee recommends toxicity screening of every agent to which there is a strong potential for human exposure. A well-designed tiered strategy could help to set priorities among environmental agents for screening and could identify end points of mechanisms of action that would trigger more in-depth testing for various end points or in various life stages" (NRC 2007). Various practical considerations should influence the creation of premarket tests to protect the public's health (Cranor 2011).

Contamination by industrial molecules is not something we can prevent. At most we can prevent contamination by *toxic* molecules, but to do this we must understand which ones are toxic. The new discoveries in biology and toxicology reveal that toxic contamination begins so early in life that there is no way that postmarket laws can address toxic contamination to prevent risks to developing children. We must find a better way, namely, by utilizing premarket testing and review of products before there is contamination and children and adults alike are put at risk.

Notes

1. Industrial Union Department, *AFL-CIO v. American Petroleum Institute*, 448 U.S. 607 (1980) (the Benzene decision).
2. U.S. Department of Health and Human Services. 2009. *Fourth National Report on Human Exposure to Environmental Chemicals.*
3. Talbot, Prudence. 2009. Department of Cell Biology & Neuroscience at the University of California, Riverside, and developmental biologist, personal communication.
4. *Natural Resources Defense Council (NRDC) v. Gorsuch,* October 1982, Civil Action No. 73-2153; *Natural Resources Defense Council (NRDC) v. Ruckelshaus,* January 1984,

Civil Action No. 75-0172; *Natural Resources Defense Council v. Thomas,* January 1985, Civil Action No. 75-1690; *Natural Resources Defense Council v. Thomas,* April 1986, Civil Action No. 75-1267.
5. Article 20 of the Helsinki Declaration; Article 1 of the Nuremberg Code.
6. Article 1 of the Nuremberg Code; Article 22 of the Helsinki Declaration.
7. Article 9 of the Nuremberg Code.
8. Article 24 of the Helsinki Declaration.
9. *Ibid.*
10. Article 13 of the Helsinki Declaration.
11. Article 3 of the Nuremberg Code; Article 11 of the Helsinki Declaration.
12. Article 7 of the Nuremberg Code.
13. Article 4 of the Helsinki Declaration.

Bibliography

Applegate, John S. 2008. "Synthesizing TSCA and REACH: Practical Principles for Chemical Regulation Reform." *Ecology Law Quarterly* 35: 721-770.

Balakrishna, Biju, Kimiora Henare, Eric B. Thorstensen, Anna P. Ponnampalam, and Murray D. Mitchell. 2010. "Transfer of Bisphenol A across the Human Placenta." *American Journal of Obstetrics and Gynecology* 202: 393.e1–393.e7.

Birnbaum, Linda S., and Daniele F. Staskal. 2003. "Brominated Flame Retardants: Cause for Concern?" *Environmental Health Perspectives* 112: 9–17.

Bohme, Susanna R., John Zorabedian, and David S. Egilman. 2005. "Maximizing Profit and Endangering Health: Corporate Strategies to Avoid Litigation and Regulation." *International Journal of Occupational and Environmental Health* 11: 338–48.

Bothwell, James. 2008. "Toy Story: Timeout for Phthalates." *McGeorge Law Review* 39: 551–63.

Bromer, Jason G., Yuping Zhou, Melissa B. Taylor, Leo Doherty, and Hugh S. Taylor. 2010. "Bisphenol-A Exposure in Utero Leads to Epigenetic Alterations in the Developmental Programming of Uterine Estrogen Response." *The FASEB Journal* 24: 2273–80.

Calafat, Antonia M., Lee-Yang Wong, Zsuzsanna Kuklenyik, John A. Reidy, and Larry L. Needham. 2007. "Polyfluoroalkyl Chemicals in the U.S. Population: Data from the National Health and Nutrition Examination Survey (NHANES) 2003-2004 and Comparisons with NHANES 1999–2000." *Environmental Health Perspectives* 115: 1596–1602.

Calafat, Antonia M., Xiaoyun Ye, Lee-Yang Wong, John A. Reidy, and Larry L. Needham. 2008. "Exposure of the U.S. Population to Bisphenol A and 4-tertiary-octylphenol: 2003–2004." *Environmental Health Perspectives* 116: 39–44.

Claxton, Larry D., Gisela de A. Umbuzeiro, and David M. DeMarini. 2010. "The *Salmonella* Mutagenicity Assay: The Stethoscope of Genetic Toxicology for the 21st Century." *Environmental Health Perspectives* 118: 1515–22.

Cohn, Barbara A., Mary S. Wolff, Piera M. Cirillo, and Robert I. Sholtz. 2007. "DDT and Breast Cancer in Young Women: New Data on the Significance of Age at Exposure." *Environmental Health Perspectives* 115:1406–14.

Cone, Maria. 2005. *Silent Snow: The Slow Poisoning of the Arctic.* New York: Grove Press.

Costa, Lucio G., and Giordanno, Gennaro. 2007. "Developmental Neurotoxicity of Polybrominated Diphenyl Ether (PBDE) Flame Retardants." *Neurotoxicology* 28: 1047–67.

Cranor, Carl F. 2011. *Legally Poisoned: How the Law Puts us at Risk from Toxicants.* Cambridge, M.A.: Harvard University Press.
Denholm, M. 2008. Cancer Agents Found in Tasmanian Devils, News.Com.AU. 22 January 2008. www.theaustralian.com.au/news/cancer-agents-in-tassie-devils/story-e6frg6ox-1111115 (last accessed 13 May 2008).
Dietert, Rodney R., and Michael S. Piepenbrink. 2006. "Perinatal Immunotoxicity: Why Adult Exposure Assessment Fails to Predict Risk." *Environmental Health Perspectives* 114: 477–83.
Ecobichon, Donald J. 2001. "Toxic Effects of Pesticides." In *Casarett and Doull's Toxicology,* ed. Curtis D. Klaassen, 6th ed., 763–810. New York: Pergamon Press.
Environmental Working Group. 2009. *Pollution in People: Cord Blood Contaminants in Minority Newborns,* www.ewg.org/minoritycordblood/fullreport (accessed 12 March 2011).
Eskenazi, Brenda, Lisa G. Rosas, Amy R. Marks, Asa Bradman, Kim Harley, Nina Holland, Caroline Johnson, Laura Fenster, and Dana B. Barr. 2008. "Pesticide Toxicity and the Developing Brain." *Basic & Clinical Pharmacology & Toxicology* 102: 228–36.
Faustman, Elaine M., and Gilbert S. Omenn. 2001. "Risk Assessment." In *Casarett and Doull's Toxicology,* ed. Curtis D. Klaassen, 6th ed. 83–104. New York: Pergamon Press.
Fimrite, Peter. 2009. "Study: Chemicals, Pollutants Found in Newborns." *San Francisco Chronicle,* 3 December 2009, http://articles.sfgate.com/2009-12-03/news/17183043_1_campaign-for-safe-cosmetics-chemicals-umbilical (accessed 12 March 2011).
Gaba, Jeffrey M. 1984. "Regulation of Toxic Pollutants Under the Clean Water Act: NPDES Toxic Control Strategies." *Journal of Air Law and Commerce* 50: 761–91.
Grandjean, Philippe, David Bellinger, Åke Bergman, Sylvaine Cordier, George Davey-Smith, Brenda Eskenazi, David Gee, Kimberly Gray, Mark Hanson, Peter Van Den Hazel, Jerrold J. Heindel, Birger Heinzow, Irva Hertz-Picciotto, Howard Hu, Terry T.-K. Huang, Tina Kold Jensen, Philip J. Landrigan, I. Caroline McMillen, Katsuyuki Murata, Beate Ritz, Greet Schoeters, Niels Erik Skakkebæk, Staffan Skerfving, and Pal Weihe. 2008. "The Faroes Statement: Human Health Effects of Developmental Exposure to Chemicals in Our Environment." *Basic and Clinical Pharmacology and Toxicology* 102: 73–75.
Grandjean, Philippe, and Philip J. Landrigan. 2006. "Developmental Neurotoxicity of Industrial Chemicals." *The Lancet* 368: 2167–78.
Guth, Joseph H., Richard A. Denison, and Jennifer Sass. 2007. "Require Comprehensive Safety Data for All Chemicals." *New Solutions* 17: 233–58.
Hansford, Dave. 2008. "Flame Retardants Found in Rare Tasmanian Devils." http://news.nationalgeographic.com/news/2008/01/080128-devils-cancer.html (last accessed 12 March 2011).
Harley, Kim. 2010. "Younger Mothers' Breast Milk Has Highest Levels of Flame Retardants." *Environmental Health News,* http://www.environmentalhealthnews.org/ehs/newscience/younger-womens-breast-milk-higher-levels-of-pbdes/ (last accessed 12 March 2011) (citing Daniels, Julie L., I-Jen Pan, Richard Jones, Sarah Anderson, Donald G. Patterson, Jr., Larry L. Needham, and Andreas Sjödin. 2009. "Individual Characteristics Associated with PBDE Levels in U.S. Human Milk Samples." *Environmental Health Perspectives,* 118: 155–60).

Harley, Kim, and Kathleen M. McCarty. 2009. "Effects of Flame Retardants on Children's Development Unclear." *Environmental Health News*, http://www.environmental healthnews.org/ehs/newscience/unclear-effects-of-flame-retardants-on-development (last accessed 12 March 2011) (citing Roze, Elise, Lisethe Meijer, Attie Bakker, Koenraad N.J.A. Van Braeckel, Pieter J.J. Sauer, and Arend F. Bos. 2009. "Prenatal Exposure to Organohalogens, Including Brominated Flame Retardants, Influences Motor, Cognitive, and Behavioral Performance at School Age." *Environmental Health Perspectives* 117: 1953–58).

Hass, Ulla, Martin Scholze, Sofie Christiansen, Majken Dalgaard, Anne Marie Vinggaard, Marta Axelstad, Stine Broeng Metzdorff, and Andreas Kortenkamp. 2007. "Combined Exposure to Anti-Androgens Exacerbates Disruption of Sexual Differentiation in the Rat." *Environmental Health Perspectives* 115: 122–28.

Heath, Clark W., Jr., Henry Falk, and John L. Creech, Jr. 1975. "Characteristics of Cases of Angiosarcoma of the Liver Among Vinyl Chorlide Workers in the United States." *Annals of the New York Academy of Sciences* 246: 231–36.

Heindel, Jerrold J. 2008. "Animal Models for Probing the Developmental Basis of Disease and Dysfunction Paradigm." *Basic and Clinical Pharmacology and Toxicology*, 102: 76–81.

Heinzow, Birger G.J. 2009. "Endocrine Disruptors in Human Breast Milk and the Health-Related Issues of Breastfeeding." In *Endocrine-Disrupting Chemicals in Food*, ed. Ian Shaw I. 322–55. Cambridge: Woodhead Publishing.

Honda, Sun'ichi, Lars Hylander, and Mineshi Sakamoto. 2006. "Recent Advances in Evaluation of Health Effects on Mercury with Special Reference to Methylmercury–A Minireview." *Environmental Health and Preventive Medicine* 11: 171–76.

Jirtle, Randy L., and Michael K. Skinner. 2007. "Environmental Epigenomics and Disease Susceptibility." *Nature Reviews Genetics* 8: 253–62.

Keeton, W. Page, Dan B. Dobbs, Robert E. Keeton, and David G. Owen, eds. 1984. *Prosser and Keeton on the Law of Torts*, 5th ed. St. Paul, M.N.: West Publishing.

Kodavanti, Prasada Rao S. 2005. "Neurotoxicity of Persistent Organic Pollutants: Possible Mode(s) of Action and Further Considerations." *Dose-Response* 3: 273–305.

Kortenkamp. Andreas. 2008. "Low Dose Mixture Effects of Endocrine Disruptors: Implications for Risk Assessment and Epidemiology." *International Journal of Andrology* 31: 233–40

Kuratsune, Masanori, Takesumi Yoshimura, Junichi Matsuzaka, and Atsuko Yamaguchi. 1972. "Epidemiologic Study on Yusho, a Poisoning Caused by Ingestion of Rice Oil Contaminated with a Commercial Brand of Polychlorinated Biphenyls." *Environmental Health Perspectives* 1:119–28.

Lee, Stephanie M. 2012. "PBDEs Linked to Delays in Development." *San Francisco Chronicle*, 15 November 2012.

Langston, Nancy. 2010. *Toxic Bodies: Hormone Disruptors and the Legacy of DES*. New Haven: Yale University Press.

Landrigan, Philip J., Carole A. Kimmel, Adolfo Correa, and Brenda Eskenazi. 2004. "Children's Health and the Environment: Public Health Issues and Challenges for Risk Assessment." *Environmental Health Perspectives* 112: 257–65.

McCurry, Justin. 2006. "Japan Remembers Minamata." *The Lancet* 367(9505): 99–100.

Miller, Mark D., Melanie A. Marty, Amy Arcus, Joseph Brown, David Morry, and Martha Sandy. 2002. "Differences Between Children and Adults: Implications for Risk Assessment at California EPA." *International Journal of Toxicology* 21: 403–18.

National Research Council (NRC). 2007. *Toxicity Testing for Assessment of Environmental Agents: Interim Report.* Washington, D.C.: National Academy Press.

Navas-Acien, Ana, Eliseo Guallar, Ellen K. Silbergeld, and Stephen J. Rothenberg. 2007. "Lead Exposure and Cardiovascular Disease—A Systematic Review." *Environmental Health Perspectives* 115: 472–82.

Needham, Larry L. 2007. Personal communication, Faroes Islands.

Perera, Frederica P., Wieslaw Jedrychowski, Virginia Rauh, and Robin M. Whyatt. 1999. "Molecular Epidemiologic Research on The Effects of Environmental Pollutants on The Fetus." *Environmental Health Perspectives* 107: 451–60.

Rawlins, Rachael. 2009. "Teething on Toxins: In Search of Regulatory Solutions for Toys and Cosmetics." *Fordham Environmental Law Review* 20: 1–50.

Schardein, James L. 2000. *Chemically Induced Birth Defects, Third Edition Revised and Expanded.* New York: Marcel Dekker.

Simon, Ted, Janice K. Britt, and Robert C. James. 2007. "Development of A Neurotoxic Equivalence Scheme of Relative Potency for Assessing the Risk of PCB Mixtures." *Regulatory Toxicology and Pharmacology* 48: 148–70.

Swan, Shanna H., Katharina M. Main, Fan Liu, Sara L. Stewart, Robin L. Kruse, Antonia M. Calafat, Catherine S. Mao, J. Bruce Redmon, Christine L. Ternand, Shannon Sullivan, J. Lynn Teague, and the Study for Future Families Research Team. 2005. "Decrease in Anogenital Distance among Male Infants with Prenatal Phthalate Exposure." *Environmental Health Perspectives* 113: 1056–61.

Titus-Ernstoff, L., E.E. Hatch, R.N. Hoover, J.R. Palmer, E.R. Greenberg, W. Ricker, R. Kaufman, K. Noller, A.L. Herbst., T. Colton, and P. Hartge. 2001. "Long-term Cancer Risk in Women Given Diethylstilbestrol (DES) During Pregnancy." *British Journal of Cancer* 84: 126–33.

U.S. Congress, Office of Technology Assessment (OTA). 1987. *Identifying and Regulating Carcinogens.* Washington, D.C.: U.S. Government Printing Office.

Weiss, Bernard. 1994. "The Developmental Neurotoxicity of Methylmercury." In *Prenatal Exposure to Toxicants: Developmental Consequences,* ed. Herbert L. Needleman & David Bellinger, 112–29. Baltimore and London: The Johns Hopkins University Press.

Wigle, Donald T., and Bruce P. Lanphear. 2005. "Human Health Risks from Low-Level Environmental Exposures." *PLoS Medicine* 2: 1232–34.

Woodruff, Tracey J., Lauren Zeise, Daniel A. Axelrad, Kathryn Z. Guyton, Sarah Janssen, Mark Miller, Gregory G. Miller, Jackie M. Schwartz, George Alexeeff, Henry Anderson, Linda Birnbaum, Frederic Bois, Vincent James Cogliano, Kevin Crofton, Susan Y. Euling, Paul M.D. Foster, Dori R. Germolec, Earl Gray, Dale B. Hattis, Amy D. Kyle, Robert W. Luebke, Michael I. Luster, Chris Portier, Deborah C. Rice, Gina Solomon, John Vandenberg, and R. Thomas Zoeller. 2008. "Meeting Report: Moving Upstream-Evaluating Adverse Upstream End Points for Improved Risk Assessment and Decision-Making." *Environmental Health Perspectives* 16: 1568–75.

 CHAPTER 10

Untangling Ignorance in Environmental Risk Assessment

Scott Frickel and Michelle Edwards

This chapter examines the regulatory response to suspected chemical hazards in New Orleans, Louisiana, following the city's catastrophic flooding from Hurricane Katrina in August 2005. For the U.S. Environmental Protection Agency, the year-long response represented an unprecedented mobilization of regulatory science, generating over 400,000 laboratory analyses of soil and flood sediment. Analysis of the resulting data, the policy frameworks that guided the collection and organization of that data, and the agency's subsequent claims about the relative absence of risk to returning city residents reveal some of the ways in which risk assessment in the U.S. environmental regulatory system is deeply structured by the production and reproduction of ignorance.

We define ignorance straightforwardly as domain-based absence of knowledge. We understand ignorance not as a cognitive condition held by individuals, but as an institutional outcome. As we argue below, in the domain of regulatory science ignorance emerges directly from within the rules, procedures, and protocols that define and structure regulatory-based risk assessment. In this context ignorance is organized within the epistemic culture of regulatory practice and can take variable forms and produce different effects. Importantly, these different forms and effects are not static. Over time, they can accumulate, combine, and stabilize within the risk regime itself. In this sense, ignorance is both an outcome of institutional processes and a dynamic feature of those processes. In New Orleans, ignorance has operated alongside knowledge as an organizing principle that guided postdisaster risk assessment work and subsequently structured what regulatory scientists came to know (and not know) about the relationship between chemical hazards, soil toxicity, and risk.

Ignorance Studies: Context and Approach

The study of ignorance has had a somewhat tortured intellectual history, one characterized by long periods of scholarly inattention punctuated by occasional flurries of short-lived enthusiasm for the subject—for reviews see Smithson (1989) and Gross (2010). One of these flurries is occurring now, as scholars have begun recently to give more sustained and systematic attention to ignorance and allied concepts as historical, philosophical, and sociological problems (Hess 2007; Sullivan and Tuana 2007; Proctor and Schiebinger 2008; Gross 2010; McGoey 2012a). In highlighting the nonproduction of knowledge, this newer body of work presents science studies with some important challenges.

As Tuana (2008) observes, Bloor's (1976) programmatic statement for a symmetrical and reflexive sociology of scientific knowledge (SSK) lays the provisional groundwork for a sociology of ignorance. SSK collapses the distinction between "true" and "false" knowledge claims, insisting that scientists' acceptance of both requires social explanations. By extension, the same logic holds for the dualism of knowledge/ignorance: both also require social explanations. Yet this has not generally come about. One of the consequences of SSK's reorientation of the field toward close investigation of knowledge practices has been that relational theories, such as Collins's (1986) theory of scientific closure and Latour's (1987) theory of actor-networks, exhibit the ironic characteristic of being both asymmetrical and nonreflexive in their inattention to ignorance. Another consequence has been missed opportunities to study ignorance empirically. For example, scientific "uncertainty" is clearly a good candidate for ignorance studies, but social scientists and science policy scholars have typically framed uncertainty as a problem of knowledge (Jamieson 1996; Shackley and Wynne 1996). This work does not directly confront the problem of ignorance because it does not theorize what remains unknown and why (Hoffman-Reim and Wynne 2002). Instead, this work investigates how scientific and policy decision making advances under conditions of *limited* knowledge, focusing on how knowledge about uncertainty and risk is produced and stabilized (Gross 2007).

Another reason for the relative dearth of published studies of ignorance is simply that it is difficult to study what by definition is not there. Historians of science have found a useful, if not wholly satisfying, way around this dilemma by framing ignorance as the process through which existing knowledge disappears over time, for example through secrecy and censorship (Galison 2008) or through deceit and suppression (Proctor 1995; Markowitz and Rosner 2002; Michaels and Monforton 2005). While politically important, these studies tend to utilize a conspiratorial logic that ties the production of ignorance to the specific political, economic, or professional interests of powerful organizations and individuals intent on keeping certain research findings private (Proctor 1995). In so doing, such studies advance an overly narrow conceptualization of ignorance, viewing it as the intentional result of purposive social action—see also McGoey (2012b). For example, several essays in Proctor and Schiebinger

(2008) exhibit this tendency. Apart from being empirically shortsighted, this narrow conceptualization is problematic because it is based on the misguided and nominally functionalist assumption that, *pace* Merton (1973), the production of ignorance results from deviant science. By extension, this implies that greater levels of public transparency will render science less deviant and thereby reduce scientific ignorance. There is little room in the narrow and conspiratorial view for theorizing ignorance as a regular feature of scientific production.

In this study, we treat ignorance more broadly as a regular outcome of risk assessment, not the result of deviant science nor one necessarily dependent on the purposive action of specific actors—although this may occur and we believe that our framework can accommodate it. Our approach is consonant with a political sociology of science perspective that highlights how distributional inequalities shape institutional and extra-institutional relations in science (Blume 1974; Frickel and Moore 2005; Moore et al. 2011). Accordingly, we identify institutional mechanisms in the regulatory response to Hurricane Katrina that produced, combined, and distributed different forms of ignorance within risk assessment frameworks and produced different effects across New Orleans neighborhoods.

This study also extends the empirical scope of ignorance studies from how existing knowledge disappears to how the absence of knowledge is generated and reproduced within regulatory agencies. The highly bureaucratized nature of the regulatory response to Katrina makes this empirical shift possible. The EPA response was delimited spatially by existing geopolitical boundaries (Orleans Parish, Louisiana) and by the topographical reach of the flood, and it was delimited temporally, with definite start and end points. And, because the response was part of a federally organized disaster response effort, U.S. law requires that all data collected within the project domain be made publicly available (United States Congress 2000). With access to these data as well as to the organizational rules, policies, and procedures that guided data collection and analysis, we can trace ignorance upstream into what Knorr Cetina (1999) has called the "epistemic machinery" of risk.

Testing Katrina's Contamination

Hurricane Katrina made landfall in Louisiana as a strong category three storm on 29 August 2005. As it moved inland, the massive storm's eye-wall grazed the southeastern edge of the New Orleans metropolitan area, causing extensive wind and storm surge damage to structures large and small. Most infamously, the storm triggered a systematic failure of the federal hurricane levee protection system. As the concrete and earthen levees surrounding the city crumbled, the storm's surge swamped New Orleans with 131 billion gallons of salt water (Smith and Rowland 2007). Floodwaters covered nearly 80 percent of the city's land area and inundated the households of over 60 percent of its population (see Figure 10.1), with standing water remaining in some neighborhoods for six weeks (Campanella 2007).

As the floodwaters gradually receded, a layer of sediment ranging in depth from several millimeters to nearly two meters blanketed the city (Nelson and Leclair 2006). Some of the sediment came laced with chemical toxicants and heavy metals, but other contaminants originated from within the city itself. Pollution point sources included gas, oil change, and auto service stations; laundries and dry cleaners; pest control companies, paint and hardware stores, hospitals, and cemeteries. Non-point pollution sources included as many as 350,000 automobiles and other vehicles submerged in the flood as well as a wide variety of hazardous substances typically stored in homes, garages, and backyard sheds.

With images of "toxic soup" driving news accounts and public concern (Goodman 2005), a dozen state and federal regulatory organizations led by the EPA and Louisiana's Department of Environmental Quality (DEQ) began collaboration on what would become a year-long effort to characterize environmental hazards in the sediment and soil of four flood-impacted parishes.

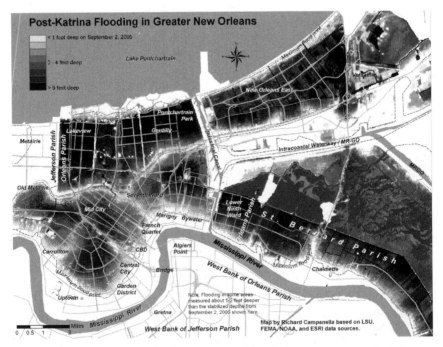

Figure 10.1. Post-Katrina Flooding in Greater New Orleans.

In line with their mission "to determine the nature and type of contamination that may have impacted residential areas due to migration of hazardous materials by flood," project organizers focused sampling efforts on flooded residential areas, largely ignoring neighborhoods that did not flood and also overlooking nonresidential (i.e., industrial and commercial) areas within the flood zone (United States Environmental Protection Agency 2006a).

From early September 2005 through July 2006, the project generated more than 400,000 chemical analyses for the presence of 195 individual contaminants from approximately 1,800 samples. Two-thirds of those samples originated in the city of New Orleans. Project scientists analyzed every sample for the presence of varying numbers of contaminants. Regulators then used those results to calculate short- and long-term human health risk.

While the suite of 195 chemical compounds that the EPA developed to guide the investigation all have industrial uses, they do not all originate in chemical laboratories. Many, such as vanadium, iron, and manganese, are also naturally occurring minerals that are common to various soil types. Synthetic chemicals targeted in the assessment include various banned or otherwise regulated pesticides and herbicides, PCBs, chlorides, and phthalates, among others. We obtained a complete list of targeted chemical compounds and calculated for each the total number of tests conducted and the number and percentage of positive results or "detects." Detections indicate contaminant presence in different sediment and soil samples resulting from laboratory analyses.

Our own analysis of the EPA test results data reveals three key points that are easily summarized. First, detections were common. Across all tests run for all compounds, lab technicians identified contaminants in more than 22,000 chemical assays accounting for just over 20 percent of total tests conducted. This amounts to an average of twenty-three different compounds detected per sample. Second, detections were also widely distributed across the range of compounds. In all, scientists identified 141 of the targeted chemicals, or nearly three-quarters of the original suite of compounds, in one or more of the collected samples. Finally, total detections for each compound concentrated in a large subset of these chemicals. Specifically, for sixty compounds, positive tests represented 10 percent[1] or more of total tests conducted for that substance. The number of samples represented by these percentages ranges from twenty-nine samples for 1,2,4-Trimethylbenzene, an aromatic hydrocarbon used as a gasoline additive, to 809 samples for lead. In between are scores of other environmental pollutants, including the banned pesticide DDT and its breakdown products DDD and DDE, detected in 262, 159, and 79 samples respectively; the recognized carcinogen and reproductive toxin bis(2-ethylhexyl) phthalate, detected in 288 samples; Indeno[1,2,3-cd]pyrene, an EPA priority pollutant and classified carcinogen, detected in 447 samples; and mercury and arsenic, both highly toxic elements with natural and industrial origins, detected in 658 and

698 samples respectively. In short, while laboratory testing identified relatively few samples containing contaminant levels high enough to automatically trigger additional investigation under existing regulatory requirements (and even fewer that required site remediation), the EPA's own data demonstrate that at lower levels heavy metals and industrial chemicals of various sorts abound in the soils of New Orleans.

For regulators vested with the responsibility to assess environmental conditions in the once-flooded city, these summary statistics mattered little. In fact, none of it was reported to the public. What did matter to regulators were affirmative answers to a single, unwavering question: Are chemical concentrations that exceed state environmental standards for human health risk evident in the data? In the vast majority of cases, regulators' answer to that question was "no." The data from hundreds of thousands of separately run tests prompted affirmative answers in fewer than 200 tests for just a handful of specific contaminants. Regulatory action motivated by the identification of those "hotspots" is summarized in the EPA's final report:

> A few localized areas were re-assessed due to elevated levels of arsenic, lead, benzo(a)pyrene, and diesel and oil range organic petroleum chemicals. The results of these re-assessments indicated that: 1) the highest concentrations of arsenic were likely associated with herbicides used at or near golf courses; 2) benzo(a)pyrene was found in a small section of the Agriculture Street Landfill Superfund site and will be addressed as the Housing Authority of New Orleans finalizes its plans for badly damaged townhomes in the area; 3) the concentrations of diesel and oil range organic chemicals are diminishing and will be monitored over time to ensure that these concentrations continue to decrease; and, 4) the elevated levels of lead detected in samples collected by EPA predate the hurricanes. The lead results from the EPA samples are comparable to the historical concentrations of lead in soil in New Orleans found in studies conducted by local university researchers before the hurricanes. The extensive sediment and soil sampling in response to Hurricane Katrina is complete. (United States Environmental Protection Agency August 2006b)

In reaching these conclusions, regulators focused on knowledge about the concentration levels of individual contaminants derived from laboratory analysis, using that knowledge in ways that progressively narrowed the scope of investigation spatially and epistemologically. From an initial concern encompassing about 55 square miles of flooded city streets,[2] investigation came to focus on a few dozen "hotspots." Similarly, the nearly two hundred analyses targeted for investigation in September 2005 had been reduced to just five by August 2006. The nightmare scenario that regulatory scientists and officials faced in the immediate aftermath of Hurricane Katrina had become scientifically certain, technically manageable, politically legitimate, and economically safe. As presented to the public, knowledge and hope had replaced ignorance and fear.

For New Orleans residents, city officials, and members of the business community, the stakes of the risk assessment had been exceptionally high. Especially for a city so dependent on tourism, it was difficult to imagine how the local economy would rebound if human health risk remained a serious concern. Amid a political atmosphere laced with apprehension, the EPA's risk assessment came as a great relief. City newspaper editors welcomed the agency's conclusions with a front page headline declaring "Final EPA Report Deems N.O. Safe" (Brown 2006). This image of a safe city stands in sharp contrast to headlines appearing in the wake of the hurricane eleven months earlier: "Entire Community Is Now a Toxic Waste Dump" (Clarren 2005), "New Orleans' Toxic Tide" (Knickerbocker and Jonsson 2005), and "Katrina Stirs Up Oily Nightmare" (Dakss 2005).

Organizing Ignorance

Rather than accept at face value this discursive shift from danger to safety, the remainder of this chapter digs beneath the EPA's formal risk statements in order to better appreciate the ways that ignorance shaped the agency's analysis of risk. To do so, we examine the institutional logic of two policy frameworks that guided regulatory scientists working in post-Katrina New Orleans. The first framework sets out the criteria governing the risk assessment process; the second develops and sets risk standards. Our investigation of the institutional logics of risk assessment reveals something other than a steady progression from ignorance to knowledge, as characterized by newspaper headlines and official reports. Instead, we find a progression from no knowledge to some knowledge and from unorganized ignorance to organized ignorance.

Risk Assessment

The Risk Evaluation/Corrective Action Program (RECAP) was adopted by DEQ in 2003 as the "primary statutory mandate for remediation activities" in the state (Louisiana Department of Environmental Quality 2003: Preamble). RECAP provides decision-making rules and procedures to assist state regulators in developing risk assessments and was the main source of information underlying the EPA and DEQ's collaborative post-Katrina response in New Orleans (and Louisiana generally). The stated function of RECAP is to establish "clear and consistent guidelines ... for the remediation of releases to air, land, and water" (Louisiana Department of Environmental Quality 2003:

Preamble). While its overriding goal is protection of "human health and the environment" in Louisiana, secondary RECAP goals include conserving institutional (i.e., DEQ) resources, protecting "the regulated community" (i.e., industry) from unnecessary oversight, and "ensuring transparency" in the risk assessment process. Toward this end, RECAP establishes risk assessment as a tool for determining, first, whether "corrective action is necessary" and, second, to "identify constituent levels in impacted media that *do not* pose unacceptable risks to human health or the environment" (Louisiana Department of Environmental Quality 2003: Preamble; our emphasis). An important, if unstated, assumption undergirding both of these institutional functions is that protection of human health and the environment does not a priori require cleanup of contaminated environments.

Read as a form of knowledge politics, the dominant institutional imperative rendered by the RECAP framework is not change, but stasis; the point is to minimize remedial activity where possible and when necessary to provide technical justification for environmental inaction. RECAP accomplishes this, in part, by setting forth a four-tier risk assessment framework in which ignorance operates as a silent organizing principle. The first tier is a "screening option" designed to establish whether contamination in residential or industrial areas pose threats and, if so, what is the specific nature of threat (Louisiana Department of Environmental Quality 2003: 1–2). At the screening stage, chemical assays of soil, air, or water samples are used to identify "constituents of concern," referring to any targeted contaminant that is "detected in at least one sample" (Louisiana Department of Environmental Quality 2003: 33, 37). While this language seems broadly encompassing, its logic plays on the inherent spatial and epistemological weaknesses of testing: testing will only identify the contaminants that regulatory scientists look for and then only in the samples they collect (Frickel and Vincent 2007).

Tiers 2–4 describe different "management options" instructing regulatory scientists in how to deal with problem areas when the results of screening warrant additional action. In step-wise fashion, each of the three management option tiers invokes progressively stricter sets of risk standards for individual contaminants, sets out additional measures to more specifically identify the boundaries of contamination, and provides increasingly intensive remediation criteria. As summarized in the official report, these "tiered Management Options allow site evaluation and corrective action efforts to be tailored to site conditions and risks. As the [management option] level increases, the approach becomes more site-specific and hence, the level of effort required to meet the objectives of the Option increases" (Louisiana Department of Environmental Quality 2003: 2).

As we interpret it, the institutional logic guiding risk assessment under RECAP is one that seeks to maximize "epistemic efficiency." Figure 10.2 il-

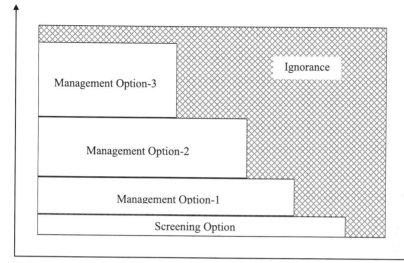

Figure 10.2. "Epistemic Efficiency" in Risk Assessment.

lustrates this process. As regulators move up the hierarchy of management options, they learn more and more about less and less. The logic of epistemic efficiency provides regulators clear organizational advantages in terms of increasing expediency, minimizing resource use, and gaining political legitimacy in response to public pressure. But there is a dark side as well. As accumulated knowledge concentrates on specific contaminants identified at specific sites, ignorance also accumulates: we learn less and less about more and more. This is "organized ignorance" and is the result of institutionalized inaction. Organized ignorance is a general form—a species or kind of ignorance—and not unique to Louisiana's own regulatory apparatus. As our next example shows, organized ignorance is characteristic of national risk standards as well.

Risk Standards

The mechanism that advances decision making in RECAP is risk standards. Risk standards are presented in DEQ and EPA documents as tables, with each contaminant assigned a numerical value. For regulatory scientists, those numbers refer to concentrations of chemicals in solution and are used to represent upper limits of "acceptable risk." For example, at the initial screening stage, such standards are used to determine whether a suspected area of concern is in fact contaminated and to identify what specific contaminants are present. Where initial screening standards are exceeded, regulatory effort under RECAP advances to Management Option 1 (and so on). Where screening standards are not exceeded, the framework does not require—and so does not compel—further action. In the New Orleans post-Katrina risk assessment most chemical concentrations derived from sediment and soil samples did not exceed initial screening standards for individual chemicals.

In combination with risk frameworks such as RECAP, risk standards are important elements in the institutional organization of ignorance in regulatory science. Embedded within each risk standard are a host of assumptions concerning, for example, a chemical's bioavailability, exposure potential, and health effects. Scientists' understanding of bioavailability rests in part on assumptions about how environmental conditions impact the chemical composition of contaminants over time or the ways that contaminants migrate through air, water, and soil. Similarly, the likelihood that contaminants will enter human bodies, directly or indirectly, depends on assumptions about the geography of contamination in relation to both natural patterns (e.g., wind, temperature, or rainfall) and social patterns (e.g., correlations between industrial pollution and urban poverty). A third set of assumptions, which we address in more detail below, concerns the health effects of chemical exposures. What do chemicals do to human bodies once inside?

Such assumptions—integral to the development of risk standards—organize three general types of ignorance. The first is ignorance that stems from scientists' limited knowledge base. For various reasons, much potential knowledge about the environmental and human health effects of chemicals remains "undone" (Hess 2009; Frickel et al. 2010). Undone science generates an absolute form of ignorance; it describes knowledge that does not exist. A second type describes ignorance in relative terms and in some ways represents the inverse of scientific uncertainty. This is ignorance that derives from the thinness of knowledge covering the health effects of certain chemicals. The knowledge is thin in part because there is not a lot of evidence with which to work, and in part because the evidence at hand has not been thoroughly assessed through expert review and thus lacks dimension or "epistemic depth." The third type of ignorance is formed in the process of extrapolation and represents a combination of absolute and relative forms. Here, regulators reach conclusions about the likely effects of a chemical on a particular endpoint based on what is known about the effects of that chemical on a different endpoint. Regulators may also extrapolate by substituting results from controlled laboratory studies onto uncontrolled events occurring outside the lab. Through extrapolation, absolute absence of knowledge is "upgraded" to a relative absence by substituting one type of knowledge or knowledge context for another. Risk standards embody all three types of ignorance.

These types of ignorance, while substantively different, are all generated by a similar institutional logic, one that we describe as maximizing "epistemic reach." Where policy frameworks guiding risk assessment tend to maximize epistemic efficiency by progressively knowing more about less, the logic guiding the construction of risk standards operates by making less knowledge count for more. The logic of epistemic reach represents a radically reductive approach to understanding the relationship between the environment and human health. We examine this argument empirically in the next section.

Human Toxicity Values

In practice, standards for human health effects occupy a central position in risk assessment. As described in the RECAP framework and as applied in New Orleans, these standards take the form of "human toxicity values" (Louisiana Department of Environmental Quality 2003: 25). These numerical values identify dose response relationships for substances and are expressed as "reference doses" or "reference concentrations" for non-carcinogenic effects and as "slope factors" for carcinogenic effects. Formulas used to calculate toxicity values contain built-in assumptions about a person's body weight (70 kg, as modeled on an "average" adult male) and their daily water intake, inhalation rate, or dermal absorption rate (Louisiana Department of Environmental Quality 2003: 67).[3]

Detailed methods for developing toxicity values illustrate the institutional logic of epistemic reach. These methods are described in the EPA's *Risk Assessment Guidance for Superfund Volume I Human Health Evaluation Manual* (1989) and were updated in 2003 in a memo from the Director of the EPA's Office of Superfund Remediation and Technology Innovation (Cook 2003). Similar to the hierarchical framework RECAP builds for risk assessment, the report and memo also establish a hierarchical framework to guide the generation and use of toxicity values. Importantly, this three-tier "hierarchy of human toxicity values" is reproduced in a document released by EPA Region 6 in November 2005, two months into its post-Katrina risk assessment project. The 2005 document establishes risk-based screening levels "to address common human health and environmental exposure pathways" and confirms that procedures drawn up in the original 1989 report to screen for site-specific hazards such as regulated landfills were employed in essentially the same form to address the urban-scale disaster in New Orleans from Hurricane Katrina sixteen years later (United States Environmental Protection Agency Region 6 2005b).

As described in all of these policy manuals, the Tier 1 source of toxicity information is the EPA's Integrated Risk Information System (IRIS). This database contains information on chemical effects that achieve "agency consensus" through an extensive review process.[4] IRIS is the richest available data source for establishing toxicity values, in terms of its scientific knowledge base as well as the level of internal and external review. Nevertheless, IRIS is far from comprehensive.

When assessment information on particular contaminants is not available in IRIS, regulators are instructed to next consult the EPA's Tier 2 database, the Provisional Peer-Reviewed Toxicity Values (PPRTV) (Cook 2003: 3).[5] Reviews of PPRTV data are developed over a shorter period of time than IRIS data, undergo fewer rounds of peer review, and are based on fewer scientific studies. In all, PPRTV represents less information based on a more limited knowledge base than IRIS.

When regulators fail to find the toxicity information they need in IRIS or PPRTV, they are instructed to consult additional "Tier 3" data sources. Tier 3 sources include the EPA Health Effects Assessment Summary Tables (HEAST), which has "not had enough review to be recognized as high quality, Agency-wide consensus information" (United States Environmental Protection Agency 1997), the EPA National Center for Environmental Assessment (NCEA), which influences agency "regulatory, enforcement, and remedial-action decisions,"[6] as well as other non-EPA sources. Thus, as regulators in search of information to use for generating toxicity values move from Tier 1 to Tier 2 and then from Tier 2 to Tier 3, the relative quantity and quality of toxicity information decreases.

When there is no toxicity information available, regulators rely on "route-to-route extrapolation." This involves substituting toxicity values of one kind for another, for example, by using an oral reference dose value in place of an inhalation reference dose value. While regulators suggest that although route-to-route extrapolation is *"a useful screening procedure,"* the Agency clearly sees this method as a measure of last resort, cautioning that *"the appropriateness of these default assumptions for specific contaminants should be verified by a toxicologist"* (United States Environmental Protection Agency Region 6 2005b: 5, original emphasis). Figure 10.3 illustrates the concept of epistemic reach in the hierarchical framework of toxicity value production.

Table 10.1 shows the frequency distributions of information for the 141 chemicals detected in New Orleans samples relative to the EPA's hierarchy of human toxicity values. This data is drawn from an electronic spreadsheet containing values used by the EPA to determine risk standards for specific contaminants in New Orleans (United States Environmental Protection Agency Region 6 n.d.). The spreadsheet also contains codes describing which database (IRIS, PPRTV, etc.) each toxicity value came from. We aggregated this data for each of the 141 contaminants detected in New Orleans to derive the distributions presented in the table. In addition to describing where particular kinds of toxicity information originate, the table also describes gaps in toxicity information. We interpret these information gaps as representing measurable

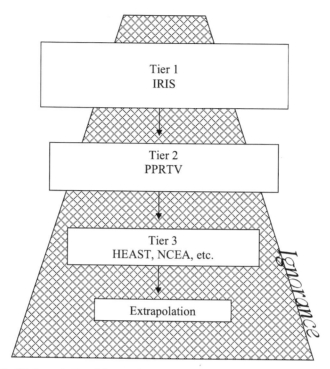

Figure 10.3. "Epistemic Reach" in Risk Standards. The EPA Hierarchy of Human Toxicity Values.

Table 10.1. Information Gaps in EPA's Hierarchy of Human Toxicity Values

Total analytes detected in New Orleans samples: N=141	Carcinogenicity Values				Non-carcinogenicity Values					
	SFo		SFi		RfDo		RfDi		RfC	
	N	%	N	%	N	%	N	%	N	%
Tier 1 IRIS	26	18.44	21	14.89	65	46.10	17	12.06	11	7.80
Tier 2 PPRTV	0	0.00	1	0.71	8	5.67	5	3.55	0	0.00
Tier 3 various	11	7.80	8	5.67	13	9.22	9	6.38	4	2.84
Route-to-Route Extrapolation	0	0.00	11	7.80	2	1.42	44	31.21	0	0.00
No Information	104	73.76	100	70.92	53	37.59	66	46.81	126	89.36
Total	141	100	141	100	141	100	141	100	141	100

Source: EPA Region VI Human Health Medium-Specific Screening Levels (HHMSSL)
Abbreviations: SFo = Slope Factor (oral), SFi = Slope Factor (inhalation), RfDo = Reference Dose (oral), RfDi = Reference Dose (inhalation), RfC = Reference Concentration.

outcomes associated with the three types of ignorance described in the previous section.

The dominant type of ignorance contributing to missing information across the various databases, represented in the bottom row in Table 10.1, is undone science. There is no information for over 70 percent of the carcinogenicity values for the contaminants found in New Orleans. For non-carcinogenicity values, the proportion of missing information ranges from 37 to 89 percent. We infer from these figures that, within the domain of EPA risk assessment—and perhaps also beyond—toxicity and related studies that regulatory scientists rely on to calculate risk standards for many of the chemical substances detected in New Orleans samples simply do not exist.

Scientists' reliance on lower quality data generates another form of ignorance underlying the EPA human toxicity values. Table 10.1 shows that the proportion of toxicity values that are produced with information from Tier 2 and Tier 3 sources relative to Tier 1 sources ranges from 24 percent to 45 percent.[7] This type of ignorance derives more from the "thinness" or limited depth of existing information than from undone science or missing knowledge. Consonant with the logic of epistemic reach, the deeper regulatory scientists go into the hierarchy of data sources in search of toxicity information, the more limited the knowledge they have to draw from becomes.

Finally, Table 10.1 also describes the distribution of ignorance as decontextualized knowledge generated through methods of extrapolation. Based on these data, regulatory scientists substituted one carcinogenicity value for another eleven times and substituted one non-carcinogenicity value for another forty-six times. For toxicity values based on inhalation reference doses, route-to-route extrapolation was by far the dominant method, used to calculate risk screening levels for 31 percent of the substances detected in New Orleans. Here too, regulators are guided in their work by a logic that makes less knowledge count for more.

The institutional logic that maximizes epistemic reach helps explain the different distributions of ignorance that are depicted schematically in Figure 10.3 and quantitatively in Table 10.1. Once institutionalized as numerical risk standards, toxicity values calculated from Tier 3 sources or based on route-to-route extrapolation become indistinguishable from toxicity values calculated from more robust data contained in Tier 1. In this way, the same institutional logic that produces ignorance in risk standards also hides it from view.

Conclusion

This study raises four key points about the institutional production of ignorance in risk assessment. First, our study describes how knowledge and ignorance are coproduced. Knowledge practices in risk assessment concentrate in ways that narrow the questions regulators ask about soil contamination and narrow the geographic scope within which those questions are treated by regulators as meaningful. Over time, knowledge about fewer contaminants in fewer places deepens. Simultaneously however, knowledge about greater numbers of contaminants and their potential distribution across larger land areas diminishes. In this way, the production of knowledge about identified risks and the production of ignorance concerning potential risks are not only mutually constituted, their coproduction within risk assessment frameworks is structured in ways that, on a policy level, helps to legitimate regulatory inaction.

Second, regulatory inaction is an outcome whose achievement in risk assessment requires considerable resources, organization, and effort. Risk standards do much of this work by signaling to regulatory scientists whether or not the type and nature of contamination identified in soil and sediment samples are meaningful as environmental hazards. In this way, risk standards perform a critical function in the production of ignorance; they operate as decision-making mechanisms that drive the risk assessment process forward by concentrating knowledge investments spatially and epistemically.

Third, as they rationalize decision making and thus help to maximize epistemic efficiency in risk assessment, risk standards are themselves produced through a different institutional logic, one that maximizes epistemic reach. The logic of epistemic reach masks the fact that the health information databases used to develop risk standards are themselves deeply structured by the lack of knowledge. The logic of epistemic reach preserves legitimacy for risk assessment decision making by rendering ignorance invisible.

Fourth, and perhaps most importantly, different forms of ignorance combine in complex ways. Within the health effects databases, ignorance that results from undone science shapes the conditions under which other types of ignorance are generated when regulators are forced to accept lower quality knowledge or to resort to extrapolation in developing risk standards. More broadly, epistemic efficiency and epistemic reach are features of the risk system that produce ignorance according to different logics but that gain force in combination. Making less knowledge from toxicity studies count for more in the production of risk standards in turn shapes risk assessment practices in ways that generate new forms of spatially ordered ignorance about contaminated soil across urban neighborhoods.

The picture of ignorance that emerges from this analysis stands in sharp contrast to more narrowly drawn depictions of ignorance as the conspiratorial result of powerful corporate and government actors purposefully manipulating or subverting scientific research. While highly problematic politically, the intentional suppression of knowledge does little to shed light on ignorance as a general feature of science. Our emphasis in this study on the institutional logics that produce ignorance in risk assessment seeks a broader understanding. In developing an institutional approach to the study of ignorance, we also hope to contribute to the political sociology of science a better understanding of how ignorance shapes and is shaped by distributions of power and resources in science. Finally, while our focus here is on regulatory science mobilized in response to urban disaster, we believe this approach can be refined and extended in application to other domains as well. For example, future studies might fruitfully develop a comparative perspective on ignorance as it is made, remade, and unmade in academic, government, industry, and civil society contexts. If, as this chapter suggests, ignorance is a both a dynamic feature of science and a regular outcome of scientific work, it stands to reason that absence of knowledge has done much to shape the history and social organization of our toxic world.

Acknowledgments

We thank Richard Campanella for producing the map that appears in Figure 10.1 and Bess Vincent for preliminary research that helped make possible the analysis presented in this chapter.

Notes

1. This is an arbitrary cutoff point, employed here to facilitate our summary. Using this metric is potentially misleading because some analyses with "detect" percentages below 10 percent actually involved more samples/sites than others with detect percentages above 10 percent.
2. Richard Campanella. Personal communication, 12 January 2010.
3. On the use of the "average male" in calculating health risks see Corburn (2005: chap. 3) and, more generally, Epstein (2007).
4. For details of the IRIS peer review process, see United States Environmental Protection Agency, "IRIS Process," *Integrated Risk Information System (IRIS),* http://www.epa.gov/iris/process.htm.
5. The data contained in PPRTV are typically developed by the Superfund Health Risk Technical Support Center (STSC) either on a chemical-by-chemical basis when requested by the EPA Superfund Offices for "contaminants lacking a relevant IRIS value" or after a "batch-wise review" is conducted for toxicity values gathered from a third database (HEAST; United States Environmental Protection Agency 1997).
6. See United States Environmental Protection Agency, "NCEA Basic Information," National Center for Environmental Assessment, http://www.epa.gov/ncea/basicinfo.htm.
7. Specifically, carcinogenicity values for SFo and SFi are both 30 percent; non-carcinogenicity values vary from 24 percent (for RfDo) to 27 percent (for RfC) to 45 percent (for RfDi).

Bibliography

Bloor, David. 1976. *Knowledge and Social Imagery.* Chicago: University of Chicago Press.

Blume, Stuart. 1974. *Toward a Political Sociology of Science.* New York: The Free Press.

Brown, Matthew. 2006. "Final EPA Report Deems N.O. Safe." *The Times-Picayune,* 19 August. http://www.nola.com/news/t-p/frontpage/index.ssf?/base/news-6/11559715 80163240.xml&coll=1 (accessed 28 September 2010).

Campanella, Richard. 2007. "An Ethnic Geography of New Orleans." *Journal of American History* 94(3): 704–16.

Clarren, Rebecca. 2005. "Entire Community Is Now a Toxic Waste Dump." Salon.com, 9 September. http://www.salon.com/news/feature/2005/09/09/wasteland/index.html (accessed 28 September 2010).

Collins, Harry. 1986. *Changing Order.* Chicago: University of Chicago Press.

Cook, Michael. 2003. "Human Health Toxicity Values in Superfund Risk Assessments [OSWER Directive 9285.7-53]." Memorandum. 5 December.

Corburn, Jason. 2005. *Street Science: Community Knowledge and Environmental Health Justice.* Cambridge, M.A.: MIT Press.

Dakss, Brian. 2005. "Katrina Stirs Up Oily Nightmare." *CBS News,* 7 September. http://www.cbsnews.com/stories/2005/09/07/earlyshow/main821663.shtml (accessed 28 September 2010).

Epstein, Steven. 2007. *Inclusion: The Politics of Difference in Medical Research.* Chicago: University of Chicago Press.

Frickel, Scott, Richard Campanella, and M. Bess Vincent. 2009. "Mapping Knowledge Investments in the Aftermath of Hurricane Katrina: A New Approach for Assessing Regulatory Agency Responses to Environmental Disaster." *Environmental Science & Policy* 12: 119–33.

Frickel, Scott, Sara Gibbon, Jeff Howard, Joanna Kempner, Gwen Ottinger, and David Hess. 2010. "Undone Science: Charting Social Movement and Civil Society Challenges to Research Agenda-setting." *Science, Technology & Human Values* 35: 444–73.

Frickel, Scott, and Kelly Moore. 2005. "Prospects and Challenges for a New Political Sociology of Science." In *The New Political Sociology of Science: Institutions, Networks, and Power,* ed. S. Frickel and K. Moore, 3–31. Madison, W.I.: University of Wisconsin Press.

Frickel, Scott, and M. Bess Vincent. 2007. "Katrina, Contamination, and the Unintended Organization of Ignorance." *Technology in Society* 29:181–88.

Galison, Peter. 2008. "Removing Knowledge: The Logic of Modern Censorship." In *Agnotology: The Making and Unmaking of Ignorance,* ed. R. N. Proctor and L. Schiebinger, 37–54. Stanford, C.A.: Stanford University Press.

Goodman, Amy. 2005. "Toxic Soup: The Deadly Floodwaters of New Orleans." Radio interview by Amy Goodman with Harold Zeliger on Democracy Now!, 8 September (accessed 28 September 2010).

Gross, Matthias. 2007. "The Unknown in Process: Dynamic Connections of Ignorance, Non-Knowledge and Related Concepts." *Current Sociology* 55: 742–59.

———. 2010. *Ignorance and Surprise.* Cambridge, M.A.: MIT Press.

Hess, David J. 2007. *Alternative Pathways in Science and Industry: Activism, Innovation, and the Environment in an Era of Globalization.* Cambridge, M.A.: MIT Press.

———. 2009. "The Limitations and Potentials of Civil Society Research: Getting Undone Science Done." *Sociological Inquiry*, 79: 306–27.
Hoffman-Reim, Holger, and Brian Wynne. 2002. "In Risk Assessment, One Has to Admit Ignorance." *Nature* 416(14 March): 123.
Jamieson, Dale. 1996. "Scientific Uncertainty and the Political Process." *Annals of the American Academy of Political and Social Science* 545(May): 35–43.
Knickerbocker, Brad, and Patrik Jonsson. 2005. "New Orleans' Toxic Tide." *Christian Science Monitor*, 8 September. http://www.csmonitor.com/2005/0908/p01s01-usgn.html (accessed 28 September 2010).
Knorr Cetina, Karin. 1999. *Epistemic Cultures: How the Sciences Make Knowledge*. Cambridge, M.A.: Harvard University Press.
Latour, Bruno. 1987. *Science in Action: How to Follow Scientists and Engineers Through Society*. Cambridge, M.A.: Harvard University Press.
Louisiana Department of Environmental Quality. 2003. "Preamble." In *Risk Evaluation/Corrective Action Program (RECAP)*. Baton Rouge, L.A.: DEQ.
———. 2005. "State, Federal Agencies Summarize Environmental Assessment of Hurricane Impacted Areas of Southeast La." Press release. 9 December. http://www.deq.louisiana.gov/portal/portals/0/news/pdf/jointenvironmentalassessmentpr.pdf (accessed 28 September 2010).
Markowitz, Gerald, and David Rosner. 2002. *Deceit and Denial: The Deadly Politics of Industrial Pollution*. Berkeley, C.A.: University of California Press.
McGoey, Linsey, ed. 2012a. "Strategic Unknowns: Towards a Sociology of Ignorance." Special issue of *Economy & Society* 41(1).
———. 2012b. "The Logic of Strategic Ignorance." *British Journal of Sociology* 63(3): 553–76.
Merton, Robert. 1973. *The Sociology of Science*. Chicago: University of Chicago Press.
Michaels, David, and Celeste Monforton. 2005. "Manufacturing Uncertainty: Contested Science and Protection of the Public's Health and Environment." *American Journal of Public Health* 95: S39–S48.
Moore, Kelly, David Hess, Daniel Kleinman, and Scott Frickel. 2011. "Science and Neoliberal Globalization: A Political Sociological Perspective." *Theory and Society* 40:505–32.
Nelson, Stephen A., and Suzanne F. Leclair. 2006. "Katrina's Unique Splay of Deposits in a New Orleans Neighborhood." *GSA Today* 16: 4–9.
Proctor, Robert. 1995. *Cancer Wars: How Politics Shapes What We Know and Don't Know About Cancer*. New York: Basic Books.
Proctor, Robert N., and Londa Schiebinger, eds. 2008. *Agnotology: The Making and Unmaking of Ignorance*. Stanford, C.A.: Stanford University Press.
Shackley, Simon, and Brian Wynne. 1996. "Representing Uncertainty in Global Climate Change Science and Policy: Boundary-Ordering Devices and Authority." *Science, Technology, & Human Values* 21(3): 275–302.
Smith, Jodie, and James Rowland. 2007. "Temporal Analysis of Floodwater Volumes in New Orleans After Hurricane Katrina." In *Science and the Storms—the USGS Response to the Hurricanes of 2005*, ed. G.S. Farris, G.J. Smith, M.P. Crane, C.R. Demas, L.L. Robbins, and D.L. Lavoie. U.S. Geological Survey Circular 1306, 57–61.
Smithson, Michael. 1989. *Ignorance and Uncertainty: Emerging Paradigms*. New York: Springer-Verlag.

Sullivan, Sharon, and Nancy Tuana, eds. 2007. *Race and Epistemologies of Ignorance*. Albany, N.Y.: SUNY Press.

Tuana, Nancy. 2008. "Coming to Understand: Orgasm and the Epistemology of Ignorance." In *Agnotology: The Making and Unmaking of Ignorance*, ed. R. N. Proctor and L. Schiebinger, 108–45. Stanford, C.A.: Stanford University Press.

United States Congress. 2000. "The Robert T. Stafford Disaster Relief and Emergency Assistance Act." *U.S. Code* vol. 42, secs. 5121 et seq.

United States Environmental Protection Agency. 1997. *Health Effects Assessment Summary Tables (HEAST), FY 1997 Update*. Washington, D.C.: EPA.

———. 2006a. "EPA Provided Quality and Timely Information on Hurricane Katrina Hazardous Material Releases and Debris Management." Office of Inspector General Evaluation Report No. 2006-P-00023. 2 May.

———. 2006b. "Summary Results of Sediment Sampling Conducted by the Environmental Protection Agency in Response to Hurricanes Katrina and Rita." 17 August. http://www.epa.gov/katrina/testresults/sediments/summary.html (accessed 28 September 2010).

———. 2006c. "Test Results." 17 August. http://www.epa.gov/katrina/testresults/index.html (accessed 28 September 2010).

———. 2009. "IRIS Process." *Integrated Risk Information System (IRIS)*. Updated 2009. http://www.epa.gov/iris/process.htm (accessed 28 September 2010).

———. 2010. "National Center for Environmental Assessment: NCEA Basic Information." 18 August. http://www.epa.gov/ncea/basicinfo.htm (accessed 28 September 2010).

United States Environmental Protection Agency Region 6. 2005a. "Emergency Response Quality Assurance Sampling Plan for Hurricane Katrina Response: Screening Level Sampling for Sediment in Areas Where Flood Water Receded, Southeast Louisiana." Unpublished document in author files. September.

———. 2005b. "Human Health Medium-Specific Screening Levels." November.

———. n.d. "Human Health Medium-Specific Screening Levels" (electronic spreadsheet file no longer publicly available).

United States Government Accounting Office. 2005. "Chemical Regulation: Options Exist to Improve EPA's Ability to Assess Health Risks and Manage Its Chemical Review Program." Washington, DC: GAO. http://www.gao.gov/products/GAO-05-458 (accessed 28 September 2010).

CHAPTER 11

Low-Dose Toxicology
Narratives from the Science-Transcience Interface

Sheldon Krimsky

Uncertainties associated with low-dose exposures to chemicals that are known to be hazardous at high doses were probably being raised at the dawn of human civilization when Homo sapiens began distinguishing among edible, near edible, and poisonous plants. The study of toxicology began around the sixteenth century with the writings of an Austrian physician and contemporary of Leonardo da Vinci, named Philip von Hohenheim, who practiced "chemical medicine." Hohenheim is more popularly known as Paracelsus, a name he adopted to elevate him above a prominent Roman physician named Celsus. Paracelsus is known to have said: "All things are poison and nothing is without poison, only the dose permits something not to be poisonous." He believed that low doses of a poison could be used to cure diseases brought about by the poison ("like cures like") (Borzelleca 2000). By the twentieth century, when toxicology had become a scientific discipline, the observations that low doses of a poison can be therapeutic became epitomized by the aphorism "the dose makes the poison."

Modern toxicology is a battleground of contested issues pertaining to low-dose exposures. The domain of published work contains a number of presuppositions, some false, some true, and some indeterminate. The core questions underlying the low-dose narratives are: (1) Are there empirical tests that can be used to evaluate the human health effects of exposures to extremely low doses of a substance? (2) Are there methods that can be used to generate a reliable dose-response curve for extremely low doses of a substance? and (3) How can we know and with what level of confidence whether a chemical below a testable dose is safe?

I approach this subject by treating the controversy over low-dose exposures as a confluence of alternative narratives. Each narrative may frame the problem differently and builds its argument or draws its conclusions from empirical evidence based on a preferred set of presuppositions, which may not be empirically verifiable or falsifiable. In this respect, rather than being in direct conflict, some of the narratives may be orthogonal to one another, like two religions whose adherents speak a different language, preach from different texts, and hold a different world view. The chapter is structured as follows. First, six guiding presuppositions that provide the framing assumptions for the low-dose narratives are identified. These framing presuppositions are themselves contested and therefore help to explain the differences in how scientists approach low-dose effects. Then, three modalities of evidence used to obtain answers to low-dose human health effects are examined. They are: direct empirical evidence; indirect empirical evidence, and theoretical evidence. Next, I explain how the gene-environment interaction model of disease complicates efforts to assess low-dose health effects.

I go on to examine the classic problem of the linear dose response curve for extremely low doses, including the reproducibility of low-dose effects. Next, a case study of Bisphenol A (BPA) reveals competing narratives of low-dose effects impeding efforts to reach a consensus position. Finally, I argue that the path of mechanistic reductionism, which largely defines the approach used by some of the narratives in toxicology, is not the best approach for regulating the health risks of low doses of toxic chemicals. Further, my approach using competing narratives helps explain why acquiring more data often fails to resolve the issue of when low-dose exposures become a health risk deserving of regulatory action.

Narrative Frames on Low-Dose Effects

One cannot read the scientific literature on low-dose toxicology without experiencing the contrasting narratives of this field. There is nothing about low-dose toxicology that is common across all chemicals, consistent across all modes of action, or predictable across all genotypes. I shall begin my inquiry by exploring several of the framing presuppositions that have become a notable part of the scientific literatures:

A. The human effects of extremely low doses of chemicals are beyond what can be learned from direct observation. One must introduce a priori (empirically unverifiable) assumptions to extrapolate from the effects of high doses to low doses of a chemical substance. These assumptions take the form of dose response curves extrapolated from points in the high-dose range or of thresholds below which there are no effects (NOEL = No Observable Effect Level).

B. There are discontinuities between the effects of chemicals at high and low doses, sometimes referred to as non-monotonic, making simple linear extrapolation unrealistic.

C. Multiple and differentiated physiological mechanisms operating in mammalian systems can make the determination of a simple dose response relationship for a single chemical, and a single outcome over low- to high-dose range, highly unrealistic.
D. Low-dose studies are difficult to replicate because they are vulnerable to sensitive stochastic effects. The analogy is the "baking effect." Even though the baker uses all the same ingredients in a precise order, with a standardized baking process, the outcome may vary significantly by virtue of a small number of stochastic effects.
E. Embryos at particular windows of development (the first trimester) are more sensitive to low-dose effects than more developed fetuses and adult organisms.
F. Epigenetics has introduced the idea that the embryo may have been affected by the mother or grandmother's exposure to low doses of chemicals. Something in the organism's external environment alters how a gene is expressed without changing the structure of the gene. This model is akin to "action at a distance." In this case it is a generational distance. The mechanisms proposed to explain such effects are attributed to a combination of "genetic switches" and "imprinting" of the genome, one of the newest and least-studied mechanisms for low-dose effects.

These presuppositions are found throughout the low-dose literature and have become subjects of intense debate and the source of policy conflicts. In some instances, these debates have helped to paralyze regulatory bodies, preventing them from reaching a conclusion on specific toxic substances. Beyond these presuppositions, there are also three epistemic modalities used in science to acquire evidence and reach conclusions on low-dose effects of chemicals. The combination of the presuppositions and the epistemic modalities contribute to a particular narrative.

Epistemic Considerations in Regulating Low-Dose Effects

Regulators have always had difficulty in determining acceptable levels of a substance at low doses. They are expected to show evidence that low doses of a substance are harmful before they can restrict or ban its use. In drug development, manufacturers are required to demonstrate both efficacy and safety before a drug is approved for consumers. For all chemicals that do not have therapeutic use, the burden is on government to show that it is unsafe. Of course, regulators can ask for data from manufacturers if they have prima facie evidence that a substance may be harmful. But the manufacturers are only expected to provide data that they can easily obtain. And here lies the problem for low doses. There are methodological problems in obtaining evidence in support of the hypothesis that the substance is harmful at low doses. And as the aphorism goes, "No evidence of a chemical risk is not evidence of no risk." This is particularly true when there are impediments to obtaining data.

A great number of claims, misunderstandings, and some myths have been raised about low doses. It is sometimes said that there can be no direct empirical evidence for low-dose effects and that all evidence must come from animals where extrapolations are made from high doses. It is also said that effects found in animals in relatively short-lived species such as the rat or mouse cannot be used to estimate the effects in a long-lived species such as humans. Believers in hormesis argue that some industrial chemicals that are known to be toxic at high doses are beneficial at low doses (Calabrese and Baldwin 2002). Others maintain that we cannot find a causal relationship at low doses because there can be no statistical verification of low-dose extrapolation (Pesch et al. 2009), where a variety of shapes of curves may all fit (Armitage 1982: 126).

There are three general methods, which I refer to as epistemic modalities, that are used to acquire information about low-dose effects of substances on humans. They are: direct evidence, mainly epidemiological studies on humans; indirect evidence by extrapolation from high and moderate doses to low doses or from animal studies to humans; and theoretical approaches that apply mechanistic modeling. Each method has its unique benefits and limitations.

Direct Evidence

Epidemiological studies of large human populations exposed to low doses of a substance can sometimes yield reliable evidence of health effects. The data can come from chemical spills or radiation exposure (Pierce and Preston 2000), where the doses of exposure are well understood or measured. To get low-dose data from animals in traditional toxicological studies that are statistically meaningful could require hundreds to thousands of animals because animals do not live that long. When oncogenic-sensitive mice were developed, low doses could be used in conjunction with smaller sample sizes. Critics of such experiments argue that the animals are so artificial that their effects cannot be used to shape policies about human disease.

Endocrine disruptors (industrial chemicals that behave like human hormones) present a different model for studying the effects of chemicals on the developing organism. Very low doses of toxicants can produce statistically reliable, observable effects with far less than 100 mice. Extrapolation is not required. Moreover, high doses of the same substances may not exhibit the same effects.

Many studies involving the impact of toxicants on the endocrine system use 6–12 mice in the experimental sample and the same in controls (Salazar et al. 2006).

Indirect Evidence

The most common method of acquiring data for low-dose effects of chemicals is through extrapolation from high doses in animal studies or from occupational data involving human exposure. This has generated debates over the shape of the dose response curves prompting some scientists to claim that the method of extrapolation introduces subjective judgments about how chemicals will respond at low doses. In addition, after deciding which extrapolation model to use, and then extrapolating down to a regulatory "safe" level, for endocrine disruptors there are effects at concentrations several orders of magnitude below the "safe" level.

Theoretical Approaches

The use of mechanistic modeling to obtain low-dose toxicity information is growing in interest (Rietjens and Alink 2006: 980). Among its benefits is that it is grounded in identifying the "mechanism of action" of a specific chemical, including its biochemical pathways, and measurable endpoint effects. This means that society does not regulate a chemical until the mechanism of action is fully determined, for each endpoint, for each chemical present, for each organ, and for each strain of animals. The bar of scientific knowledge required for mechanistic modeling can be much higher than that for indirect evidence and, where available, direct evidence.

One of the most complex problems in public health is how to regulate substances that are known to cause cancer in animals, or at least some species of animals, but for which there is not direct evidence that they cause cancer in humans. For many years regulators assumed that there was no safe dose of a carcinogen. That was the premise behind the U.S. Delaney Clause of the 1958 Food, Drug, and Cosmetic Act. The complexity of regulating carcinogens mirrors the complexity of the mechanisms underlying cancer. It has been often said that cancer is not one disease but many diseases under the same name. Some cancers have a long latency period from the point of exposure. Moreover, the etiology of cancer is a multistage process and each stage provides necessary but not sufficient conditions for a particular cancer to develop. Other theories of cancer etiology influence the way we think about low-dose carcinogens. These include the idea that cancer is a breakdown of the immune system or a malfunctioning of the signaling that takes place between cells of different tissues. One of the most important new theories is based on the link between genes and the environment.

Gene-Environment Interactions: The No-Effect Outcome

While many assumed there was no safe dose of a carcinogen, it is difficult to understand how, in some cases, low doses of a carcinogen may be more lethal than high doses. Some studies have shown that the risk of lung cancer associated with smoking and a particular polymorphism is greater at lower doses of cigarette smoke than at high doses. It is hypothesized that the genetic susceptibility to cancer (in a gene-environment interaction) may be more responsive at low doses because "from a metabolic point of view... at high dose levels the relevant enzyme is saturated both in rapid and in slow metabolizers, while this does not happen at low doses" (Vineis 1997: 1). Thus, at high dose levels of a cigarette carcinogen, critical enzymes that are part of the cancer etiology are disabled, but become active at low levels. Another hypothesis is that one must look at the "carcinogen" in its context of use. For example, at low doses, mixtures, including chemicals in one's diet, "might prove to be more important than exposure to single agents" (Vineis 1997: 3). Variations in peoples' genetic susceptibility to chemical diseases may skew linear dose response curves, making the risk of contracting cancer higher than expected.

At very low doses, the chemicals may reside in certain regions of the body that are more susceptible to organ damage or oncogenic effects. Gerde (2005: 145) notes: "While the overall dose of inhaled substances can be reasonably measured and assessed, the local dose to disease-prone regions of the respiratory tract is often impossible to measure directly."

Another complexity in low-dose assessment of chemical hazards is that some individuals have different metabolic genes (polymorphisms) that encode enzymes that are involved in the metabolism of carcinogenic agents. People with different metabolic polymorphisms may have higher or lower risk of cancer when exposed to certain chemicals. Not everyone reacts to low doses in the same way. There are gene-environment interactions at work. Without understanding the genetic factors, studies may obscure the low-dose effects of chemicals (Taioli and Garte 1999).

Consider an experimental design involving two groups of people as in a case control study. Study Group I is the experimental group and Study Group II is the control (see Figure 11.1).

Suppose we cluster all the people in whom we see an effect into Group I. People with similar characteristics but without an effect are clustered into Group II. We then examine the individuals in Group I to see if there is a cor-

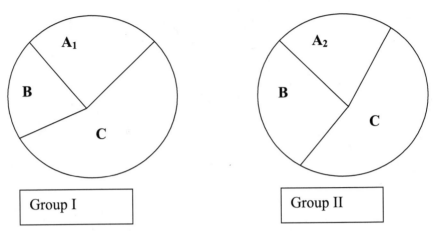

Figure 11.1. Case Control Study. Hidden Genetic Effects.

relation between environmental contaminant C and the people who have the effect. Then we determine whether the contaminant correlates with the people in Group II. The true cause of the effect is: $A_1 + B + C \rightarrow$ Effect. A_1 is a genetic factor (polymorphism) and B is a social component, constant for both groups. In Group II we have: $A_2 + B + C \rightarrow$ No Effect. The environmental factor C does not show up any more strongly for Group I than it does for Group II. Without knowing about the existence of the polymorphisms, one might conclude that C is not the cause of the effect. In fact, C is a necessary condition, but requires both A_1 and B for its effect. Without understanding gene-environment interactions, case control studies may not reveal low-dose effects. This could help explain the negative results in case control studies of adult women with breast cancer on whether DDT or PCBs could be a contributing cause.[1]

Another explanation that can account for a no-effect outcome in case control studies of low-dose exposure of persons to endocrine-active chemicals is that such studies neglect the stage of development during which the exposure took place—the "window of exposure." Chemicals may affect embryos and fetuses differently than they do adults. "Data from studies with adult animals thus cannot be used to predict the pharmacokinetics of chemicals in pregnant females and fetuses" (Welshons et al. 2003: 1001). One study compared women exposed to DDT before the age of fourteen with women exposed after that age. The investigators found a fivefold increased risk of breast cancer among women who were first exposed to DDT before the age of fourteen in around 1945, when DDT came into widespread use. Women who were not exposed to DDT before age fourteen did not have a higher risk of breast cancer (Cohn et al. 2007). Given the uncertainties around low-dose exposures, when no direct evidence is available, scientists have made two bold assumptions: that the dose response curve is linear and that there are no thresholds at low doses.

The Linear, Non-Threshold Default Position

The default model for chemical carcinogens and radiation has been the assumption of a linear non-threshold (LNT) effect. When first adopted, it was based on the assumption that a single mutation can launch the cell into becoming a cancer cell. But today the onset of cellular carcinogenesis is considered more complex than the "one-hit" hypothesis. There are cell repair mechanisms that can respond to mutagenesis before carcinogenesis takes hold. Moreover, mutations are required in more than a single cell. Traditionally, regulatory bodies adopted a two-tiered approach to low-dose extrapolation. The LNT was applied to carcinogens and the NLT approach was used for chemicals that exhibited noncancer effects. However, the dichotomy is losing its force among toxicologists in favor of more reductionist approaches that look at "mechanisms of toxicity" that are purported to reveal more information about low-dose effects. Mechanism of Action (MOA) after all requires a more detailed understanding of biological events at the molecular level. Once the toxicological approach for low-dose extrapolation turns to mechanism of action, you must have models that require validation with more levels of complexity than simply linear extrapolation. And when there is a dearth of good data to validate the mechanistic models, regulatory decision making is put on hold. Good data are exactly what is missing in the low-dose range. Some scientists advocate using the LNT assumption unless there is sufficient justification to accept the MOA model. However, once MOA is sought as the gold standard, commercial interests may hold it up as the desired standard for regulation, possibly slowing down any progress in regulating new chemicals.

Increasingly scientists are questioning the dichotomy between cancer and noncancer outcomes in low-dose extrapolations of exposures of chemicals and radiation. One of the findings of a 2007 EPA and Johns Hopkins Workshop was: "The historical dichotomy between low dose response extrapolation methods (typically applied to cancer and non cancer outcomes) should be set aside."[2] Their findings state that the emphasis should be placed on low-dose extrapolation models informed by the mechanisms of toxicity.

The Adaptive Response

Health physicists have been studying low-dose radiation since the aftermath of World War II. One of the unexpected outcomes of these studies is the adaptive response to low-level radiation, which confounds the conventional wisdom that has embraced the LNT view of radiation effects. Scientists studying the effects of ionizing radiation on human lymphocytes found that, compared to nonirradiated cells, low-level radiation provided more protection to high doses of ionizing radiation and chemical mutagens (S. Wolff et al. 1988). Cai (1999) noted that "Adaptive response (AR) induced by low-dose radiation (LDR) ... is the induction of cellular resistance to genotoxic effects caused by subsequently high-dose radiation (HDR)." It is hypothesized that the low-level ionizing radiation boosts the repair mechanism of cells (antioxidant activity) preparing them for mutagens (radiation or chemicals). The system of cells is being viewed as analogous to an immune system, which, by being exposed to certain proteins, can be activated to fight against viruses and bacteria. The visualization of the cells and DNA as an "immune-like" system could revolutionize health physics and toxicology and open the door for a hormesis-like theory of radiation (hormesis is the theory that low doses of substances that are toxic at high doses may be beneficial to human health). Some scientists are applying the same idea to low doses of chemicals without using the term "hormesis." In addition, they are using the "drug framework" for industrial chemicals. Recognized as having both positive and negative effects, drugs are approved when it is found that the positive effects outweigh the negative effects. Here's how one group of scientists views the assessment of low-dose exposures of industrial chemicals under the "drug framework": "the biological effects at low levels of exposure not only may be adverse but also can be beneficial depending on the target organ, the actual endpoint studied, the receptors activated, and/or the gene expression, protein and metabolite patterns affected" (Rietjens and Alink 2006: 977). They argue that toxicologists "should redirect their focus from looking at adverse effects only to also characterizing the beneficial effects, including even the beneficial effects of supposed adverse effects" (Rietjens and Alink 2006: 980). A recent example of a claim of "adaptive response" is the report that cell phone radiation reduces Alzheimer's disease in mice.[3] A good example of an emerging narrative framework for endocrine disruptors can be found in the case of Bisphenol A.

Low-Dose Exposures to Bisphenol A

Bisphenol A (BPA) was first reported to be synthesized in 1891 by the Russian chemist Aleksandr P. Dianin (1851–1918) (Dianin 1891; 1914; Rubin and Soto 2009). He prepared BPA from a condensation of acetone, which is how it got the suffix "A." Its estrogenic properties were discovered by Dodds and Lawson in 1938 by tests on ovariectomized rats. BPA was manufactured in the late 1930s, when it had been introduced extensively in consumer products. Toxicological data were reported decades ago and a No Observed Effect Level (NOEL) was established in animal studies. But since the discovery of endocrine-disrupting chemicals in the mid 1990s, BPA was studied at much lower concentrations. The reason it was studied at concentrations far below what was considered an acceptable dose was that a new mechanism of interaction was introduced. The mechanism involved estrogen receptors. Chemicals can bind to the receptors, which are either inside or on the cell surface, and disturb the normal endocrine system by "mimicking, modulating, or antagonizing" (McLachlan 2001) the pathway of an endogenous hormone. This mechanism was first discovered for estrogen receptors, but soon was extended to many other hormone systems.

Most NOELs are determined by adult exposures. But scientists have recently distinguished between the effects of chemicals on embryos and fetuses (in pregnant women) and adults. Hormones in development operate at particular time windows. Very small changes in hormone levels at a particular time of development may have dramatic effects on the organism, possibly at some later time. A two-tier system of toxicology is in the making, reflecting independent operating mechanisms. The traditional toxicological range was seldom fifty times below the Maximum Tolerable Dose (MTD) in animals. The MTD for BPA is 1,000 mg/kg/day. The EPA's lowest observed effect level (LOEL) is 50 mg/kg/day. The Reference Dose (RfD) of Bisphenol A based on a safety factor of 1,000 was calculated to be 50 µg/kg/day.

Low doses of endocrine-disrupting chemicals (EDCs) were tested on another set of toxicological assumptions—namely, that EDCs have the greatest impact when exposure occurs during development. During embryonic and fetal development, "endogenous hormones regulate the differentiation and growth of cells, and developmental processes appear to have evolved to be exquisitely sensitive to changes in hormone concentrations.... Even in animals that are genetically identical, small fluctuations in endogenous hormonal signals during development provide the basis for significant variability in phenotype" (Welshons et al. 2003: 995). Because EDC compounds fall into different mechanistic models than traditional toxicants, high-to-low-dose extrapolations cannot be used. The assumptions of threshold values, monotonic dose response curves, and singular dose response curves do not map reality. The mechanism of action of most toxicants is unknown (Welshons et al. 2003: 995). The endpoints, such as tumors or liver toxicity, are measured without understanding the pathways leading to the pathology (Hanahan and Weinberg 2000). With EDCs, scientists are continuing to work out the mechanistic pathways. Three main problems with traditional toxicological approaches applied to EDCs are: (1) they operate with only one macro-endpoint and assume a single mechanism of action; (2) they do not take into consideration latency effects; and (3) they neglect windows of vulnerability in the development of an organism.

Extrapolating from high to low doses takes for granted a single mechanism and neglects a second or third mechanism that may not express abnormalities in the organisms for years after exposure (fetal to adult latency). Because of the mechanism of EDCs and receptors, and the notion of receptor occupancy, nonlinearities in effects are quite plausible. When all receptors are occupied, additional doses of the endocrine disruptor will not induce a hormonal effect; it can only induce secondary effects not mediated by the estrogen receptor. The new generation of endocrine toxicologists has learned that the saturation of response can occur before the saturation of receptor occupancy (see Figure 11.2).

Another complexity of the endocrine system is that the EDC-receptor ligand may activate different genes, wherein "the activation of different genes requires different numbers of receptors to be occupied" (Welshons et al. 2003: 998). Those studying endocrine disruptors have identified at least two levels in toxicological studies. Level 1 involves high doses and acute toxicity (cytotoxicity or cell death) and does not depend on receptors for the dose response. Level 2 involves low-dose activation of hormone receptors. The dose range of Level 1 is up to 100 million times greater than the dose range of Level 2 (see Figure 11.3).

In 2000, the National Toxicology Program of the National Institute of Environmental Health Sciences (NIEHS) conducted a peer review to evaluate the scientific evidence for reported low-dose effects and dose response relationships for endocrine-disrupting chemicals in mammalian species. Thirty-six scientists made up the subcommittee for the review. They used the following operational definition for low-dose effects: "Low-dose effects were considered to be occurring when a non-monotonic dose response resulted in significant effects below the presumed NOEL expected by the traditional testing programs" (Melnick et al. 2002: 429). The subpanel concluded that "there is credible evidence that low doses of BPA (bisphenol A) can cause effects on specific endpoints" (Melnick et al. 2002: 428). The subpanel also noted that "it is not persuaded that a low-dose effect of BPA has been conclusively established as a general or reproducible finding" (Melnick et al. 2002: 429). The workshop participants drew up a formidable research agenda to narrow the uncertainties about low-dose effects of BPA. One of these proposals could keep a number of research teams occupied for generations, namely, to fully elaborate the mechanism at the molecular level of low-dose interactions.

The American Plastic Council has opposed studies on the reproductive and developmental effects of chemicals claiming that the low-dose effects of BPA

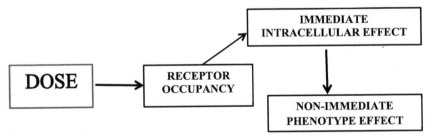

Figure 11.2. Causal Chain of Endocrine Receptor Mediated Effects.

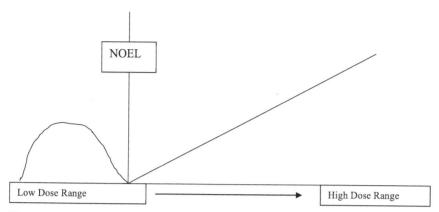

Figure 11.3. A Two-Range Dose Response Curve Reflecting Two Mechanisms of Action.

have not been demonstrated.[4] The Council funded a study by a group of scientists in 2003 who applied a "weight of evidence" evaluation of BPA, which included studies published through 2002 on the potential reproductive and developmental toxicity of BPA. The published report stated: "The panel found no consistent affirmative evidence of low-dose BPA effects for any endpoint. Inconsistent responses across rodent species and strains made generalizability of low-dose BPA effects questionable" (Gray et al. 2004: 875). Witorsch (2002) argues that the physiology of the gestation of the mouse differs markedly from that of a human, and therefore low-dose results on mice of endocrine disruptors cannot tell us anything about humans.

Reproducibility of Low-Dose Experiments

Low-dose experiments can be difficult to replicate. Epidemiologic experiments are typically opportunistic and are sometimes carried out when a major chemical spill occurs. Some of the animal studies involve tens of thousands of animals and are almost never replicated because of expense. One of the largest reported tumor studies in a rodent model used 24,000 animals. In a study of the carcinogenic effects of dibenzopyrene (DBP) 42,000 trout were used. The trout were fed as little as 0.45 ppm doses of DBP for four weeks to detect one additional cancer in 1,000 trout (Williams et al. 2003). Even with experiments that involve a small number of animals, replication can be confounded because the strains of the animals are different, the feed is not uniform, or the ambient environment varies between the experiments.

A peer review report from an NIEHS panel wrote: "The major problem with regard to the issue of low-dose effects of BPA and related compounds pertains to the consistency of results from study to study."[5] The subpanel concluded: "There is credible evidence that low doses of BPA (bisphenol A) can cause effects on specific endpoints. However, due to the inability of other credible studies in several different laboratories to observe low dose effects of BPA and the consistency of those negative studies, the subpanel is not persuaded that a low dose effect of BPA has been conclusively established as a general and reproducible finding."[6]

A group of scientists published a letter in *Toxicological Science* in response to a previously published research article (Ryan et al. 2010) where rats were fed BPA during pregnancy and lactation and showed no effects on either the male or female offspring. These effects were found in other experiments where the same doses were administered. The authors noted that Ryan et al. used a strain of rats that were quite insensitive to ethinyl estradiol (EE) and therefore they should have used a positive control. They noted: "It is unacceptable in any research with experimental animals to not include both a negative control and an appropriate positive control." Even when the same strain of mice is used and efforts to repeat an experiment are made, the outcomes may be different. One of the first studies linking BPA to prostate enlargement was performed by vom Saal et al. in 1998.

Two separate studies from other laboratories were conducted in an effort to replicate low-dose effects of BPA using the same strain of mice and following the same research design as the 1998 study. Neither of the follow-up studies showed effects on prostatic weights or daily sperm production. Responding to the failure of replication of their results, vom Saal reported: "A critical issue in experiments concerning effects of low doses of estrogenic chemicals is that a common rodent feed used in toxicological studies has been reported by investigators at the National Institute of Environmental Health Sciences (Thigpen et al. 2003) to be highly variable in estrogenic activity ... raising the possibility that endocrine-disrupting components in this feed played a role in the failure of these studies to show low-dose effects of BPA". (vom Saal and Hughes 2005: 929).

John Ashby (2001) wrote in *Toxicology Letters* that different strains of mice yield different effects of BPA. He said that this explains why he was unable to confirm the mouse prostate effects of BPA reported by Nagel et al. (1997). Richard Sharpe of Edinburgh University showed that rats exposed in the womb to octylphenol and butylbenzyl phthalate experienced reductions in testicular weight (Sharpe et al. 1995). The results could not be replicated when Sharpe repeated the phthalate experiment and others repeated the octylphenol experiment (Sharpe et al. 1998).

Vom Saal and Hughes reported a biasing effect of industry-funded papers published on BPA: "As of the end of 2004, we are aware of 21 studies that report no harm in response to low doses of BPA. Source of funding is highly correlated with positive or negative findings in published studies, 94 of 104 (90 percent) report significant effects at doses of BPA, 50 mg/kg/day. No industry-funded studies (0 of 11, or 0 percent) report significant effects at these same doses" (vom Saal and Hughes 2005: 928).

Vom Saal spoke about how companies were interested in striking a deal. After his early BPA prostate studies, he reported: "Dow chemical sent a guy down here and he said we can arrive at a mutually beneficial outcome, where you don't publish this work on bisphenol A until the chemical industry has replicated your study, and approval for publication was received by all the plastic manufacturers" (Krimsky 2000).

The "funding effect" in science means that the source of funding affects the outcome of a study. The "funding effect" has been demonstrated in a number of studies in biomedical science (Krimsky 2003). It has also been cited in toxicology (Michaels 2008), public health (McGarity and Wagner 2008), global warming (Gelbspan 1997), nutrition (Nestle 2001; Levine et al. 2003) and almost any academic discipline with strong commercial ties. Because of the sensitivity of low-dose experiments, the "funding effect" can be a determining factor in whether low-dose effects become recognized within the scientific community.

Conclusion: Mechanistic Reductionism and Its Role in Policy Stasis

Discussion within the scientific community about low-dose exposures has not changed much in fifty years. It is all about obtaining better data, discovering the biochemical and now genetic mechanisms of foreign chemicals on the human physiology, identifying the uncertainties and proposing a new experiment that will be analyzed, reanalyzed, and meta-analyzed. Ironically, as toxicological science progresses, the uncertainties over the health effects of low doses are not narrowed but broadened because each new experiment raises new questions. The relevant metaphor is the "peeling onion" where for each discovery we reach new depths of uncertainty. It is somewhat paradoxical that more science results in more uncertainty. What we have is a scientific Ponzi scheme, where each payoff (testing a hypothesis) results in new questions, and the payoff, if it ever comes, awaits new experiments that lead to new questions involving new uncertainties.

If the goal of regulatory agencies is to seek closure on the uncertainties before they can regulate a substance, they will forever be grasping for straws. Ana Soto once remarked: "If you are going to study in detail for each chemical, its absorption, degradation and storage, we will never end up with an answer, not in fifty years ... no one can tell you for sure about the risks until we run all the experiments ... even if we had all this knowledge about the fate of individual chemicals, this might still not be enough" (Cadbury 1997: 180).

If mechanistic reductionism is not the answer to addressing the health and environmental effects of low-dose exposures of chemicals and radiation, then what is? One approach has been comparative risk assessment. If you know that a person receives 100 units of natural radiation a year and a technological device exposes one to the same modality of radiation (ionizing or non-ionizing at the same frequencies) at 0.01 units per year, it can be reasonably argued that the added radiation, ceteris paribus, will not be significant. Current debates in mammography, cell phones, and whole body scans in airports are about the added risks of cancer to incremental exposures or continuous exposures impacting large populations.

There has been a change in perspective and scientific breakthroughs regarding low-level exposure of endocrine-modulating chemicals. Because the endocrine system can be affected by very low doses of hormones, especially during specific windows of embryogenesis, scientists have been able to obtain results using small numbers of animals and thus have not had to depend on linear extrapolations from high doses or large animal populations. While these studies have challenged the assumption that low-dose effects of chemicals are beyond direct human observation, they have had little immediate effect on regulation of the chemicals because industrial lobbyists ask for mechanistic results, replicated studies, and consistency in every experimental outcome. And when we add to the demands the study of combinatorial effects of chemicals, the complexity rises exponentially. Carpy et al. (2000) note: "Despite a large body of knowledge in the field of risk assessment methodologies for exposure to chemical pesticide mixtures, there is no single methodological approach in 'combination toxicology' and health risk assessment of chemical mixtures, and therefore professional judgment is still required."

Alternatives to low-dose toxicology that are not rooted in mechanistic and reductionist models are based on a set of principles that seek to minimize regret and engage the "precautionary principle." They produce a different set of narratives. Some examples of basic verifiable knowledge claims of potential risk and possible approaches to be taken in response are: (1) chemicals that bioaccumulate in the body; (2) synthetic chemicals that attach to hormone receptors; (3) synthetic chemicals that leach into food in quantities that are hazardous to test animals; (4) synthetic chemicals that interact with important human biochemical pathways; and (5) synthetic chemicals that cross the placenta and expose the fetus.

Suppose we know that some synthetic chemical V found in our food in low doses bioaccumulates in the human body. That is, the human body does not have the enzymes necessary to metabolize the chemical; instead the chemical accumulates in our fat tissue. A reasonable person might ask: Why would I want a synthetic chemical of no known contribution to my health or nutrition to bioaccumulate? Why would I want to be a waste receptacle for a chemical that is not necessary for my health and well being? Do we need to know the exact mechanism of action of the chemical on my organs or on my genes? Do we need scores of animal tests to determine what the chemical does at high doses and then to extrapolate that to doses that are most common in human tissue? Do we need a series of reproducible tests on multiple endpoints in animals that are proven to model human physiology before a regulatory decision can be made?

In another example, suppose chemical W attaches to hormone receptors in human cells and either blocks or activates the hormone receptor. Do we want to play Russian roulette with our bodies by permitting our exposure to chemicals that bind to our cellular hormone receptors? The xenobiotic hormones mimic the body's own hormones and may either block or activate genetic mechanisms for hormone production. Unless we have chosen to introduce the xenobiotics for medical therapy, it is reasonable to assume that the chemicals are not likely to benefit the individual and may create harm. A reasonable person would not want to expose themselves to synthetic organic xenobiotics that could be biologically active in unpredictable ways. Once again, do we need to work out all the details of the biochemical pathways with evidence of their pathology to bodily organs or cells before we take prudent steps of precaution? For certain chemicals the effects at high doses may not be the same as the effects at low doses. Extrapolation from high to low doses in these situations will not yield reliable outcomes. Low doses must be studied sui generis despite the difficulty of acquiring reliable data. By virtue of their sensitivity, low-dose experiments are less likely to deliver unambiguous results. Consequently, as a public health precautionary measure, we should find surrogate models of decision making that will not impose imponderable burdens of evidence for demonstrating a risk.

In a third case synthetic chemical X is found in low quantities in fresh and prepared food. Animal studies indicate that the quantities of the chemicals in the food when fed to animals exhibit pathologies. Taking account of safety factors in animal to human extrapolation, is this sufficient to establish a precautionary response to the allowable concentrations of the chemical X in the food supply?

For the fourth case let us assume there is strong evidence in animal studies that a synthetic chemical Y or one of its metabolites interferes with an important biochemical pathway, which is also found in humans. Do we need to demonstrate the effect in human subjects before we take precautionary approaches in limiting human exposure? One such example was discussed by scientists at the University of Lausanne, Federal Polytechnic School and the National Cancer Institute. They described a pathway that involves the pollutant diethylhexyl phthalate (DEHP) and concluded that "exposure to the environmental pollutant DEHP has far reaching metabolic consequences" (Feige et al. 2010: 240).

Finally, in the fifth case, chemical Z is found to transfer from a pregnant mother to her developing fetus across the placenta. Moreover, small quantities of chemical Z are known to have an adverse effect on fetal development. One such case is the transfer of thyroxine (T4) from maternal blood to the embryo. If a xenobiotic chemical Z increases maternal thyroxine (T4), *then some of that thyroxine will enter the fetus*. With no more information than the importance of a proper balance of T4 to healthy fetal development, that may be sufficient to prevent pregnant women from being exposed to chemical Z (Contempré et al. 1993).

The take-home message of these cases is that the grounds for substituting, banning, or regulating a chemical need not await a complete reductionist analysis of its biochemical and genetic pathways that demand reproducibility and validated animal models that predict human effects. Instead it may be reasonable to act on some commonsense principles that provide precautionary early warning signals.

Notes

1. See, for example, the study by Hunter et al. (1997) published in the *New England Journal of Medicine*, which some observers believed put an end to speculations that DDT and PCBs could be a cause of breast cancer.
2. Workshop, U.S. Environmental Protection Agency and Johns Hopkins Risk Sciences and Public Policy Institute, "State of the Science Workshop: Issues and Approaches in Low Dose Response Extrapolation for Environmental Health Risk Assessment," 23–24 April 23–24 2007, Baltimore, M.D.
3. Katherine Noyes, "Cell Phone Radiation May Thwart Alzheimers," *TechNewsWorld*, 7 January 2010. http://www.technewsworld.com/story/69052.html. "After years of controversy over whether cell phone radiation might cause cancer, scientists have reached the startling conclusion that it might actually cure Alzheimer's disease. Young mice exposed to long-term radiation equivalent to human cell phone use of a couple of hours a day were protected from Alzheimer's, and memory function was restored in old mice already afflicted."
4. Neil Franz, "Industry Hopes to Avoid Low-Dose Testing." *Chemical Week*, 7 November 2001, 38.
5. National Institute of Environmental Health Sciences (NIEHS), National Toxicology Program (NTP). Endocrine Disruptor Low Dose Peer Review Report, August 2001, 910. Panel met October 10–12, 2000. http://ntp-server.niehs.nih.gov/htdocs/liaison/lowdosewebpage.html (accessed 13 January 2010).
6. Ibid. iv.

Bibliography

Armitage, Peter. 1982. "The Assessment in Low-Dose Carcinogenicity." *Biometrics Supplement* 38: 119–29.

Ashby, John. 2001. "Testing for Endocrine Disruption Post-EDSTAC: Extrapolation of Low-Dose Rodent Effects to Humans." *Toxicology Letters* 120: 233–43.

Borzellecca, Joseph F. 2000. "Paracelsus: Herald of Modern Toxicology." *Toxicological Sciences* 53: 2–4.

Cadbury, Deborah. 1997. *The Feminization of Nature*. London: Hamish Hamilton.

Cai, Lu. 1999. "Research of the Adaptive Response Induced by Low-Dose Radiation: Where We Have Been and Where We Should Go?" *Human Experimental Toxicology*, 18: 419–25.

Calabrese, Edward J., and Linda A Baldwin. 2002. "Defining Hormesis." *Human and Experimental Toxicology*, 21: 91–97.

Carpy, Serge A., Werner Kobel, and John Doe. 2000. "A Review of the 1985-1998 Literature on Combination Toxicology and Health Risk Assessment." *Journal of Toxicology and Environmental Health Part B* 3: 1–25.

Cohn, Barbara A., Mary S. Wolff, Piera M. Cirillo, and Robert I. Sholtz. 2007. "DDT and Breast Cancer in Young Women: New Data on the Significance of the Age at Exposure." *Environmental Health Perspectives*, 115: 1406–14.

Contempré, Bernard, Eric Jauniaux, Rosa Calvo, D. Jurkovic, S. Campbell, and Gabriella Morreale de Escobar. 1993. "Detection of Thyroid Hormones in Human Embryonic Cavities During the First Trimester of Pregnancy," *Journal of Endocrinology Metabolism* 77: 1719–22.

Dianin, Aleksandr P. 1891. *Zhurnal russkogo fiziko-khimicheskogo obshchestva*, 23: 492.

———. 1914. *J. Russ. Phys. Chem. Soc.* 36: 1310.

Feige, Jérôme N., Alan Gerber, Cristina Casals-Casas, Qian Yiang, Carine Winkler, Elodie Bedu, Manuel Bueno, Laurent Geiman, Johan Auwerx, Frank J. Gonzalez, and Béatrice Desvergne. 2010. "The Pollutant Diethylhexyl Phthalate Regulates Hepatic Energy Metabolism via Species-Specific PPARα-Dependent Mechanisms." *Environmental Health Perspectives* 118: 234–41.

Gelbspan, Ross. 1997. *The Heat is On*. New York: Perseus.

Gerde, Per. 2005. "Animal Models and their Limitations: On the Problem of High-to-Low Dose Extrapolations Following Inhalation Exposures." *Experimental Toxicological Pathology* 57: 143–46.

Gray, George M., Joshua T. Cohen, Gerald Cunha, Claude Hughes, Ernest E. McConnell, Lorenz Rhomberg, I. Glenn Sipes, and Donald Mattison. 2004. "Weight of the Evidence Evaluation of Low-Dose Reproductive and Developmental Effects of Bisphenol A." *Human and Ecological Risk Assessment* 10: 875–921.

Hanahan, Douglas, and Robert A. Weinberg. 2000. "The Hallmarks of Cancer." *Cell* 100: 57–70.

Hunter, David J., Susan E. Hankinson, Francine Laden, Graham A. Colditz, JoAnn E. Manson, Walter C. Willett, Frank E. Speiser, and Mary S. Wolff. "Plasma Organochlorine Levels and the Risk of Breast Cancer." *New England Journal of Medicine* 337: 1253–58.

Krimsky, Sheldon. 2000. *Hormonal Chaos*. Lanham, M.D.: Johns Hopkins University Press.

———. 2003. *Science in the Private Interest*. Lanham, M.D.: Rowman & Littlefield.

Levine, Jane, Joan Dye Gussow, Diane Hastings, and Amy Eccher. 2003. "Authors' Financial Relationships with the Food and Beverage Industry and their Published Positions on the Fat Substitute Olestra." *American Journal of Public Health* 93: 664–69.

McGarity, Thomas O., and Wendy E. Wagner. 2008. *Bending Science*. Cambridge, M.A.: Harvard University Press.

McLachlan, John A. 2001. Environmental Signaling: What Embryos and Evolution Teach Us about Endocrine Disrupting Chemicals. *Endocrine Reviews* 22:319–341.

Melnick, Ronald, George Lucier, Mary Wolfe, Roxanne Hall, George Stancel, Gail Prins, Michael Gallo, Kenneth Reuhl, Shuk-Mei Ho, Terry Brown, John Moore, Julian Leakey, Joseph Haseman, and Michael Kohn. 2002. "Summary of the National Toxicology Program's Report of the Endocrine Disruptors Low-Dose Peer Review." *Environmental Health Perspectives* 110: 427–31.

Michaels, David. 2008. *Doubt is Their Product*. Oxford, U.K.: Oxford University Press.

Nagel S. C., F.S. vom Saal, K.A. Thayer, M.G. Dhar, M. Boechler, W.V.Welshons. 1997. Relative Binding Affinity-Serum Modified Access (RBA-SMA) Assay Predicts the Relative *in vivo* Bio-activity of the Xenoestrogens Bisphenol A and Octylphenol. *Environmental Health Perspec*tives 105:70–76.

Nestle, Marion. 2001. "Food Company Sponsorship of Nutrition Research and Professional Activities: A Conflict of Interest?" *Public Health Nutrition* 4: 1015–22.

Pesch, Beate, Anne Spickenheuer, Dirk Taeger, and Thomas Brüning. 2009. "Low-dose Extrapolation in Toxicology: An Old Controversy Revisited." *Archives in Toxicology* 83: 639–40.

Pierce, Donald A., and Dale L. Preston. 2000. "Radiation-Related Cancer Risks at Low Doses Among Atomic Bomb Survivors." *Radiation Research* 154: 178–86.

Rietjens, Ivonne M.C.M., and Gerrit M. Alink. 2006. "Future of Toxicology-Low-Dose Toxicology and Risk-Benefit Analysis." *Chemical Research in Toxicology*, 19: 977–81.

Rubin, Beverly S., and Ana. M. Soto. 2009. "Bisphenol A: Perinatal Exposure and Body Weight." *Molecular and Cellular Endocrinology* 304: 55–62.

Ryan, Bryce C., Andrew K. Hotchkiss, Kevin M. Crofton, and L. Earl Gray Jr. 2010. "In Utero and Lactational Exposure to Bisphenol A, in Contrast to Ethinyl Estradiol, Does Not Alter Sexually Dimorphic Behavior, Puberty, Fertility, and Anatomy of Female LE Rats." *Toxicological Sciences* 114: 133–48.

Salazar, Keith D., Michael R. Miller, John B. Barnett, and Rosana Schafer. 2006. "Evidence for a Novel Endocrine Disruptor: The Pesticide Propanil Requires the Ovaries and Steroid Synthesis to Enhance Humoral Immunity." *Toxicological Sciences* 93: 62–74.

Sharpe, Richard M., Jane S. Fisher, Mike M. Millar, Susan Jobling, and John P. Sumpter. 1995. "Gestational Lactational Exposure of Rats to Xenoestrogen Results in Reduced Testicular Size and Sperm Reduction." *Environmental Health Perspectives* 103: 1136–43.

Sharpe, Richard M., Katie J. Turner, and John P. Sumpter. 1998. "Endocrine Disruptors and Testis Development." *Environmental Health Perspectives* 106: A220–A221.

Taioli, Emmanuela, and Seymour Garte. 1999. "Low Dose Exposure to Carcinogens and Metabolic Gene Polymorphisms." *Advances in Nutrition and Cancer 2*, ed. Vincenzo Zappia, Fulvio Della Ragione, Alfonso Barbarisi, Gian Luigi Russo, and Rossano Dello Iacovo. New York: Klewer Academic/Plenum Pub.

Vineis, Paolo. 1997. "Molecular Epidemiology: Low-Dose Carcinogens and Genetic Susceptibility." *International Journal of Cancer* 71: 1–3.

vom Saal, Frederick S., Paul S. Cooke, David L. Buchanan, Paolo Palanza, Kristina A. Thayer, Susan C. Nagel, Stefano Parmigiani, and Wade W. Welshons. 1998. "A Physiologically Based Approach to the Study of Bisphenol A and Other Estrogenic Chemicals on the

Size of Reproductive Organs, Daily Sperm Production, and Behavior." *Toxicology and Industrial Health* 14: 239–60.

vom Saal, Frederick S., and Claude Hughes. 2005. "An Extensive New Literature Concerning Low-Dose Effects of Bisphenol A Shows the Need for a New Risk Assessment." *Environmental Health Perspectives* 113: 926–33.

Welshons, Wade W., Kristina A. Thayer, Barbara M. Judy, Julia A. Taylor, Edward M. Curran, and Frederick S. vom Saal. 2003. "Large Effects from Small Exposures. I. Mechanisms for Endocrine-Disrupting Chemicals with Estrogenic Activity." *Environmental Health Perspectives* 111: 994–1006.

Williams, David E., George S. Bailey, Ashok Reddy, Jerry D. Hendricks, Aram Oganesian, Gayle A. Orner, Cliff B. Pereira, and James A. Swenberg. 2003. "The Rainbow Trout (Oncorhynchus mykiss) Tumor Model: Recent Applications in Low-Dose Exposures to Tumor Initiators and Promoters." *Toxicological Pathology* 31, Supplement: 58–61.

Witorsch, Raphael J. 2002. "Low-Dose In Utero Effects of Xenoestrogens in Mice and their Relevance to Humans: An Analytical Review of the Literature." *Food and Chemical Toxicology* 40: 905–12.

Wolff, Sheldon, Veena Afzal, John K. Wiencke, Gregorio Olivieri, and A. Michaeli. 1988. "Human Lymphocytes Exposed to Low Doses of Ionizing Radiation Become Refractory to High Doses of Radiation as Well as Chemical Mutagens that Induce Double-Strand Breaks in DNA." *International Journal of Radiation Biology* 53: 39–48.

 CHAPTER 12

Unruly Technologies and Fractured Oversight
Toward a Model for Chemical Control for the Twenty-First Century

Jody A. Roberts

The story of chemical control in the twentieth century boils down to a single paradox: the more "innovative" chemists have proven to be in manufacturing and manipulating matter, the more unpredictable their chemistries became. Standard histories of chemistry recount the evolution of tools—physical and conceptual—that allowed chemists (broadly speaking) to continue an uninterrupted progression in their abilities to control matter at the molecular level leading from early efforts to mix, combine, and purify the elements of nature and leading to the eventual synthesis of wholly new materials previously unknown or seemingly impossible.[1] Our world is now largely a product of these efforts, providing the material basis for anything from textiles to electronics to drugs. Nowhere is this more evident than in the synthetic world of plastics, a bland label for molecules that provide the foundation for building materials, office equipment, vehicles, kitchenware, clothing, medical equipment, and nearly every other product used in a modern, Western, everyday life.

Indeed, behind every industrial and technological revolution lies another more hidden, less discussed revolution in the chemistry of materials that made those changes possible. The information technology revolution serves as an exemplar of this phenomenon. Underlying the development of new plugged-in and networked societies sits years of research into advanced materials that made possible the manufacturing of silicon chips, the establishment of Silicon Valley, and the fulfillment of Moore's Law (Lécuyer 2006; Lécuyer and Brock 2010).

While these traditional accounts of the history of chemistry in the twentieth century celebrate the evolution of the chemist's ability to manipulate, create, and control matter, it might be argued that a more proper telling of the story of chemistry in the previous decades would feature the ways in which chemicals, both new and old, continued to resist every tool of control the chemist developed. Smelter smoke filled agricultural and urban valleys (Wirth 2000). Pesticide residues remained on food and in the food chain. Lead—from paint and automobile exhaust—became the scourge of urban centers (Markowitz and Rosner 2002: chaps. 1–3). Other heavy metals filled lakes and streams. Plastics, designed for durability, filled the ocean with materials that can ride global tides and expose our bodies to inescapable materials. By the end of the century, chemists, toxicologists, developmental biologists, endocrinologists, and others continued to develop new tools and new languages to describe the ways in which chemicals continued to outwit us: mass spectrometry, gas chromatography, HPLC, FTIR, NMR; bioaccumulation, biopersistence, endocrine disruption, mutagenesis. The products of these processes (intentional and otherwise) have made these molecular marvels truly "unruly" technologies. Our social and technoscientific tools seem incapable of ruling over them, controlling them, or keeping them in place. With every advance in chemical and material sciences, or every demonstration of human cleverness, also seems to emerge an additional manifestation of the ways in which chemicals continue to outsmart us.

The result: a Silicon Valley filled with advanced materials, and also with persistent pollutants. We have computers in millions of homes, and millions more in waste heaps. Computers connect the world through information networks, and their disposal connects us in a network of waste transfer. The same precious metals that make the machines such marvels leach into the soil around electronic waste centers. And while cities around the United States strive to protect themselves from this toxic second life through recycling initiatives, "recycling" typically results in the exposure of low-wage workers in economically exploited areas of the world to these very same chemicals (Pellow and Park 2002; Grossman 2006; Pellow 2007).

It is in this way that we have entered what Ulrich Beck terms the risk society (Beck 1992). In his oft-cited text, Beck speaks of risk not as regulators, policy makers, and technoscientists do (that is, as a hazard that must be controlled in order to limit our exposure, or as a calculated manifestation of economic and health possibilities), but rather as a cultural state of being that arises when our abilities to alter the world outpace our understanding of what those alterations might mean. It signals a moment when our technoscientific and sociopolitical mechanisms of control (developed through and in parallel to the very same technologies that have created this situation) prove inadequate, outdated, and useless for moving beyond a constant state of risk. The technosciences fail because, as Beck points out, they are intimately connected to the system that has created the problem in the first place and that is largely responsible for its perpetuation. "As they are constituted," Beck says, "the sciences are *entirely incapable* of reacting adequately to civilizational risks, since they are prominently involved in the origin and growth of those very risks. Instead … the sciences become the *legitimating patrons* of a global industrial pollution and contamination ... as well as the related generalized sickness and death of plants, animals, and people" (Beck 1992: 59). The sociopolitical tools fail as well because the tools developed during the nineteenth century to handle such problems, "increased production, redistribution or expansion of social protection," contribute to rather than redress the project of modernization that has produced our risk society (Beck 1992: 52). In the context of governing chemicals, the thesis goes something like this: in the process of our creations outwitting us, we have had revealed the inadequacies of the regulatory state—which relies on the use and distribution of technoscientific fixes—that had (ostensibly) been designed to protect us.

This paradox of innovation in the molecular technosciences has itself led to innovations in other fields—from the technologies of tracking, detecting, and cleaning up molecular messes to the tools of governance designed to command, control, and prevent disruptions to ecological systems while preventing disruptions to and perturbations in market systems. The combination of the failures of the regulatory state coupled with the search for alternatives to prevent or mitigate these risks has also prompted action by non-state actors and collaborations between various actor groups.

In what follows, I outline the ways in which some of these various types of actors—the state, industry, technoscientists, communities, and NGOs—have evolved in recent decades in their attempt to keep pace with these changes in our molecular environment. In so doing, I hope to bring to the forefront the ways in which these mechanisms have failed, but also some of the innovations that may be worth saving. Building on these innovations, I offer some closing thoughts on how we might start building a system of oversight for the twenty-first century.

Government Control of Chemicals

Over the course of the twentieth century, the U.S. regulatory model for controlling chemical exposures has evolved from a focus on foodstuffs, drugs, and personal care items toward a more complete (if piecemeal) approach to industrial chemicals. While the first half of the century was dominated by the Food, Drug, and Cosmetic Act, the second half saw expansion of regulatory controls in response to the rapid development of the chemical industry in the wake of World War II. As new chemicals developed during wartime found their way into commercial civilian markets, a patchwork of regulations and regulatory agencies was developed to address emerging public concerns. The Federal Insecticide, Fungicide, and Rodenticide Act (or FIFRA) became law in 1947 and laid the foundation for how to manage new chemicals that began entering the market in the wake of two World Wars. By the 1960s, growing public awareness about the potential impacts related to the widespread use of these chemicals (due in no small part to Rachel Carson's *Silent Spring* published in 1962) combined with a budding environmental movement helped to highlight the inadequacies of laws such as FIFRA for serving as comprehensive approaches to chemical assessment and management.

As the 1960s ended and the 1970s began, the U.S. Congress had been primed for serious regulatory reforms. Images of the Cuyahoga River on fire; forests damaged by acid rain; smog resting over urban centers in steel towns in the East and Los Angeles in the West; PCBs throughout the Hudson River Valley—all these helped to create the context for new environmental laws that would address targeted pollutants in air and water. The Clean Air and Clean Water Acts became legal manifestations of broad-based and bipartisan support for reform at the federal level. Within the newly created Environmental Protection Agency, the offices that oversaw the implementation of these statutes became visible representatives of a new commitment to cleaning up chemical pollutants from our environment. However, the chemicals addressed through these more prominent environmental statutes covered a minuscule fraction of the overall number of chemicals traveling through commerce. To address this wider base of industrial chemicals, the Nixon administration submitted the Toxic Substances Control Act (TSCA) to Congress in 1971 (Council on Environmental Quality 1971). Unlike the media-based statutes (i.e., those focused on air and water), however, TSCA was not greeted with bipartisanship in the Congress; instead, it languished in committee for the next five years as key provisions were negotiated.

By addressing industrial chemicals writ large, TSCA had the potential to touch nearly every facet of the economy, a fact that intensified the debates surrounding passage of the law.[2] When TSCA did finally pass in 1976, most parties involved—including Congress, environmental NGOs, and EPA staff—had written the statute off as unworkable and likely suffering from fatal flaws.[3] Over the course of the next three decades, EPA staff worked to implement the various provisions of TSCA including the development of an inventory of chemicals in commerce, a program for reviewing new chemicals, and the exercise of regulatory controls on existing chemicals. In an attempt to test the might of TSCA, the EPA spent a decade developing a rule to restrict nearly all uses of asbestos in commercial applications. In 1991, when the Fifth Circuit Court of Appeals delivered its decision that the EPA had not met its statutory obligations in developing the rule, TSCA became for all intents and purposes an empty statute.[4] The weaknesses inherent in the patchwork system of chemical regulations supposedly held together by TSCA had been exposed.

But for all of the failings of TSCA, the programs and projects developed along the way created useful tools. The Toxics Release Inventory (created as part of the Emergency Planning and Community Right to Know Act, or EP-CRA, in 1986) developed a new picture of the relationship between chemicals in production, commerce, and the environment.[5] Several programs created under the Pollution Prevention Act of 1990, which was passed between the filing of the asbestos rule and the subsequent ruling, provided opportunities to create voluntary programs that would focus on the development of alternatives.[6] And state-based initiatives, such as California's Proposition 65, drew attention to the potential hazards of chemicals and consumer products.[7] None, of these, however, provided an adequate substitute for comprehensive chemical regulation.[8]

Industrial Control of Chemicals

Industry efforts similarly have evolved over the past century. A short history of industrial involvement in the oversight of chemicals would highlight one main theme: stall attempts at the construction of new regulations for as long as possible. This general project has typically taken two specific forms: push for voluntary regulations where possible and stress the need for additional research when science looks uncertain (Michaels 2008; Ross and Amter 2010). While these specific tactics have been outlined in great detail, there is much more nuance to industry involvement than what is revealed in these typical histories. Additionally, in the process of stalling, the chemical industry at times developed programs that succeeded in changing the ways in which chemicals are produced and distributed, which had significant impacts on local and global environments.

Early voluntary efforts to address chemical control are characterized by attempts to make processing and production more efficient. Escaping effluent isn't so much an environmental and health hazard as it is lost dollars, which brings a financial incentive to clean up operations. This is typically manifested in proper maintenance of the facility, more efficient operations, and finding marketable uses for "waste" products. This argument has been successfully deployed across the century and across business sectors—from refining and smelting to processing specialty chemicals and producing pharmaceuticals. But increased efficiency as pollution prevention only takes a company so far (since at some point it is cheaper to be inefficient than it is to fix the problem) (Gorman 2001). But as high-profile incidents (such as Bhopal, India, and Nitro, West Virginia, to name two) made their way into public media, and government regulation likewise evolved, two new tactics were developed by industry. The first, aimed at government interactions, involves the development of voluntary standards. The development of Responsible Care by the Chemical Manufacturer's Association (now the American Chemistry Council) was designed to demonstrate that the industry could move without government intervention to address issues that arose during the disaster at Bhopal. The trade association interpreted the disaster as requiring not only tighter controls on plant operations, but also needing better relationships between plants and their neighbors. Programs such as the Community Advisory Panel, which brings together community representatives with plant managers, emerged out of this system (Lynn et al. 2000; American Chemistry Council 2001; 2004).

For all of the energy spent on stalling regulations, the absence of a regulatory framework can prove problematic as well. The case of nanomaterials highlights these tensions. In the absence of solid regulatory guidance from the U.S. EPA, DuPont decided to create its own oversight mechanism through an experimental collaboration with the Environmental Defense Fund (EDF).[9] The partnership provided each with a unique opportunity to address internal concerns while also setting an agenda for what oversight might look like in the future. DuPont gained a partnership with a recognized and trusted leader in the environmental advocacy community. And they also created an agenda that would protect their own developing work in nanomaterials, which could perhaps insulate them from future scrutiny. EDF had the opportunity to develop an image of working with, not simply against, corporations in support of responsible research and environmental protection—especially in a situation where the government appeared to have abdicated its role in oversight. While the partnership and its activities have been short-lived, the experiment demonstrated a willingness by both parties to explore new modes of governance in the twenty-first century.

Communities Take Action

Communities have come to play a crucial role in the development of strategies to limit and control chemical exposures. Over the course of the twentieth century, the nature of the communities involved in these processes and the strategies developed have changed considerably. Early "community" action was rooted in workplace/occupational exposures with the sphere of control limited largely to that space of production. Other communities, too, were involved as the scope and scale of production increased. Farmers downwind of smelters complained of the damage to their crops.

Our perception of which communities count as exposed communities and what actions they take have also expanded over the past century. Concerns about worker health and the workplace as a site for exposure dominated concerns about the hazards associated with the chemical enterprise. Over recent decades, our perception of an exposed community has grown to include those brought close to hazard through geographical proximity, temporally bounded events, and everyday practices. In particular, three transitions have helped to expand the scope of our understanding of communities of exposure and their ability to become primary actors. First, the development and deployment of the concept of environmental justice (EJ) has provided communities with a means for calling attention to the ways in which chemical hazards are disproportionately experienced. By highlighting these injustices of exposure, the EJ community has succeeded in bringing a new voice to activism and advocacy in local, national, and global discussions about the impacts of chemical exposures.[10]

Second, communities have developed tools that have allowed them access to scientific debates about health and exposure. Citizen science projects have helped to challenge traditional notions of expertise and authority when speaking about exposure and health. Tools such as those used in the Bucket Brigade give local residents power and voice while producing alternative data sets. While most of these tools have failed to fully infiltrate the system of regulatory science (in courts and government agencies) that constructs standards and asserts authority, they have provided opportunities for community voices to be heard.[11]

Third, consumers as a community have been mobilized to think more critically about the products that surround them in everyday life. Stirred by information campaigns by organizations such as the Environmental Working Group and disappointed by the lack of action being taken at the federal level, communities of consumers have pushed for chemical bans based on exposure to consumers. From action against Bisphenol A in children's products to concern over phthalates in cosmetics and personal care products, consumers are using the market and local governments to take action. These communities have largely replaced the unions and workers in pushing for regulatory reform. Between online activism and the development of social media tools, communities of consumers have become increasingly powerful even as their modes of action have been critiqued (Galusky 2004; Szasz 2007).

Alternatives Innovation

The contributions of science and engineering to a system of chemical governance have largely gone unnoticed. Yet, scientific and technological developments in the past century have changed dramatically the risks associated with chemicals. Perhaps one reason this group is ignored is that it is often perceived to be synonymous with the actions of industry. To pass over the changes within these technoscientific practices without notice, however, ignores some of the more dramatic internal changes that have taken place.

Advancements in these areas might be divided into two groups: increased analytical capability and changing conceptual tools. The second half of the century introduced new and far more powerful analytical instruments and techniques that evolved out of many of the wartime physics projects and birth of physical instrumental analysis in chemical laboratories. In particular, the introduction of new detectors and separation techniques allowed traces of molecules previously unfathomable to become visible traces in streams and rivers, urban smog, and upper atmospheres. Out of such analytical experiences emerged changing conceptions of our environment. Molecules persist and accumulate. Molecules travel through the global still, riding warm currents from sites of production only to settle in the cold air of the poles. This information has dramatically changed the way we think about exposure because we can, for the first time, get a sense of the aggregate exposures that we all carry in our bodies. The Centers for Disease Control and Prevention's Biomonitoring Program is just one instance of the power of these new techniques.[12] And the results are already influencing the ways in which governments think about chemical control.[13]

Seeing the environment linked through these travels has led to new thinking on these connections yielding such conceptual frameworks as Gaia, earth systems, and deep ecology. Analytical innovations connect intimately with these conceptual revolutions. Likewise, conceptual reframing away from limiting exposures to hazardous materials toward a redesign of molecules to be nonhazardous yields green chemistry. Green chemistry (and other tools developed concurrently) provides not just a way of thinking about chemical risk, but a new way of thinking about the way chemistry is practiced in its most base forms. These developments importantly bring scientists and engineers into the mix of a large system of chemical governance.

The Rising Presence of NGOs

NGOs (nongovernmental organizations) have become more prominent players in our system of chemical governance. NGOs found a voice in the 1970s that has gained strength over the succeeding decades. Likewise, the diversity of perspectives offered has expanded to include not just litigation, but increasingly a more nuanced and focused look at the various ways in which to intervene in the system. As states and the federal government moved toward more neoliberal approaches to environmental governance in the latter parts of the century, NGOs positioned themselves as important arbiters between the actors already outlined. Thus, their most prominent position on the playing field seems to have become one of filling voids and making connections. The result has been increased power for these groups and the development of innovative ways for adapting and developing new tools for governance.

The general failure of centralized chemical regulations over the course of the last thirty plus years has resulted in a patchwork system of programs and initiatives that work "over, under, and around" these chemical regulations.[14] That is not an indictment of these particular programs, many of which have resulted in significant new protections, better understandings of exposure routes, and serious reductions of specific pollutants and toxicants in the environment. It is, however, recognition of the limited scale and scope of these initiatives when they are not backed up by a formalized and enforceable regulatory system. For all of the power of NGOs, they cannot make partnerships legally binding. For all of the industry's talk of responsibility, it means little if there is no system of accountability. For all of the advances of citizen science, there remains a role for experts in the system. And for all of the market-based changes consumer communities can bring, we can't shop our way to safety (Szasz 2007). The failures of these projects in isolation and the resulting heterogeneous protections have resulted in calls for an international system of governance that would provide uniformity in global protections that mirror the global travel and presence of chemicals (Selin 2010). Whatever shapes these new national and global regulations take, there are important lessons that can be learned from the products of our current fractured system of oversight. In particular, while a globalized system of control might remove some of the heterogeneity in protections, it should be careful not to homogenize the world. Keeping the local and the global in balance will be the crucial test for constructing a global system of chemical governance.

Robustness and Redundancy: Building a System for Twenty-First Century Chemical Control

What key features could a twenty-first century system of governance possess, and what challenges will test that system? I begin here with two conundrums left over from our current modes of governance—dealing with uncertainty and defining a "vulnerable" population—as instructive for both pointing out weaknesses in the current system and also potentially benefiting from the incorporation of some of the regulatory innovations that have taken place.

Uncertainty and Precaution

The politics of uncertainty might be *the* key artifact of the twentieth century. Somehow, this seemingly simple concept has become the prime battleground for debates concerning safety, risk, and regulation. If the problem of uncertainty serves as the key characteristic of the past century, perhaps our ways of dealing with it will define our new century.

Already we have seen the development of more sophisticated approaches for unpacking, dismantling, and exposing the problems of uncertainty, political as well as epistemological. We have seen the development of new major policy initiatives in the European Union that attempt to manage head on, rather than sidestep, the difficulties of dealing practically with uncertainty. And, perhaps as a result of the failings of the last century, we have seen a robust system of participation by communities of activists, scientists, and bureaucrats develop to track and monitor unexpected and uncertain effects from chemical development, production, use, and disposal. Most of these approaches have been successful in finding ways to battle through the muddled mess that ensues around issues of certainty and scientific fact. Fortunately, perhaps, this is territory that science and technology studies have spent a great deal of time exploring. Finding ways to translate this academic literature into real world application is the challenge given to those of us straddling the world of STS academics and a more engaged scholarship.

In previous work, I argued that one avenue for navigating around this debate might be to sidestep the issue of scientific fact by incorporating some variant of what Bruno Latour suggests might be a more appropriate focus, on matters of concern.[15] The idea suggests that actors might be mobilized to engage with an issue because of its political importance. Here, science speaks, but it is not the final arbiter.[16] Applying this to the U.S. context, however, seems perhaps impossible given the state of the position science has come to occupy in these sorts of broader debates. As just one example, arguments for precaution (mostly, but not always discussed in concert with the precautionary principle) are treated as taboo topics in the U.S. context.

"Vulnerable" Populations

One inroad to chemical reform in the United States has been through the highlighting of the excessive burden placed on vulnerable populations. In most cases, this has had two results. First, we've made children (age three and under) the exemplar of a vulnerable population—and quite rightly so. This age group is exposed to more potentially toxic chemicals, pound for pound, than an adult through interactions with food, water, objects, and the environment. Their systems—hormonal, neurological, skeletal—are still under development, leaving them potentially at risk of disruption of their development. And, in large part, little if any research exists on the effects of chemical exposures on this population. For these reasons, researchers and advocates alike have seized on this population to demonstrate the potentially catastrophic problems of exposing this group to daily doses of dozen if not hundreds or thousands of environmental chemicals. The result has been the development of new research programs (funded in large part by the NIEHS), the instantiation of a new popular consciousness about potential chemical exposures, and the passage of the Kids Safe Chemical Act, which goes further than any other current U.S. law to remove certain substances from commercial products. One might even credit this approach for instigating and maintaining momentum for broad chemical reforms in the United States.

"Vulnerable population" has also come to have a more clinical and research-focused use that worries less about "populations" in a more traditional sense, but builds on the -omics revolutions of the past decade(s) to redefine populations in terms of those with similar genomic sensitivities to specific chemical exposures. The hope is that by identifying inherent sensitivities we can limit exposures to those chemicals by those populations. That is, rather than seeing hazard or risk as something to be generalized, it assumes that exposure can be controlled in such a way that a chemical of concern can continue to be used. A look at current research programs underway at federal research institutes confirms this direction, which is in keeping with a more general trend towards personalized genomic medicine. The two, of course, go well together. This institutionalized approach to identifying and protecting vulnerable populations, however, is of concern for at least two reasons. First, it takes for granted much of the history of understanding how we've come to know about hazards and health risks associated with chemical exposures via communities disproportionately exposed to specific chemicals. We know much of what we know because of a century plus of action and activism by workers and environmental justice communities demanding research into the effects of exposure to specific chemicals and classes of chemicals. These vulnerable (because disproportionately exposed) communities will disappear in a top-down system that seeks to erase communities defined by geography and occupation and replace them with individuals placed into categories constructed based on personal biological data. The danger is the same one presented by the nonspecific biomonitoring data being collected by the Centers for Disease Control and Prevention through NHANES (the National Health and Nutrition Examination Survey). While the accumulated data has on the one hand provided a boon to public health research, its lack of much demographic information (e.g., geographic) has created the potential for a scenario that says we are all exposed (perhaps even equally so), potentially silencing the most important tenant of the EJ movement: that their exposures are disproportionately high. The same might be said for workers.

A second group also disappears in a top-down approach such as this: communities disproportionately exposed because of some other (e.g., medical) necessity. The work of the group Health Care Without Harm, for example, is designed to define a different sort of vulnerable community, those in healthcare situations. Indeed, for anyone who ever passes through a hospital Intensive Care Unit (ICU) or Neonatal Intensive Care Unit (NICU), the shear volume of synthetic materials and the intimate ways in which they are connected to patients is truly remarkable. Vulnerability in this situation is not a matter of inherent susceptibility but results from intimate and constant contact—and therefore constant dosing.

The knowledge generated by and because of the actions of these groups has been crucial for furthering our more general understanding of the hazards posed by chemical exposures. These cases speak more broadly to the important role that citizen action, NGOs, and broad public participation have played (and continue to play) in creating this moment in which we are rethinking how we conceptualize and deal with the risks associated with chemicals. In any new system, these communities will continue to play a pivotal role.

Moving Forward

The situations presented above present certain challenges that can't be met by simply developing a larger, more centralized, global regulatory system. Instead, both situations could benefit from innovations developed more recently that have helped to provide voice, agency, and alternatives in thinking through and acting in response to chemical exposures. More importantly, perhaps, the decentralized and overlapping elements of these innovations could make a system of governance more robust and more redundant, which would alleviate pressure on the regulatory agencies and ensure exposure gaps are covered.

The systems of cooperation that develop between stakeholders—particularly those between industry and neighboring communities, industry and NGOs, and industry and regulators—have created precisely the sorts of elements that ensure a robust system. As the U.S. government has struggled to find a way to include the public in decision-making processes, industry has found its own way (National Research Council 2008). The CAP system may be flawed, but many of those flaws are attributable to the lack of a presence of government agencies. A stronger regulatory effort should support these interactions without squelching them. Likewise, the voluntary agreements between, for example, DuPont and EDF, provided a creative starting point for thinking about limits on nanoscale research. But those interactions should be beginnings, not ends.

In a decentralized system, data flows from many places; a more centralized system of governance should work to maintain these flows and to find ways to integrate them. The original TSCA was designed to incorporate data from labs around the globe. That never happened. In the meantime, data began flowing from other locations as well—citizen science projects, epidemiological studies, genomic studies, and the like. As they stand, these data points appear to be disconnected and incommensurable. A twenty-first century system finds meaningful data in untraditional places and finds ways to make them contribute to a more complete picture.

Programs such as Design for the Environment and green chemistry found roots in the rocky soil of the EPA following the asbestos ruling. Their emphasis wasn't on regulatory action, but on the promotion of alternatives. Their successes should be noted, their reach expanded, and their power increased by giving them a more prominent place in a system of chemical governance. But this framework will have to keep careful watch: calling something green doesn't make it so. A smarter and more flexible program recognizes its limits, and realizes that perceptions of risk and safety are epistemologically and temporally bounded. At any moment new data may change our views on a chemical. The process of substitution and innovation, then, is a continuous one.

All of these changes would be for naught if we didn't take as a cornerstone a sense of justice in developing this system. Traditional risk analyses and cost-benefit ratios have failed not only because of the limits on our ability to use the right information to make this calculations, but because they erase the contours of risk and benefit and exposure that characterize our globalized network of production and consumption.

Notes

1. See general histories of chemistry such as Brock (1992) and Levere (2001). Not surprisingly, the stories told from within the chemical sciences highlight these features even more. The stories that come closest to highlighting both of these features typically come from reformists, such as those involved in marginalized activities like green chemistry. See, for example, Anastas and Warner (1998). But even in the latter cases, greater control is sought to make up for previous lapses in control.
2. The Toxic Substances Control Act: From the Perspective of J. Clarence Davies, interview by Jody A. Roberts and Kavita D. Hardy (Philadelphia: Chemical Heritage Foundation, Oral History Transcript, 2009).
3. The Toxic Substances Control Act: From the Perspective of Steven D. Jellinek, interview by Jody A. Roberts and Kavita D. Hardy (Philadelphia: Chemical Heritage Foundation, Oral History Transcript, 2010).
4. *Corrosion Proof Fittings v. EPA*, 947 F.2d 1201 (5th Cir. 1991).
5. The Toxic Substances Control Act: From the Perspective of Charles L. Elkins, interview by Jody A. Roberts and Kavita D. Hardy (Chemical Heritage Foundation, Oral History Transcript, 2010).
6. The Toxic Substances Control Act: From the Perspective of Mark A. Greenwood, interview by Jody A. Roberts and Kavita D. Hardy (Chemical Heritage Foundation, Oral History Transcript, 2010).
7. Safe Drinking Water and Toxic Enforcement Act of 1986.
8. The Toxic Substances Control Act: From the Perspective of James Aidala, interview by Jody A. Roberts and Kavita D. Hardy (Chemical Heritage Foundation, Oral History Transcript, 2010).
9. Environmental Defense—DuPont Nano Partnership, *Nanorisk Framework* (2007).
10. Important work here includes Bullard (1983 and 1990), Pellow (2002), and United Church of Christ Commission for Racial Justice (1987).
11. See, for example, Corburn (2005) and Ottinger (2009, 2010).

12. See, for example, Centers for Disease Control and Prevention. 2009. *Fourth National Report on Human Exposure to Environmental Chemicals.*
13. Provisions for biomonitoring were present in both draft chemical reform bills submitted to the U.S. House and Senate TSCA reform. See Safe Chemicals Act, S.3209. 111th Congr., 2d Sess. (2010) and Toxic Chemicals Safety Act, H.R. 5820. 111th Congr., 2d Sess. (2010).
14. The Toxic Substances Control Act: from the Perspective of James Aidala, interview by Jody A. Roberts and Kavita D. Hardy, (Chemical Heritage Foundation, Oral History Transcript).
15. See Roberts (2010).
16. Both describing and adjusting this balance has been a topic of inquiry for decades, from Alvin Weinberg's treatment of "science and trans-science" (1970) to Sheila Jasanoff's exploration of the role of science advisors (1994) to the work of Roger Pielke (Pielke 2007; Pielke and Klein 2010).

Bibliography

American Chemistry Council. 2001. *Guide to Community Advisory Panels.* Washington D.C.: American Chemistry Council.
———. 2004. "American Chemistry Council on the Twentieth Anniversary of the Bhopal Tragedy: Lessons Learned Lead to Safer Operations." Washington D.C.: American Chemistry Council.
Anastas, Paul T., and John C. Warner. 1998. *Green Chemistry: Theory and Practice.* New York: Oxford University Press.
Beck, Ulrich. 1992. *Risk Society: Towards a New Modernity.* London: Sage.
Brock, William H. 1992. *The Chemical Tree.* New York: W.W. Norton & Company.
Bullard, R. 1983. "Solid Waste Sites and the Black Houston Community." *Sociological Inquiry* 53: 273–88.
Bullard, Robert D. 1990. *Dumping in Dixie: Race, Class, and Environmental Quality.* New York: Westview.
Carson, Rachel. 1962. *Silent Spring.* Boston: Houghton Mifflin.
Corburn, Jason. 2005. *Street Science: Community Knowledge and Environmental Health Justice.* Cambridge, M.A.: MIT Press.
Council on Environmental Quality. 1971. *The President's 1971 Environmental Program.* Washington, D.C.: U.S. Government Printing Office.
Galusky, Wyatt. 2004. "Virtually Uninhabitable: A Critical Analysis of Digital Environmental Anti-toxics Activism." Ph.D. diss., Virginia Polytechnic Institute and State University.
Gorman, Hugh S. 2001. *Redefining Efficiency: Pollution Concerns, Regulatory Mechanisms, and Technological Change in the U.S. Petroleum Industry.* Akron: University of Akron Press.
Grossman Elizabeth. 2006. *High Tech Trash: Digital Devices, Hidden Toxics, and Human Health.* Washington, D.C.: Island Press.
Jasanoff, Sheila. 1994. *The Fifth Branch: Science Advisors as Policy Makers.* Cambridge, M.A.: Harvard University Press.

Lécuyer, Christophe. 2006. *Making Silicon Valley: Innovation and the Growth of High Tech, 1930-1970*. Cambridge: MIT Press.
Lécuyer, Christophe, and David C. Brock. 2010. *Makers of the Microchip: A Documentary History of Fairchild Semiconductor*. Cambridge, M.A.: MIT Press.
Levere, Trevor H. 2001. *Transforming Matter: A History of Chemistry from Alchemy to the Buckyball*. Baltimore: Johns Hopkins University Press.
Lynn, Frances M., George Busenberg, Nevin Cohen, and Caron Chess. 2000. "Chemical Industry's Community Advisory Panels: What has been Their Impact?" *Environmental Science and Technology*, 34(10): 1881–86.
Markowitz, Gerald, and David Rosner. 2002. *Deceit and Denial: The Deadly Politics of Industrial Pollution*. Berkeley, C.A.: University of California Press.
Michaels, David. 2008. *Doubt is Their Product: How Industry's Assault on Science Threatens Your Health*. Oxford: Oxford University Press.
National Research Council. 2008. *Public Participation in Environmental Assessment and Decision Making*. Washington: National Academies Press.
Ottinger, Gwen. 2009. "Epistemic Fencelines: Air Monitoring Instruments and Expert-Resident Boundaries." *Spontaneous Generations*, 3:1, 55–67.
———. 2010. "Buckets of Resistance: Standards and the Effectiveness of Citizen Science." *Science, Technology, and Human Values*, 35:2, 244–70.
Pellow, David N. 2002. *Garbage Wars: The Struggle for Environmental Justice in Chicago*. Cambridge, M.A.: MIT Press.
———. 2007. *Resisting Global Toxics: Transnational Movements for Environmental Justice*. Cambridge: MIT Press.
Pellow, David N., and Lisa Sun-Hee Park. 2002. *The Silicon Valley of Dreams: Environmental Injustice, Immigrant Workers, and the High-Tech Global Economy*. New York: New York University Press.
Pielke, Roger A., Jr. 2007. *The Honest Broker: Making Sense of Science in Policy and Politics*. Cambridge: Cambridge University Press.
Pielke, Roger, and Roberta A. Klein, eds. 2010. *Presidential Science Advisors: Perspectives and Reflections on Science, Policy and Politics*. New York: Springer.
Roberts, Jody A. 2010. "Reflections of an Unrepentant Plastiphobe: Plasticity and the STS Life." *Science and Culture* 19(1): 101–20.
Ross, Benjamin, and Steven Amter. 2010. *The Polluters: The Making of Our Chemically Altered Environment*. Oxford: Oxford University Press.
Selin, Henrik. 2010. *Global Governance of Hazardous Chemicals*. Cambridge: MIT Press.
Szasz, Andrew. 2007. *Shopping Our Way to Safety: How We Changed from Protecting the Environment to Protecting Ourselves*. Minneapolis: University of Minnesota Press.
United Church of Christ Commission for Racial Justice. 1987. *Toxic Wastes and Race in The United States: A National Report on the Racial and Socio-Economic Characteristics of Communities with Hazardous Waste Sites*. New York: United Church of Christ.
Weinberg, Alvin M. 1970. "Science and Trans-Science." *Minerva* 10(2): 209–22.
Wirth, John D. 2000. *Smelter Smoke in North America: The Politics of Transborder Pollution*. Lawrence: University Press of Kansas.

Contributors

Barbara L. Allen is professor and director of the science and technology studies graduate program at Virginia Tech's Washington, D.C. area campus. Her books include: *Uneasy Alchemy: Citizens and Experts in Louisiana's Chemical Corridor Disputes* (2003) and *Dynamics of Disaster: Lessons, on Risk, Response, and Recovery* (2011). She has published many articles on the public shaping of technical and scientific knowledge in journals such as: *Social Studies of Science; Science, Technology and Human Values;* and *Technology in Society*. Allen's current research, funded by the U.S. National Science Foundation, comparatively examines participatory science and regulatory change in polluted communities in the United States, Italy, Germany, and France.

Stefania Barca is a senior researcher at the Center for Social Studies of the University of Coimbra, Portugal, where she coordinates the research group on "Social policies, labour and inequalities" and the Ph.D. program "Democracy in the 21st century." She obtained her Ph.D. in Economic History from the University of Bari (Italy) in 1997. She has published a number of articles in Italian and international history journals, and three books. Her last publication, *Enclosing Water: Nature and Political Economy in a Mediterranean Valley* (Cambridge, U.K.: White Horse Press 2010), has been awarded the Turku Prize for best book in European environmental history. Her new research project deals with industrial hazards and the relationships between labor and the environment in a transnational perspective. She has been recently elected vice-president of the European Society for Environmental History (ESEH).

Soraya Boudia is professor of science, technology, and innovation studies at the University of Paris-Est Marne-la-Vallée. Her scholarly work focuses on the transnational government of technological and health environmental risks. She is the author of *Marie Curie et son laboratoire. Science et industrie de la radioactivité en France* (Paris: Editions des Archives Contemporaines 2001). Among the volumes she has coedited are "Risk and Risk Society in Historical Perspective" (*History and Technology,* 2007), and *Toxicants, Health and Regulations* (2013), both with Nathalie Jas, and "Transnational History of Science" (*British Journal of History of Science,* 2012) with Nestor Herran and Simone

Turchetti. She is currently writing a forthcoming book on scientific expertise and nuclear risks.

Laura Centemeri is researcher in environmental sociology at the CNRS (France). She holds a Ph.D. in Economic Sociology from the University of Brescia (Italy). She has been a postdoctoral research fellow at the EHESS in Paris and at the University of Milan, and senior researcher at the Centre for Social Studies (CES) of the University of Coimbra (Portugal). She is the author of *Ritorno a Seveso. Il danno ambientale, il suo riconoscimento, la sua riparazione* (Milano: Bruno Mondadori 2006). Among her recent publications are "Retour à Seveso. La complexité morale et politique du dommage à l'environnement," *Annales. Histoire, Sciences Sociales*, 2011, 66(1).

Carl F. Cranor is Distinguished Professor of Philosophy and member of the faculty of the environmental toxicology graduate program at the University of California, Riverside. For twenty-five years his research has focused on philosophical issues concerning risks, science, and the law. He is the author of *Regulating Toxic Substances: A Philosophy of Science and the Law* (New York: Oxford University Press 1993), *Toxic Torts: Science, Law and the Possibility of Justice* (Cambridge, U.K.: Cambridge University Press 2006), and *Legally Poisoned: How the Law Puts Us at Risk from Toxicants* (Cambridge, U.S.: Harvard University Press, 2011) as well as coauthor of *Identifying and Regulating Carcinogens* (U.S. Congress, Office of Technology Assessment, 1987), and *Valuing Health: Cost Effectiveness Analysis for Regulation* (Institute of Medicine, 2006). He has served on science advisory panels as well as on Institute of Medicine and National Academy of Sciences Committees. He is an elected fellow of the American Association for the Advancement of Science and the Collegium Ramazzini.

Angela N.H. Creager is the Philip and Beulah Rollins Professor of History at Princeton University. She is the author of *The Life of a Virus: Tobacco Mosaic Virus as an Experimental Model, 1930-1965* (2002) and *Life Atomic: A History of Radioisotopes in Science and Medicine* (2013). Her article with Gregory Morgan, "After the Double Helix: Rosalind Franklin's Research on Tobacco Mosaic Virus," *Isis* 99 (2008): 239–72, was awarded the *Price/Webster Prize* from the History of Science Society in 2009.

Michelle Edwards is a Ph.D. candidate in the Department of Sociology at Washington State University (WSU). Her research on topics ranging from social disorganization theory to community organizing to survey methodology has appeared in *Deviant Behavior* (2010), *Organization & Environment* (2011), and *Survey Practice* (2012). Her dissertation examines resident perceptions of drought risk and adaptive capacity to drought at different spatial scales across two states.

Scott Frickel is associate professor of sociology at Washington State University. He is the author of *Chemical Consequences: Environmental Mutagens, Scientist Activism and the Rise of Genetic Toxicology*, winner of the 2005 Robert K. Merton Award, and coeditor of *The New Political Sociology of Science: Institutions, Networks, and Power*. His current book projects are *Ground Truth*, a study of the knowledge politics that conditioned regulatory and civil society responses to Hurricane Katrina in New Orleans, and a coedited volume, *Utopian Knowledge? Critical Perspectives on Interdisciplinary Research*.

Jean-Paul Gaudillière is a senior researcher at Institut National de la Santé et de la Recherche Médicale and director of the Center for Research on Science, Medicine, Health and Society (Cermes3) in Paris. His work addresses many aspects of the history and sociology of the biomedical sciences during the twentieth century. His present research focuses on the history of pharmaceutical drugs—the ways in which they are invented, commercialized, used, and regulated. On these subjects, he has recently edited two special issues: "Drug Trajectories," *Studies in the History and Philosophy of the Biological and Biomedical Sciences*, 36:4 (2005) and "How Pharmaceuticals Became Patentable in the Twentieth Century," *History and Technology*, 24:2 (2008), and with Volker Hess, the volume *Ways of Regulating Drugs in the Nineteenth and Twentieth Century* (London: Palgrave, 2012).

Nathalie Jas is a senior researcher at the French National Institute for Agricultural Research (INRA). A historian and a STS scholar, she analyses the intensification of agriculture and its social, environmental, and health effects. She is the author of *Au carrefour de la chimie et de l'agriculture : les sciences agronomiques en France et en Allemagne 1840-1914* (Paris: Les Belles Lettres 2001). Among the volumes she has coedited are "Risk and Risk Society in Historical Perspective" (*History and Technology*, 2007) and *Toxicants, Health and Regulations* (London: Pickering and Chatto, 2013), both with Soraya Boudia. She is writing her forthcoming book, a history of the government of sanitary risks posed the agricultural pesticides in twentieth century France.

Paul Jobin is currently associate professor in the Department of East Asian Studies at the University of Paris Diderot and Director of CEFC Taipei, the Taiwan branch of the French Centre for Research on Contemporary China. His research focuses on industrial hazards in Taiwan and Japan. His Ph.D. dissertation (*Maladies industrielles et renouveau syndical au Japon*) received the Shibusawa-Claudel Prize and was published at EHESS in 2006. He has recently coordinated a three-year collective research project on silicosis and asbestos in France and Japan. Another important fieldwork study deals with nuclear plant workers.

Sheldon Krimsky is the Lenore Stern Professor of Humanities and Social Sciences in the Department of Urban & Environmental Policy & Planning in the School of Arts & Sciences and adjunct professor in Public Health and Community Medicine in the School of Medicine at Tufts University. He has authored or coedited eleven books, including *Genetic Justice: DNA Databanking, Criminal Investigations and Civil Liberties*, 2011 with Tania Simoncelli and *Race and the Genetic Revolution* with Kathleen Sloan. His forthcoming book *Genetic Explanations: Sense and Nonsense*, edited with Jeremy Gruber, is to be published by Harvard University Press. He has served on several expert committees. Currently, he serves on the Board of Directors for the Council for Responsible Genetics as a Fellow of the Hastings Center on Bioethics and on Committee A of the American Association of University Professors. Professor Krimsky has been elected fellow of the American Association for the Advancement of Science.

Nancy Langston is professor of environmental history at Michigan Technological University. In 2012, she was the King Carl XVI Gustaf Professor of Environmental Science at Umeå University in Sweden. She has served as president of the American Society for Environmental History and editor of *Environmental History*. Her most recent book, *Toxic Bodies: Hormone Disruptors and the Legacy of DES* (Yale: Yale University Press 2010) examines the environmental history of endocrine-disrupting chemicals in the United States. Her other books include *Forest Dreams, Forest Nightmares* (Washington: University of Washington Press, 1995) and *Where Land and Water Meet* (Washington: University of Washington Press, 2003).

Jody A. Roberts's work explores the intersections of emerging molecular sciences and public policy and the ways in which tensions brought about between the two get resolved. He received advanced degrees in science and technology studies from Virginia Tech, where he cultivated an interest in the practice of the molecular sciences and the politics of the broader world. Those interests became the basis for the projects that formed the Chemical Heritage Foundation's Environmental History and Policy Program, which explores social, technical, and policy innovations for governing molecules. Before becoming the first manager of the Environmental History and Policy Program, he was the Charles C. Price Fellow and Gordon Cain Fellow at CHF. Roberts is a senior fellow in the Environmental Leadership Program. He also lectures in the history and sociology of science department at the University of Pennsylvania and in the Center for Public Policy at Drexel University.

Yu-hwei Tseng is a Ph.D. candidate in the Institute of Health Policy and Management, National Taiwan University. Her publications in peer-reviewed journals include "Cuban Paradox from the Perspective of Society and Health" in *Taiwan Journal of Public Health* (2010) and "Impact of Microcredit and International Aid on NGOs in Developing Countries—A Bangladesh Observation" in *Community Development Quarterly* (2011). She won the Best Conference Paper Award in the 12th Asia-Pacific Rim Union Doctoral Student Conference for her research "Health for All or Market for All?" (2011).

Index

A
Acceptable Daily Intake (ADI), 10
activism, 8, 10–12, 14–15, 53, 59, 69, 91, 137, 152, 162–64, 260, 264
activists, 10, 12, 15, 17, 22, 39, 57, 82, 97, 126, 128, 136, 141, 146, 147n, 172, 182, 263
aflatoxin, 51, 56
American Cyanamid, 55
American Industrial Health Council (AIHC), 95, 103, 105
American Medical Association (AMA), 31
Amendola, Gianfranco, 125
Ames, Bruce, 8, 47, 50, 56, 59, 60n
Ames test, 9, 46–47, 50–51, 54–56, 58–59, 60n4, 60n8, 60n12
Amoco Chemicals Corporation, 95
ANIC Petrochemical Plant, 118, 220
Arapahoe Chemicals of Colorado, 34
arsenic, 154, 195, 204, 205, 219–20
Ashby, John, 246
Atomic Bomb Casualty Commission, 50, 60n
Atomic Energy Commission (AEC), 49, 75

B
Beadle, George, 47
Behrens, Edwin, 95, 104, 109n
benzene, 116, 154, 155, 178, 179, 196
Berg, Paul, 95, 104, 109n
Bertani, Guiseppe, 52
Bettini, Virginio, 125, 127–28, 131nn12–13
Bichler, Joyce, 65, 78
Bisphenol A (BPA), 89, 235, 242–44, 246–47, 260

Blum, Arlene, 19, 54, 60n
Bortollozzo, Gabriele, 15
brominated fire retardants, 195, 197, 202, 203, 204
Brooks, Harvey, 7, 99, 109n
Brower, David, 12
Burroughs, Wise, 35, 69

C
caesium 137, 1
Canadian Department of National Health and Welfare, 33
cancer, 29, 31, 36, 39–40, 43n, 46–47, 49, 53–59, 59n, 65, 69, 71, 73–74, 77–78, 80–82, 88, 92, 92n, 96–97, 101, 108, 118, 134–35, 137, 143, 149n, 154, 159–61, 166, 170–74, 176–77, 182, 185–86, 188, 195–96, 200–203, 238–41, 245, 248, 250n1, 250n3
  causing, 47, 53
  incidence, 56–57, 137, 173
  risk(s), 31, 43n, 46–47, 70, 91, 134, 173
  specialist, 12, 58, 76, 88
carcinogen, 9, 10, 29, 31, 36, 39, 46, 51, 53, 55–59, 60n, 66, 69, 70, 73–74, 76, 96, 103–4, 106, 109n5, 109n7, 134, 173, 178–79, 204, 219
Carcinogens, Mutagens, Reproductive Toxicants (CMR), 9, 59
carcinogenesis, 8, 34, 50, 57, 60n13, 72, 75–77, 81, 88, 90, 143, 146, 177, 241
carcinogenic, 9, 30–31, 50, 52, 55–56, 65, 70–71, 73–74, 76, 88, 95, 97, 106, 119, 134, 142–43, 177–79, 196, 225, 239, 245

carcinogenicity, 8, 46, 47, 52, 53, 55, 58, 75–77, 89, 92, 107, 134–35, 143, 145–46, 160, 177, 179, 227–28, 230n
Carson, Rachel, 1, 12–13, 49, 70, 96, 117, 124, 257
Centers for Disease Control and Prevention (CDC), 198, 261, 264
Chernobyl, 1
China Petroleum Development Company (CPDC), 181, 185, 186
Ching-Chang, Lee, 15, 181–82, 184, 187
Cini, Marcello, 127
Clean Water Act, 207, 257
Colborn, Theo, 88
Coleman, James, 103
Committee on Public Engineering Policy (COPEP), 98
Committee on Risk and Decision-Making (CORADM), 102–3
Committee on Science and Public Policy (COSPUP), 99
Commoner, Barry, 13, 56, 117, 119, 125, 127–28, 129n2, 130n3, 130n9, 131nn12–13
consumers, 4, 18, 33–34, 43, 65, 67, 90, 207, 236, 260
contaminants, 2, 8, 10, 11, 37, 38, 134, 171, 197, 200, 218, 219, 220, 222, 223, 226, 228, 229, 230n5, 240
contamination(s), 1, 3, 7, 12–18, 20–21, 23–24, 49, 118, 120, 125, 128, 130n, 135–45, 171–72, 182, 195, 197, 200, 209–10, 217, 219, 222, 224, 229, 256
Conti, Laura, 14, 118–28, 130nn4–5, 130n8, 130n14, 139
Covello, Vincent, 102
Crick, Francis, 48
Cumming, Robert B., 102

D
DDT, 12, 69, 88, 96, 122–23, 202, 219, 240, 250n
Deisler, Paul F., 102, 104, 109n
Delaney Clause, 65, 69, 73, 77, 97, 238
Delaney, James, 69

Demerec, Milislav, 52–53
DES (diethylstilbestrol), 6, 9–10, 29–37, 40–43, 65–83, 88–92, 92nn1–2, 92n7, 195, 201, 204
dioxin, 16, 118–20, 122, 134–46, 147n, 148n7, 148n12, 170, 180–86, 203, 208, 210
DNA, 46, 49–50, 53–54, 56, 58–59, 200, 242
Dodds, Charles, 30, 68, 242
Dow Chemical (Company), 95, 247
DuPont, 55, 259, 265, 266n

E
Earth Day, 49
ecology, 57, 60n, 83, 92n, 100, 115, 117, 121–22, 124–29, 188, 261
ecotoxicology, 8, 14, 72, 83
effects, 9, 17, 21, 33, 82, 135, 139, 142, 204, 249
Ehrenberg, Lars, 102
Electric Power Research Institute, 98–99
endocrine disruption, 38, 67, 81–83, 88, 92, 255
endocrine disruptors (EDs/EDCs), 10, 22, 29, 30, 37–40, 42, 65, 67–68, 80–83, 88–89, 92nn10–11, 122, 237–38, 242–46, 250n, 255
ENIChem, 165
Environmental Defense Fund, 12, 259
environmental health science, 121, 152, 163
environmental justice, 20, 22, 23, 155, 164, 260, 264
environmental law. See law, environmental
environmental groups/movement/ organizations, 29, 46, 56, 162, 69, 116, 125, 159, 207, 257
Environmental Mutagen Society, 53, 102
Environmental Protection Agency (EPA), 7–8, 103, 154, 172, 215, 219, 220, 225–26, 230n4, 230n6, 250n2, 256
environmentalism, 12, 14, 47, 58, 116–17, 121, 125–26, 130n6

environmentalists, 8, 39, 46, 47, 56, 117, 125, 130n, 163, 166
epidemiology, 67, 83, 91, 120, 130, 134, 145, 155, 160, 170, 173–74, 177, 178, 180, 187–88
epidemiology, popular, 15, 152, 165, 170–71, 174, 182
estrogen (synthetic, residues), 30–34, 41, 65, 68–69, 74, 76, 79, 83n, 83, 89, 203, 243–44
European Community (EEC), 39
European Union (EU), 39–40, 263
experiments (animal, data), 32, 38–40, 54–55, 69, 74, 76, 208–10, 237, 245–47, 249
expertise, 1–6, 8–9, 11, 13, 15–23, 27, 36, 65–66, 68, 70, 73, 75–76, 80–81, 83, 88–89, 91–92, 116, 156, 260
experts, 1, 7–9, 15–16, 18, 66, 70, 73–75, 96–97, 100–102, 105, 108, 127–28, 138–40, 143, 147, 152, 157, 161, 164, 170, 177, 188, 199, 262
exposure, 7, 10–11, 15, 17, 19, 29, 35–37, 40–43, 46, 51, 53, 56–57, 59, 73, 75–77, 79–81, 83, 88, 89, 92, 97, 101, 103, 106–7, 117, 134–35, 143–44, 155–61, 163, 165, 167n, 173–74, 177–78, 186, 195–96, 198–204, 206–7, 209–10, 224–25, 234–43, 247–49, 255–56, 259–66, 267n
extrapolation, 74, 152, 224, 226–29, 235, 237–38, 241, 243, 248, 249
Exxon and Campbell Wells, 154, 155, 165, 166

F
Federal Insecticide, Fungicide, and Rodenticide Act (FIFRA), 256–57
Flamm, Gary, 102
Food and Drug Administration (FDA), 6, 8, 30, 37–38, 43n3, 43n5, 43n7, 65
Food, Drug, and Cosmetic Act (U.S.), 6, 30, 42, 68, 70, 238, 256
Foundation for Taxpayers and Consumer Rights, 78
Freeman, Chris, 99

Friends of the Earth, 12
Frontali, Nora, 120
Fukushima, 1
furofuramide, 54, 56

G
Gass, G. H., 75–76, 89
General Electric (GE), 171, 173, 176, 177
genetic toxicology, 53, 58
Goslin, David, 103–4, 109n
grassroots organizations, 129
Greenpeace, 12, 159, 162

H
Hamilton, Alice, 121
Handler, Philip, 104–5, 109nn2–3, 109n9
Hartman, Philip, 47–48
health
   environmental, 2, 4, 7–8, 10, 12, 19, 43, 53–54, 83, 88, 102, 105, 121, 123, 127, 129, 135–37, 141, 145–47, 149n, 152, 153, 159, 163, 205, 250n2, 250n5
   occupational, 103, 116–17, 127, 129, 129n
   public, 4, 6, 21, 36–39, 42, 47, 50, 53, 67, 69, 71, 73, 77, 88–89, 91, 104, 106–7, 116–18, 120, 126, 129, 130n, 135–37, 140, 142, 143–46, 157–58, 161, 166, 168n, 170, 174, 183, 188, 195–97, 206–8, 238, 247, 249, 264
Henry, Hunter, 95, 250n
Herbst, Arthur, 71
Hollaender, Alexander, 50, 60n7
Horecker, Bernard, 47
hormones, 14, 37, 59n1, 68, 75, 122, 195, 197, 237, 243, 248
   sex, 66, 70, 74, 77, 92
   steroid, 41, 55
   synthetic, 13, 32, 34, 36, 40, 66, 74
   xeno-, 83, 249
House Committee on Science and Technology, 101
Hwan-Jang, Hwang, 181, 185, 188

I
iatrogenicity, 80
ICMESA chemical plant, 118, 128, 134, 135, 137–39, 144–46, 147n3
ignorance, 3, 17, 21, 23, 40, 70, 180, 189n, 193, 208, 215–17, 221–25, 228–29
Instituto Ramazzini (Italian Foundation), 13, 165
International Agency for Research on Cancer (IARC), 8, 96, 134
International Institute for Applied Systems Analysis (IIASA), 100
International Labour Organization (ILO), 97
International Programme on Chemical Safety (IPCS), 96
iodine 131, 1

J
Jacob, François, 48
Jensen, Soren, 14
Jones, Hardin, 75

K
Karnaky, Karl John, 31, 40–41
Kates, Robert, 99, 100
Katrina (hurricane), 21, 166, 215, 217–18, 220–21, 224–25
knowledge, 19, 20, 21, 40–41, 49, 54, 57, 70, 76, 78, 134, 143–47, 152–54, 157, 163, 166, 215–17, 220–25, 228–30, 247–48, 265
    alternative, 14–16, 18, 113, 117, 118, 163
    expert, 4, 16, 142, 165
    regulatory, 72–73
    scientific, 1–6, 13, 16, 19, 21, 23, 80, 107, 108, 135, 136, 138, 141, 188, 226, 238

L
labor movement, 116–17, 164
latency, 100–101, 108, 176, 187, 202, 238, 243
law, environmental, 184, 205, 257
Lederberg, Joshua, 52–53, 60n
Leet, Richard, 95
Lega per l'Ambiente, 125
Lilly, Eli, 35, 65, 69, 78
Litton Bionetics, 54
Louisiana's Department of Environmental Quality (DEQ), 155, 218, 221, 222, 223
low dose, 9–11, 22, 37, 59, 72–77, 88–92, 96–97, 101, 119, 123, 234–49, 250n2, 250nn4–5

M
Martin, Malcom, 48, 56
Maximum Allowed Concentration (MAC), 10, 119–120, 127–128
McCarville, William J., 104
Medical Research Council (MRC), 68
Menkes, Joshua, 102
mercury, 116, 123–24, 180–82, 186, 195–96, 219
methylmercury, 199, 201
Minamata, 182, 201
Mirer, Franklin E., 105, 107
Mitchell, Herschel, 47, 59n
Monsanto, 14, 103–4
Montedison company, 15, 117, 160–61
Morlay, Barclay, 95
Muller, Herman, 53, 60n
mutagens, 46–53, 55–56, 58–59. *See also* Carcinogens, Mutagens, Reproductive Toxicants (CMR)
mutagenesis, 8, 14, 46, 48–49, 52–53, 55, 58, 60n
mutagenicity, 8–9, 46–47, 49, 52–58, 60n, 119
mutation, 8, 46–53, 55–56, 58–59, 60n, 76, 200, 241

N
Nader, Ralph, 12, 78, 90
National Academy of Engineering (NAE), 98
National Academy of Sciences (NAS), 98, 106
National Cancer Institute, 43n, 49, 55, 71, 81, 249

National Center for Toxicological Research, 76
National Institute of Environmental Health Sciences (NIEHS), 244, 246, 250
National Institutes of Health (NIH), 47, 56, 59, 82
National Research Council (NRC), 53, 95, 98, 106, 265
National Science Foundation (NSF), 101-2, 104
National Toxicological Program, 178, 179, 244
Nebbia, Giorgio, 125
Neel, James, 56, 60n
Newell, Allen, 102, 165
No Observable Effect Level (NOEL), 235, 242, 243, 244
nongovernemental organizations (NGOs), 108, 152, 256, 257, 261-62, 265

O
Oak Ridge Laboratory, 50, 58, 102
Occupational Safety and Health Administration (OSHA), 55, 103
Office of Science and Technology Policy (OSTP), 97, 104, 106
Omenn, Gilbert, 97, 106, 198
Organisation for Economic Co-operation and Development, 97
organophosphates, 123, 200
O'Riordan, Tim, 102

P
Paccino, Dario, 126
perchlorate, 197, 204
pesticides, 5, 6, 12, 49, 57, 81-82, 88, 90, 195-200, 204, 206-10, 219
pharmaceuticals, 43, 67, 196-97, 199, 207-9, 258
pharmacology, 44n, 74, 91
phthalates, 195, 197, 204, 219, 260
pollutants, 5, 11, 40, 57, 34, 58, 88, 95, 97, 122-25, 127-28, 130n11, 136, 138, 148n7, 207-8, 219, 249, 255, 257, 262
pollution, 4-5, 7, 12-13, 38-39, 58-59, 82, 88, 96, 100, 101, 117, 137, 153, 155, 159, 161-62, 177, 181, 186, 196, 205, 218, 224, 256, 258
polybrominated diphenyl ethers (PBDEs), 197, 198, 202, 203, 20TB
polychlorinated biphenyls (PCBs), 14, 17, 130n11, 198-99, 202, 203, 219, 240, 250n1, 257
polycyclic aromatic hydrocarbons (PAHs), 51, 200, 219
precaution, 6, 8, 22, 29, 30, 32, 34-35, 37-39, 42-43, 44n, 74, 249, 262-63
precautionary (principle, action, approach, framework), 22, 29, 30-31, 33, 37-40, 42-43, 44n, 67, 91, 139, 148n, 248-50, 263
Press, Frank, 105, 109n1, 109n13, 110
Proctor and Gamble, 103-4
public use of science (PUS), 152-53
polyvinyl chloride (PVC/CVM), 122, 159, 160-63, 167n

R
radiation, 47, 49-50, 52-53, 58, 96, 117, 195, 202, 204, 237, 241-42, 248
Radio Corporation of America (RCA) 15, 170-79, 186-88
Raiffa, Howard, 100
RAND Corporation, 103
Red Book (NRC's), 11, 95, 96, 103, 106-8
Registration, Evaluation, Authorisation and Restriction of Chemical substances (REACH), 23
regulatory
    agencies, 3, 8, 23, 34, 37, 44, 72, 73, 101, 104, 106, 108, 217, 247, 256, 265
    institutions, 11, 97, 98
    policy, 47, 102, 104
    tools, 67, 72, 89, 91
risk
    analysis, 101-6
    assessment, 21, 90, 91, 95-98, 100, 102-6, 108, 172, 204, 207-8, 215, 217, 221-25, 228-30, 248
    management, 11, 67, 90, 91, 101, 106
Rodricks, Joseph, 102

Rowe, Wallace, 56
Rubin, Harry, 55
Ruckelshaus, William, 103

S

safety, 30, 32–33, 36, 41, 42, 47, 53, 55–56, 59, 65, 68, 75, 99, 120, 140, 161, 164, 185, 196, 209, 221, 236, 249, 262, 266
Safiotti, Umberto, 76
Salmonella [or *Salmonella typhimurium*], 46–48, 51, 55–56
science, regulatory, 72–73, 77, 81, 90, 92, 215, 224, 230, 260
Scientific Committee on Problems of the Environment (SCOPE), 99
screening, 8, 46, 50, 52, 54–55, 60n5, 83, 88, 89, 210, 222–28
Seveso, 3, 16, 17, 101, 118–21, 125, 127–28, 130n5, 130n11, 131n14, 134–51
Shell, 103–4
Sharpe, Richard, 246
Shubik, Philipp, 96
Sinclair, Graig, 99
Singer, Maxine 56
Sitton, Paul, 105–6, 109n1
Sloan, Miner Joe, 104
Smith, George, 32
Smith, Olive, 32
Social Science Research Council (U.S.), 99
Soong, Derkau, 181, 187
Soto, Ana, 88, 247
Stallones, Reuel A., 105
Starr, Chauncey, 98–99
Stilbosol, 35
Stockholm Conference, 7, 12, 96, 99
Sugimura, Takashi, 56
Szybalski, Waclaw, 52, 60n8

T

Tardiff, Robert, 102
Technology Assessment and Risk Analysis (TARA), 102
tests (toxicological) 8, 9, 41, 50, 51, 54–56, 60n12, 69, 73, 138, 139, 155, 209, 210, 219–20, 234, 242, 249

tetrachlorodibenzodioxin (TCDD), 119, 134, 137, 138, 147n1, 148n6, 186, 187
threshold, 5, 10–11, 14, 37, 41, 73–74, 76–77, 101, 122, 177, 235, 240, 241, 243
Throdahl, Monte, 104
Tomatis, Laurenzo, 96
Toxic Substances Control Act (TSCA), 20, 257, 265, 266
toxicants, 1–24, 97, 135, 136, 143, 145, 147, 152, 166, 177–79, 195, 199–205, 218, 237, 243, 262
toxicity, 6, 8, 36, 76, 88, 107, 115, 117, 119, 126, 127, 135, 138, 145, 155, 160, 173, 177, 195, 200, 203, 207, 208, 209, 210, 215, 225–28, 238, 241, 243, 244, 245
toxicology, 5, 9, 10, 36, 53, 58, 72, 74, 83, 134, 170, 177, 180, 187, 197, 210, 234–35, 242–43, 248
toxins, 37, 124, 154, 170, 171, 178, 187, 188, 219
trichlorophenol (TCP), 122, 138, 148n6
tris(2,3-dibromopropyl) phosphate (Tris), 54, 56, 60n11
Tumor, Horace, 80
Tversky, Amos, 103

U

uncertainty, 8, 15, 18–19, 21, 31, 37, 39–41, 57, 60n15, 70, 80, 97–98, 108, 119, 136, 138–43, 166, 170, 180, 189n10, 216, 224, 247, 262–63
unions, 3, 12, 107, 117, 126, 129, 152, 161–65, 172, 210n1, 260
United Nations Conference on the Human Environment (Stockholm Summit). *See* Stockholm Conference
United Nations Environment Programme (UNEP), 7, 97, 99, 106, 268

V

vom Saal, Frederic S., 88, 246, 247

W

Weyerhaeuser, 103

White, Priscilla, 31
Whitfield, Harvey, 48
Williams, Patricia, 15, 155, 157–58
Wingspread Conference, 10, 38, 44n18, 81, 88, 92n9
Witkin, Evelyn, 52, 59
women's health movement, 66, 67, 72, 81, 90

World Health Organization (WHO), 8, 96, 97, 106, 185

X
xenoestrogens, 88, 89, 203

Z
Zinder, Norton, 56